The Besterman World Bibliographies

The Besterman World Bibliographies

Physical Sciences

A BIBLIOGRAPHY OF BIBLIOGRAPHIES

By Theodore Besterman

VOLUME TWO

TOTOWA, N. J.
Rowman and Littlefield
1971

Published by
Rowman and Littlefield
A Division of
Littlefield, Adams & Co.
81 Adams Drive
Totowa, N. J. 07512

★

★

Typography by George Hornby
and Theodore Besterman

ISBN 0-87471-054-5

Contents

Geology.

1. *Bibliographies*

EMM[ANUEL] DE MARGERIE, Catalogue des bibliographies géologiques. Congrès géologique international: 1896. pp.xx.733. [3925.]

— — [supplement]. Edward B[ennett] Mathews, Catalogue of published bibliographies in geology, 1896–1920. National research council: Bulletin (no.36): Washington 1923. pp.228. [3312.]

HOMER P[AYSON] LITTLE, List of manuscript bibliographies in geology and geography. National research council: Reprint and circular series (no.27): Washington 1922. pp.17. [144.]

2. *Periodicals*

SPIS wydawnictw perjodycznych znajdujących się w bibljotece P[ań每stwowy] I[nstytut] G[eologiczny]. . . . Catalogue &c. [Warszawa 1928]. pp.22. [300.]

— Uzupełnienie. 1931. pp.8. [150.]
printed on one side of the leaf.

ENGLISH periodicals on geology and mining in the Science library. Science library: Bibliographical series (no.528): 1940. single leaf. [18.]*

[BRUNO BALBIS], Riviste de geodesia, geofisica, sismologia, meteorologia, ecc. i cui articoli possono essere forniti dal servizio fotoriproduzione documentaria. Consiglio nazionale della ricerche: Centro nazionale di documentazione scientifico-tecnico: Roma 1954. pp.[ii].17. [250.]*

GEOLOGY. A select list of british scientific periodicals. Royal society [&c.]: 1960. pp.31. [27.]

3. *General*

CARL FRIEDRICH WILHELM SCHALL, Oryktologische bibliothek. Weimar 1787. pp.[xviii].286. [1500.]

CHRISTIAN KEFERSTEIN, Geschichte und litteratur der geognosie. Halle 1840. pp.xiv.282. [1000.]

CATALOGUE of the books and maps in the library of the Geological society. 1846. pp.xv.226. [5000.]
— [another edition]. Catalogue of the library [&c.]. By James Dallas. 1881. pp.[v].618. [15,000.]

Geology

LOUIS [JEAN RODOLPHE] AGASSIZ, Bibliographia zoologiæ et geologiæ. . . . Edited by H[ugh] E[dwin] Strickland [vol.iv: and sir William Jardine]. Ray society: 1848–1854. pp.xxiii.506+[iii].492+[iii].658+[iii].604. [40,000.]

CHARLES GIRARD, American zoological, botanical, and geological bibliography for the year 1851. [*s.l.* 1852]. pp.19. [300.]

PAUL PARTSCH, Katalog der bibliothek des K.K. hof-mineralien-cabinetes. Wien 1851. pp.xii.233. [3000.]

—— Zweite . . . auflage, neu geordnet . . . von dr. Albrecht Schrauf. 1864. pp.xi.340. [6750.]

MONTANISTISCHE bibliothek. Verzeichniss der in Deutschland und im auslande in den jahren 1866–1870 auf den gebieten des berg-, hütten- und salinenwesens, der mineralogie, geognosie, geologie und paläontologie erschienenen bücher, zeitschriften und karten. Leipzig 1871. pp.xvi.72. [1000.]

THE GEOLOGICAL record. . . . An account of works on geology, mineralogy, and palæontology published during the year.

1874. Edited by William Whitaker. 1875. pp.xvi.397. [2000.]

1875. 1877. pp.xx.443. [2000.]
1876. 1878. pp.xxii.416. [2500.]
1877. 1880. pp.xxvi.432. [3000.]
1878. Edited by W. Whitaker and W. H. Dalton 1882. pp.xxxi.496. [3500.]
1879. 1887. pp.xxxiv.418. [3500.]
1880–1884. Edited by William Topley and Charles Davies Sherborn. 1888–1889. pp. xl.544+viii.563. [15,000.]

no more published.

JOHN PLANT, Catalogue of the library of the Manchester geological society. [Manchester] 1875. pp.38. [350.]

—— List of additions to the library. Manchester.

1904–1906. pp.xvi. [200.]
1906–1907. pp.xiv. [175.]
1907–1908. pp.xvi. [200.]

KATALOG der bibliothek der Königlichen geologischen landesanstalt und bergakademie. Berlin 1876. pp.xvii.661. [10,000.]

BERNHARD VON COTTA, Geologisches repertorium. Beiträge zur geschichte der geologie (vol.i): Leipzig 1877. pp.ix.400. [4500.]

HENRY WHITE and THOMAS W. NEWTON, A

catalogue of the library of the Museum of practical geology and Geological survey. 1878. pp. [iv].602. [10,000.]

[BERNHARD BARHAM WOODWARD], Catalogue of the library of the Geologists' association. Compiled by the Honorary librarian. 1879. pp.[iv].79. [1250.]

— — Addenda et corrigenda. [1880]. pp.12. [100.]

WILLIAM AMBROSE TAYLOR, A catalogue of the library of the Royal geological society of Cornwall. Plymouth 1882. pp.iv.160. [1500.]

G[USTAVE] DEWALQUE, Catalogue des ouvrages de géologie, de minéralogie et de paléontologie, ainsi que des cartes géologiques, qui se trouvent dans les principales bibliothèques de Belgique. Liége 1884. pp.xiii.393. [3750.]

RÓBERT FARKASS, A Magy. kir. földtani intézet könyv- és térképtárának czimjegyzéke. Budapest 1884. pp.[ii].ix.192. [3000.]

— — 1. pót-czímjegyzéke. . . . Összeállјtotta Bruck József. 1886. pp.60. [600.]

[—] — [another edition]. 1911. pp.xiii.488. [8500.]

WILLIAM ROBERT BION, Catalogue of the library

of the Geological survey of India. Calcutta 1884. pp.[ii].iii.336. [5000.]

C. EKAMA, Catalogue de la Bibliothèque. . . . Cinquième livraison: Paléontologie, géologie, minéralogie. Fondation Teyler: Harlem 1886. pp.[iii].459–551. [500.]

CHARLES LAPWORTH, Books on historical geology. Birmingham reference library: Lectures [1886]. pp.[ii].83–110.vii. [200.]

KATALOG der bibliothek der Deutschen geologischen gesellschaft. Berlin 1887. pp.166. [5000.]

— [another edition]. Bearbeitet von P[aul] Dienst. [1930]. pp.xi.1161. [30,000.]

CATALOGUE of the library of the Edinburgh geological society. Edinburgh 1887. pp.31. [500.]

BIBLIOGRAPHIA hopkinsiensis, 1876–1893. Parts II and III. Chemistry, geology and mineralogy. Johns Hopkins university: Baltimore 1893. pp.vi. 50. [1000.]

CATALOGO della biblioteca dell'Ufficio geologico. R. corpo delle miniere: Roma 1895. pp.viii. 283. [8000.]

GEOLOGICAL literature added to the Geological society's library.

[i]. July–December 1894. 1895. pp.58. [1250.]

[ii]. 1895. 1896. pp.157. [3000.]
[iii]. 1896. 1897. pp.207. [3500.]
[iv]. 1897. 1898. pp.196. [3500.]
[v]. 1898. 1899. pp.185. [3500.]
[vi]. 1899. 1900. pp.176. [3500.]
[vii]. 1900. 1901. pp.182. [3500.]
[viii]. 1901. 1902. pp.195. [3500.]
[ix]. 1902. 1903. pp.200. [3500.]
[x]. 1903. 1904. pp.186. [3500.]
[xi]. 1904. 1905. pp.180. [3500.]
[xii]. 1905. 1906. pp.201. [3500.]
[xiii]. 1906. 1907. pp.198. [3500.]
[xiv]. 1907. 1908. pp.227. [3500.]
[xv]. 1908. 1909. pp.208. [3500.]
[xvi]. 1909. 1910. pp.215. [3500.]
[xvii]. 1910. 1911. pp.218. [3500.]
[xviii]. 1911. 1912. pp.164. [2500.]
[xix]. 1912. 1914. pp.266. [4500.]
[xx]. 1913. 1920. pp.247. [3500.]
[xxi]. 1914. By Arthur Greig. 1922. pp.[ii]. iii.194. [3000.]
[xxii]. 1915–1919. By T. H. Digges La Touche. 1923. pp.546. [9000.]
[xxiii]. 1920. 1921. pp.[ii].iii.124. [3500.]
[xxiv]. 1921. 1922. pp.[ii].iii.116. [3500.]
[xxv]. 1922. 1923. pp.[ii].iii.136. [4000.]

— Subject index . . . 1920, 1921 and 1922. 1937. pp.[ii].220.

[xxvi]. 1923. 1924. pp.[ii].iii.131. [4000.]

[xxvii]. 1924. 1925. pp.[ii].iii.132. [4000.]

— Subject index . . . 1923 & 1924. 1933. pp.[ii].144.

[xxviii]. 1925. 1926. pp.[ii].iii.220. [5000.]

[xxix]. 1926. 1927. pp.[ii].iii.236. [4500.]

[xxx]. 1927. 1928. pp.[ii].iii.233. [4500.]

[xxxi]. 1928. 1929. pp.[ii].iii.237. [4500.]

[xxxii]. 1929. 1930. pp.[ii].iii.245. [4500.]

[xxxiii]. 1930. 1931. pp.[ii].iii.274. [5000.]

[xxxiv]. 1931. 1932. pp.[ii].iii.304. [5500.]

xxxv. 1932. 1933. pp.[ii].iii.308. [5500.]

xxxvi. 1933. 1934. pp.[ii].iii.304. [5500.]

xxxvii. 1934. 1935. pp.[ii].iii.328. [6000.]

no more published; except as shown a subject index forms part of each volume.

MICHEL [FÉLIX] MOURLON, Bibliographia geologica. Répertoire des travaux concernant les sciences géologiques d'après la classification décimale. Service géologique de Belgique: Bruxelles.

Série A, se rapportant aux publications antérieures à 1896.

i. 1899. ff.[iii].351.pp.353–367.ff.368–371. [3000.]

ii. 1900. pp.[vi].ff.400.pp.401–414.ff.416–420. [3000.]

iii. 1901. pp.[vii].ff.388.pp.389–400.ff.401–404. [3000.]

iv. 1902. pp.[vi].ff.386.pp.387–392.ff.393–394. [3000.]

v. 1902. pp.[vi].ff.356.pp.357–368.ff.369–370. [3000.]

vi. 1903. pp.[vi].ff.356.pp.357–366.f.367. [3000.]

vii. 1903. pp.[vi].ff.377.pp.379–391.f.392. [3000.]

viii. 1904. pp.[v].ff.357.pp.359–373.f.374. [3008.]

ix. 1906. pp.[v].ff.356.pp.357–374. [3009.]

Série B, se rapportant aux travaux parus à partir du 1er janvier 1896.

i. 1897 [1898]. pp.[viii].ff.216. [2000.]

ii. 1899. pp.[vi].ff.341.pp.343–361.ff.362–366. [3000.]

iii. 1900. pp.[vi].ff.374.pp.375–388.ff.389–390. [3000.]

iv. 1901. pp.[viii].ff.382.pp.383–397.ff.398–401. [3004.]

v. 1902. pp.[vi].ff.356.pp.357–371.pp.372–373. [3000.]

vi. 1903. pp.[vi].ff.377.pp.379–395.ff.396–398. [3000.]

vii. 1904. pp.[vi].ff.367.pp.369–387. [3008.]

no more published.

LIST of publications by members of the Department of geology of the Leland Stanford junior university. [Palo Alto, Cal.] 1903. pp.25. [250.]

GEOLOGY. International catalogue of scientific literature (section H): Royal society.

i. 1903. pp.xiv.220. [1517.]
ii. 1904. pp.viii.256. [1702.]
iii. 1905. pp.viii.248. [1713.]
iv. 1906. pp.viii.250. [1670.]
v. 1907. pp.viii.235. [1568.]
vi. 1908. pp.viii.299. [2167.]
vii. 1909. pp.viii.377. [3146.]
viii. 1910. pp.viii.283. [2555.]
ix. 1912. pp.viii.245. [2177.]
x. 1912. pp.viii.282. [2736.]
xi. 1914. pp.viii.342. [4388.]
xii. 1916. pp.viii.375. [3778.]
xiii. 1918. pp.viii.279. [2468.]
xiv. 1920. pp.viii.280. [2393.]

no more published.

LIST of additions to the library, 1904–1905, 1905–1906. Manchester geological & mining

society: Manchester [1906]. pp.xvi. [200.]
— 1907–7. pp.xiv. [150.]

EDWARD B[ENNETT] MATHEWS and G. E. REED, Bibliography of the faculty and students of the Department of geology of the Johns Hopkins university, 1883–1913. Baltimore 1913. pp.143. [2750.]

CATALOGUE of books in the library of the Mining & geological institute of India. Calcutta 1916. pp.22. [250.]

CATALOGUS van de boeken en tijdschriften geplaatst in de Laboratoria voor scheikunde . . . en voor mineralogie en geologie. Universiteit: Bibliotheek: Amsterdam 1919. pp.[iii].95. [geology: 300.]

A. BORN, Allgemeine geologie und stratigraphie. Wissenschaftliche forschungsberichte: Naturwissenschaftliche reihe (vol.ii): Dresden &c. 1921. pp.[x].146. [1767.]

BIBLIOGRAPHIE des sciences géologiques. Publiée par la Société géologique de France avec le concours de la Société française de minéralogie. Fédération des sociétés françaises de sciences naturelles.

[1923]. pp.36.[ii].37.64.39.45. [4000.]
1924. pp.303. [4000.]
1925. pp.321. [4000.]
1926. pp.347. [5000.]
1927. pp.312. [4000.]
1928. pp.299. [4000.]
1929. pp.269. [4000.]

2nd ser.

i. 1930. pp.319. [3917.]
ii. 1931. pp.432. [5620.]
iii. 1932. pp.356. [4606.]
iv. 1933. pp.388. [5202.]
v. 1934. pp.342. [4769.]
vi. 1935. pp.294. [3875.]
vii. 1936. pp.371. [4697.]
viii. 1937. pp.339. [4753.]
ix. 1938. pp.344. [6000.]
x. 1939. pp.221. [4000.]
xi. 1940. pp.107. [2000.]
xii. 1941. pp.91. [1500.]
xiii. 1942. pp.87. [1500.]
xiv. 1943. pp.93. [1500.]
xv. 1944. pp.64. [1000.]
xvi. 1945. pp.201. [4000.]
xvii. 1946. pp.525. [10,000.]
xviii. 1947. pp.430. [6000.]

continued in the form of extracts from the Bulletin

analytique [signalétique] du Centre national de la recherche scientifique.

GEOLOGIE und paläontologie, bearbeitet von Hans Praesent. . . . Mineralogie, bearbeitet von Hans Hentschel. Jahresberichte des Literarischen zentralblattes über die wichtigsten wissenschaftlichen neuerscheinungen des gesamten deutschen sprachgebietes (1924, vol.xxi):Leipzig 1925. pp. 99. [750.]

W[ILFRED] BONSER, Catalogue of the geological books in the library of University college, London. Geologists' association: 1927. pp.viii.102. [1225.]

THE GEOLOGICAL review. Brief abstracts of the important current literature related to general geology and the geology of petroleum. [Edited by Esther S. Johnston]. Number 1[-12]. Denver 1928-1929. pp.96. [300.]
no more published.

BIBLIOGRAPHY and index of geology exclusive of north America. Geological society of America: [Washington].

i. 1933. By John M[ilton] Nickles and Robert B. Miller. pp.xxii.405. [3500.]
ii. 1934. pp.xxiii.420. [3500.]

iii. 1935. [By] J. M. Nickles, Marie Siegrist and Eleanor Tatge. pp.xx.408. [3500.]
iv. 1936. pp.xxi.480. [4500.]
v. 1937. pp.xxi.510. [5000.]
vi. 1938. pp.xxi.471. [5000.]
vii. 1939. pp.xxi.522. [5500.]
viii. 1940. pp.xix.386. [4000.]
ix. 1941–1942. pp.xix.450. [4500.]
x. 1943–1944. pp.xiv.203. [2000.]
xi. 1945–1946. [By] Marie Siegrist and Eleanor Tatge. pp.xviii.474. [4500.]
xii. 1947. pp.x.359. [4000.]
xiii. 1948. pp.xi.386. [4000.]
xiv. 1949. [By] Marie Siegrist. pp.xii.375. [4000.]
xv. 1950. [By] Marie Siegrist and Marcia Lakeman. pp.xiii.429. [4500.]
xvi. 1951. [By] Marie Siegrist, Mary C. Grier, Marcia Lakeman. pp.xii.462. [4500.]
xvii. 1952. pp.xiv.621. [6000.]
in progress.

GEOLOGY and geography. Library association: County libraries section: Readers' guide (no.30): 1939. pp.32. [750.]

LUDWIG RÜGER [*and others*], Geology and pelaeontology. Office of military government for

Germany: Field information agencies technical: Fiat review of german science, 1939–1946: Wiesbaden 1948. pp.[viii].247. [1000.]
the text is in german.

HARRIET K. LONG, Manual of reference sources in geology. Washington university: Libraries: Reference manual (no.5): St. Louis 1950. pp.[ii]. ff.ii.16. [56.]

[CH. EXNER and J. WINDBRECHTINGER], Geologische literatur in Österreich, 1945–1950. Geologische bundesanstalt: Verhandlungen: Sonderheft (B): Wien 1951. pp.44. [1300.]

RICHARD M[AXWELL] PEARL, Guide to geologic literature. New York &c. 1951. pp.xi.239. [500.]

BILTEN sadržaja stranih časopisa. Dodatak "Zelenom biltenu". C: Geologija, rudarstvo, metalurgija. Savezna uprava za unapredenje proizvodnje: [Belgrade].
i. 1952. pp.[312]. [1000.]
no more published?

BRIAN MASON, The literature of geology. American museum of natural history: New York 1953. pp.vii.155. [750.]

GEOLOGICKÝ průzkum. A[–B]. (Seznam litera-

tury). Ústřední technická knihovna ČSR: Praha 1954. pp.[ix].146+[vii].68. [1000.]*
no more published?

РЕФЕРАТИВНЫЙ журнал. Геология и география [from 1956: Геология]. Академия наук СССР: Институт научной информации: Москва.

1954. pp.339.viii + 192.viii + 184.viii + 195. viii+232.viii+237.x. [7567.]

1955. pp.266.xii + 262.x + 260.xii + 263.xii + 266.x + 314.xi + 280.xii + 293.xi + 306.xiv+337.xiv+402.xvi+484. [20,492.]

1956. pp.208 + 216 + 206 + 208 + 188 + 185 + 168 + 164 + 163 + 192 + 187 + 204+184. [13,763.]*

1957. pp.192 + 217 + 220 + 228 + 224 + 260 + 260 + 308 + 272 + 296 + 305 + 284. [18,394.]

1958. pp.338+336+311+363+365.8+404

1959. pp.400 + 356 + 368 + 360 + 368 + 336 + 340 + 380 + 336 + 368 + 344 + 326+442. [26,874.]

1960. pp.308 + 294 + 316 + 304 + 320+ 320 + 326 + 328 + 433 + 452 + 508 + 464+492+754. [28,342.]

1961. pp.[360] + [368] + [367] + [368] +

[376] + [352] + [336] + [328] + [328] + [288] + [288] + [314] + 180*. [30,000.]

1962. pp.[340] + [351] + [368] + [384] + [420] + [444] + [428] + [420] + [384] + [420]+

in progress; from 1956 a separate series was devoted to geography.

[DANIEL S. TURNER], Bibliography of geology theses, colleges and universities of the United States. Denver [1954]. pp.[iii].483. [5750.]*

[—] — [another edition]. Bibliography of theses written for advanced degrees in geology and related sciences at universities and colleges in the United States and Canada through 1957. By [Byron] John Chronic and Halka Chronic. 1958. pp.[428]. [11,091.]*

LISTE mensuelle des traductions du Service d'information géologique . . . du B. R. G. M. [Bureau des recherches géologiques et minières]. 1954 &c. *in progress.*

[EDWARD PASSENDORFER *and others*], Polska bibliografia analityczna. Geologia. Polska akademia nauk: Ośrodek bibliografii i dokumentacji naukowej: Warszawa.

i. 1955. pp.   . [80.]

ii. 1956. pp.63. [112.]
iii. 1957. pp.48.44. [213.]
no more published.

KOMPENDIEN-KATALOG. Geologie, paläontologie, petrographie, bodenkunde, mineralogie, lagerstättenkunde, hydrologie, geophysik, meteorologie. Köln &c. [1956]. pp.vii.166. [3500.]
limited to german works published 1945–1955.

L[OUIS] GLANGEAUD, Géologie. Association pour la diffusion de la pensée française: Bibliographie française établie à l'intention des lecteurs étrangers: 1957. pp.90. [400.]
limited to works published since 1948.

ABSTRACTS of bulgarian scientific literature. Geology and geography. Bulgarian academy of sciences: Department for scientific and technical information and documentation: Sofia.*
i. 1957–1958. pp.[ii].52. [76.]
ii. 1959.
iii. 1960. pp.72. [106.]
iv. 1961. pp.40. [132.]
in progress.

GEOMORPHOLOGICAL abstracts. London School of economics and political science: Department of geography.*

1–3. 1960. pp.23.36.34. [223.]
4–7. 1961. pp.40.36.36.36. [424.]
8–11. 1962. pp.40.36.40.48. [447.]
12–15. 1963. pp.192. [549.]
in progress.

4. *Cartography*
[*see also section 6, below*]

BERNHARD [VON] COTTA, Geognostische karten unseres jahrhunderts. Freiburg 1850. pp.60. [571.]

SPIS map znajdujących się w bibljotece P. I[nstytut] G[eologiczny]. [Warszawa 1931]. pp. 35. [750.]
printed on one side of the leaf.

WALTER H[ERMAN] BUCHER [*and others*], Catalogue of small-scale geologic maps useful for broader regional studies (with chief emphasis on modern maps). (Preliminary edition). National research council: Committee on tectonics: Washington [1933]. pp.[ii].vi.132. [1000.]*
limited to north America.

5. *Economic geology*

ANNOTATED bibliography of economic geology. Lancaster, Pa. [printed].

[i. 1928]. John M. Nickles, editor, pp.x.380. [1756.]
[ii. 1929]. pp.vi.208+ix.209–574. [2480.]
[iii. 1930]. pp.ix.248+ix.249–477. [2160.]
[iv. 1931]. pp.ix.189+viii.191–390. [2224.]
[v. 1932]. pp.viii.228+viii.229–431. [2505.]
[vi. 1933]. pp.viii.165+viii.167–372. [2085.]
[vii. 1934]. pp.viii.191+viii.193–382. [3166.]
viii. [1935]. pp.viii.243+viii.245–429. [2617.]
ix. [1936]. Editor, Robert B. Miller. pp.viii. 226+viii.227–433. [2460.]
x. [1937]. pp.viii.234+viii.235–446. [2537.]
— General index, volumes 1–x. 1939. pp.[ii]. 496.
xi. [1938]. pp.viii.208+viii.209–391. [2247.]
xii. [1939]. pp.viii.340. [2034.]
xiii. [1940]. pp.viii.165+viii.166–337. [1995.]
xiv. [1941]. pp.viii.167+vii.168–328. [1978.]
xv. [1942]. pp.vii.287. [1744.]
xvi. [1943]. pp.vii.423. [2511.]
xvii. [1944]. 1950. pp.viii.423. [2118.]
xviii. [1945]. 1954. pp.viii.358. [1931.]
xix. [1946]. [*not yet published.*]
xx. [1947]. Editor: Helen G. Townsend. pp. viii.196. [1248.]
xxi. [1948]. pp.viii.191. [1279.]
xxii. [1949]. pp.viii.252. [1518.]

xxiii. [1950]. [Edited by] Alan M[ara] Bateman. pp.viii.264. [1599.]
xxiv. [1951]. pp.viii.278. [1779.]
xxv. [1952]. pp.viii.167+xi.213. [2303.]
xxvi. [1953]. pp.viii.355. [1263.]
xxvii. [1954]. pp.viii.362. [1153.]
xxviii. [1955]. pp.xi.350. [1030.]
xxix. [1956]. pp.xi.371. [1146.]
xxx. [1957]. pp.viii.368. [1102.]
xxxi. [1958]. pp.xi.337. [1027.]

in progress.

IMM abstracts. A survey of world literature on economic geology, mining, mineral dressing, extraction metallurgy, allied subjects. Institution of mining and metallurgy.

i. 1950–1951. pp.341. [2000.]
ii. 1951–1952. pp.388. [2500.]
iii. 1952–1953. pp.341. [2000.]
iv. 1953–1954. pp.331. [2000.]
v. 1954–1955. pp.280. [1250.]*
vi. 1955–1956. pp.369. [1500.]*
vii. 1956–1957. pp.355. [1588.]*
viii. 1957–1958. pp.421. [1905.]*
ix. 1958–1959. pp.195. [2113.]
x. 1959–1960. pp.219. [2256.]
xi. 1960–1961. pp.203. [2174.]

xii. 1961–1962. pp.204. [2318.]
xiii. 1962–1963. pp.224. [2183.]
in progress; the first volume formed part of the Institution's Bulletin.

6. *Countries &c.*

Africa

H[ARRY] P. SAUNDERS, Bibliography of south african geology. Geological commission: Cape Town 1897. pp.[iii].56. [592.]

A[RTHUR] L[EWIS] HALL, A bibliography of south african geology to the end of 1920: authors' index. Union of South Africa: Department of mines: Geological survey: Memoir (no.18): Pretoria 1922. pp.376. [5794.]

— — A subject index . . . (no.22): 1924. pp.384.

— — 1921 to 1925 (inclusive). Authors' index . . . (no.25): 1927. pp.117. [1594.]

— — 1926 to 1930 (inclusive). Authors' index . . . (no.27): 1931. pp.160. [2052.]

— — 1931 to 1935 (inclusive). Authors' index . . . (no.30): 1937. pp.168. [2233.]

BIBLIOGRAFIA geologica italiana per gli anni 1915–1933. . . . Africa orientale italiana. Roma [1936]. pp.116. [100.]

Geology

BIBLIOGRAPHIE géologique de l'Afrique centrale. (Édition provisoire). Congrès géologique international: Association des services géologiques africains *and* Société géologique de Belgique: Liége 1937. pp.298. [4000.]

— Compléments. . . . Par A. Jamotte. Bureau d'études géologiques et minières coloniales: Paris 1948. pp.37. [300.]

ARDITO DESIO, Bibliografia geologica italiana dell'Africa. Ministero dell'Africa italiana: Collezione scientifica e documentaria (vol.xv): Roma 1950. pp.viii.83. [1250.]

GEOLOGICAL bibliography of Africa south of the Sahara. Commission for technical co-operation in Africa south of the Sahara.

i. Bibliography of the Karroo system. [1956]. pp.iii.111. [1400.]

ii. Bibliography of the jurassic and cretaceous systems. 1959. pp.[ii].83. [1006.]

Albania

[MARIO G. MAGNANI], Bibliografia geologica e geografico-fisica della regione albanese. R. ufficio geologico: Roma 1939. pp.66. [822.]

with a loose leaf of additions.

Geology

America

a. *General*

JULES MARCOU and JOHN BELKNAP MARCOU, Mapoteca geologica americana. A catalogue of geological maps of America (north and south), 1752–1881. Geological survey: Bulletin (no7.): Washington 1884. pp.184. [924.]

b. *North*

[JAMES D. DANA], Catalogue of official reports upon the geological surveys of the United States and british provinces. [*s.l.*] 1867. pp.14. [250.]

BIBLIOGRAPHY of north american geology. Geological survey: Bulletin (no.44 &c.): Washington.

1886. By Nelson H. Darton . . . (no.44): 1887. pp.35. [300.]

[*continued as:*]

Record of north american geology.

1887–1889. By Nelson Horatio Darton . . . (no.75): 1891. pp.173. [1000.]

1890 . . . (no.91): 1891. pp.88. [500.]

1891 . . . (no.99): 1892. pp.73. [500.]

[*continued as:*]

Bibliography and index of north american

geology, palaeontology, petrology, and mineralogy.

1892–1893. By Fred Boughton Weeks . . . (no.130): 1896. pp.210. [1121.]

1894 . . . (no.135): 1896. pp.141. [677.]

1895 . . . (no.146): 1896. pp.130. [575.]

1896 . . . (no.149): 1897. pp.152. [788.]

1897 . . . (no.156): 1898. pp.130. [742.]

1898 . . . (no.162): 1899. pp.163. [941.]

1899 . . . (no.172): 1900. pp.141. [799.]

1892–1900 . . . (no.188–189): 1901–1902. pp. 717+337. [6548.]

1901 . . . (no.203): 1902. pp.144. [866.]

1902 . . . (no.221): 1903. pp.200. [1193.]

1903 . . . (no.240): 1904. pp.243. [1367.]

1904 . . . (no.271): 1905. pp.218. [1500.]

1901–1905 . . . (no.301): 1906. pp.770. [5750.]

[*continued as:*]

Bibliography of north american geology.

1906–1907. By F. B. Weeks and J. M. Nickles . . . (no.372): 1909. pp.317. [2672.]

1908. By John M. Nickles . . . (no.409): 1909. pp.148. [1215.]

1909 . . . (no.444): 1910. pp.174. [1305.]

1910 . . . (no.495): 1911. pp.179. [1410.]

1911 . . . (no.524): 1912. pp.162. [1266.]

1912 . . . (no.545): 1913. pp.192. [1257.]
1913 . . . (no.584): 1914. pp.183. [1357.]
1914 . . . (no.617): 1915. pp.167. [1357.]
1915 . . . (no.645): 1916. pp.144. [1127.]
1916 . . . (no.665): 1917. pp.172. [1281.]
1917 . . . (no.684): 1918. pp.154. [1194.]
1918 . . . (no.698): 1919. pp.148. [1099.]
1919–1920 . . . (no.731): 1922. pp.[ii].232. [2108.]
1921–1922 . . . (no.758): 1924. pp.[iii].273. [2241.]
1923–1924 . . . (no.784): 1927. pp.iii.280. [2577.]
1925–1926 . . . (no.802): 1928. pp.iii.286. [2927.]
1919–1928 . . . (no.823): 1931. pp.iii.1005. [13,000.]
1929–1930 . . . (no.834): 1931. pp.ii.280. [2960.]
1931–1932 . . . (no.858): 1934. pp.ii.300. [3454.]
1933–1934. By Emma M[ertins] Thom . . . (no.869): 1935. pp.ii.389. [3836.]
1934–1935 . . . (no.892): 1937. pp.ii.504. [4716.]
1929–1939 . . . (no.937): 1944. pp.[ii].1063+[ii].1065–1546. [20,000.]

1940–1941 . . . (no.938): 1942. pp.ii.479. [6000.]

1942–1943 . . . (no.949): 1945. pp.viii.460. [4000.]

1944–1945 . . . (no.952): 1947. pp.ii.496. [4000.]

1946–1947. By E. M. Thom, Marjorie Hooker and Ruth Reece Dunaven . . . (no.958): 1949. pp.iii.658. [5000.]

1948 . . . (no.968): 1950. pp.iii.309. [3500.]

1949 . . . (no.977): 1951. pp.273. [2500.]

1950. By M. Hooker . . . (no.985): 1952. pp. iii.394. [3000.]

1951. . . . (no.1025): 1955. pp.iii.378. [3500.]

1952–1953. By Ruth Reece King [*and others*]. . . . (no.1035): 1956. pp.iii.714. [5500.]

1954. . . . (no.1054): 1957. pp.iii.484. [4000.]

1955 . . . (no.1065): 1958. pp.iii.511. [4000.]

1956. . . . (no.1075): 1959. pp.iii.520. [4000.]

1957. . . . (no.1095): 1960. pp.iii.531. [4000.]

1958. . . . (no.1115): 1961. pp.iii.592. [4800.]

1959 . . . (no.1145): 1961. pp.iii.605. [4500.]

1960.

in progress.

NELSON HORATIO DARTON, Catalogue and index of contributions to north american geology, 1732–1891. Geological survey: Bulletin (no.127):

Washington 1896. pp.1045. [10,000.]

FRED BOUGHTON WEEKS, North american geologic formation names. Bibliography, synonymy and distribution. Geological survey: Bulletin (no.191): Washington 1902. pp.448. [10,000.]

JOHN M. NICKLES, Geologic literature on north America, 1785–1918. Geological survey: Bulletin (nos.746–747): Washington 1923–1924. pp.[ii]. 1167+[ii].658. [45,000.]

c. *South and central*

GEOLOGY of latin America ... in the Columbus memorial library of the Pan american union. Washington [1917]. ff.13. [200.]*

HENRY B. SULLIVAN, A catalogue of geological maps of south America. American geographical society: Research series (no.9): New York 1922. pp.[ii].iv.191. [209.]
printed on one side of the leaf.

JOSEPH T[HEOPHILUS] SINGEWALD, Bibliography of economic geology of south America. Geological society of America: Special papers (no.50): [New York] 1943. pp.v.159. [2750.]

M[ANUEL] MALDONADO-KOERDELL, Bibliografía geológica y paleontológica de America central.

Geology

Instituto panamericano de geografía e historia: Publicación (no.204): México 1958. pp.288. [2000.]

Antarctic

GEORGE LESLIE ADKIN [*and others*], Preliminary bibliography of the geology of Ross dependency. Geological society of New Zealand: Antarctic research subcommittee: [1956]. ff.[7]. [150.]*

— — Supplement. [1956]. ff.[3]. [60.]*

Argentina

ENRIQUE SPARN, Bibliografía de la geología, mineralogía y paleontología de la república Argentina hasta el año 1899. Academia nacional de ciencias: Miscelánea (no.3): Córdoba 1921. pp.93. [937.]

— — [supplements].

i. 1900–1914. . . . (no.2): 1920. pp.58. [519.]

ii. 1899–1914.

iii. 1915–1921, con dos suplementos para los años hasta 1914. 1922. pp.51.vii.xviii. [635.]

iv. 1922–1924, con un suplemento para los años 1915–1921. pp.37.viii. [394.]

v. 1925–1927. [1928]. pp.44. [440.]

vi. 1928–1931. 1935. pp.52. [514.]

vii. 1932–1935. 1936. pp.62. [600.]

Geology

PUBLICACIONES de la Dirección de minas y geología. Ministerio de agricultura de la nación: Buenos Aires 1940. pp.31. [400.]

— Supplemento. [1942]. single sheet. [20.]

CATÁLOGO de publicaciones de la Dirección nacional de geología y minría, incluyendo los informes inéditos. Buenos Aires 1960. pp.113.

Asia and central

A. N. ASATKIN and V. A. SAMOILOV, Библиография Дальневосточного края 1890–1931. . . . Том II, Геология, полезные ископаемые, палеонтология. Дальневосточный краевой исполнительный комитет: Москва 1935. pp.xl.415. [2000.]

Australia

ROBERT ETHERIDGE and ROBERT LOGAN JACK, Catalogue of works . . . and maps on the geology, palæontology, mineralogy, mining and metallurgy, etc. of the australian continent and Tasmania. 1881. pp.[ii].ii.196. [1250.]

— — [another edition]. [Department of mines: Sidney 1882]. pp.[ii].ii.131. [2500.]

A[NDREW] GIBB MAITLAND, Bibliography of the geology of Western Australia. Geological survey: Bulletin (no.1): Perth 1898. pp.31. [400.]

Geology

CATALOGUE of the publications of the Geological survey of South Australia. [Adelaide 1915]. pp.12. [150.]

GENERAL index to reports, 1870–1910. Western Australia: Geological survey: Bulletin (no.60): Perth 1916. pp.230. [15,000.]

LIST of publications, 1879–1946. Queensland geological survey: Brisbane [1946]. pp.24. [278.]

E. N. TEESDALE-SMITH, Bibliography of south australian geology. South australian department of mines and geological survey: [Adelaide] 1959. pp.240. [1500.]*

Austria

FRANZ HERITSCH, Verzeichnis der geologischen literatur der österreichischen Alpenländer. Leoben 1914. pp.[v].69. [2037.]

Balkans

V. P. PETKOVIĆ, Bibliographie géologique de la péninsule balkanique, 1904–1909. Геолошки анали балканскога полуострва (vol.vi, part 2): Београд 1911. pp.lviii. [500.]

Belgium

L. WALSCHOT, Kwartairgeologische bibliografie van België. Bibliographia belgica (no.69): Brussel 1962. pp.[ii].xiii.318. [1419.]*

Brazil

JOHN C[ASPER] BRANNER, A bibliography of the geology, mineralogy and paleontology of Brasil. Rio de Janeiro 1903. pp.115. [1250.]

ALPHEU DINIZ GONSALVES, Bibliographia da geologia, mineralogia e paleontologia do Brasil. Serviço geologico e mineralogico do Brasil: Boletim (no.27): Rio de Janeiro 1928. pp.[iii].iii. 205. [3000.]

DOLORES IGLESIAS, Bibliografia e indice da geologia do Brasil, 1640–1940. Departamento nacional da produção mineral: Divisão de geologia e mineralogia: Boletim (no.111): Rio de Janeiro 1943. pp.323.

JOSUÉ CAMARGO MENDES, Bibliografia geológica, mineralógica, petrográfica e paleontológica da estado de São Paulo. Secretaria da agricultura, industria e comercio: Instituto geográfico e geológico: São Paulo 1944. pp.57. [560.]

DOMÍCIO DE FIGUEIREDO MURTA, Literatura geológica do estado de Minas Gerais. Departamento de fomento industrial: Belo Horizonte 1946. pp. 3–183. [3000.]

Canada

LIST of publications of the Geological survey of Canada. Montreal 1873. pp.7. [40.]

—— [another edition]. W[alter] F[rederick] Ferrier, Annotated catalogue of and guide to the publications of the Geological survey, Canada, 1845–1917. Geological survey: [Publication] (no. 1723): Ottawa 1920. pp.[iv].544. [7500.]

—— [supplement]. Lorne B. Leafloor, Publications of the Geological survey of Canada (1917–1952). 1952. pp.[v].82. [2000.]

there are numerous intermediate indexes, lists and catalogues.

BIBLIOGRAPHY of canadian geology and palaeontology. Royal society of Canada: Ottawa.

1905. By H[enry] M[arc] Ami. pp.301–326. [250.]
1906. pp.143–156. [175.]
1907. pp.191–204. [150.]
1908–1911. By L[eopold] Reinecke. pp.88. [1250.]
1912. By Wyatt Malcolm. pp.287–345. [300.]
1913. pp.317–350. [400.]
1914. pp.279–305. [300.]

1915. pp.131–168. [400.]

P. J. MORAN, Published maps (1917–1930 inclusive). Geological survey of Canada: Ottawa 1931. pp.[ii].16. [375.]

LIST of geological maps published by the Quebec Department of mines. [Quebec 1944]. pp.4.4. [131.]*

LILLIAN B. KERR, Bibliography of geology, coal, oil, natural gas, and industrial minerals in the post-cambrian region of southern Manitoba to 1945. Department of mines and natural resources: Mines branch: Winnipeg 1949. pp.[ii].25. [450.]*

WALTER O[SCAR] KUPSCH, Annotated bibliography of Saskatchewan geology, 1823–1951, incl. Saskatchewan geological survey: Report (no.9): Regina 1952. pp.106. [525.]

A LIST of recently published memoirs, preliminary papers and maps of Ontario. Geological survey: Ottawa 1954. pp.[i].28. [350.]*

GEOLOGICAL reports and maps of Quebec, 1917–1959. Geological survey: Ottawa 1959. pp.34.

GEOLOGICAL reports and maps of the Atlantic provinces, 1917–1959. Geological survey: Ottawa 1959. pp.31.

Geology

GEOLOGICAL reports and maps of Ontario, 1917–1959. Geological survey: Ottawa 1959. pp.30.

GEOLOGICAL reports and maps of Yukon territory, 1917–1959. Geological survey: Ottawa 1959. pp.16.

GEOLOGICAL reports and maps of British Columbia, 1917–1960. Geological survey: Ottawa 1960. pp.35.

GEOLOGICAL reports and maps of Alberta, 1917–1960. Geological survey: Ottawa 1960. pp.27.

GEOLOGICAL reports and maps of Manitoba, 1917–1960. Geological survey: Ottawa 1960. pp.23.

Chile

J[OHANNES OTTO] BRÜGGEN, Bibliografía minera i jeolójica de Chile. Ministerio de industria i obras publicas [: Servicio jeolójico: Publicaciones (folleto 4):] Santiago 1919. pp.[ii].142.

China

CHUNG YU WANG, Bibliography of the mineral wealth and geology of China. 1912. pp.[vii].63. [750.]

CATALOGUE of the library of the National geological survey of China. Peking 1927. pp.iii.45. [400.]

C[HI] C[HÊN] LIU, Catalogue of the library of the National geological survey of China. Maps and atlas, series A. [Peking] 1928. pp.viii.89. [500.]

LIST of publications of the National geological survey of China. Peiping 1930. pp.[ii].16. [150.]

— [another edition]. Publication list. 1934. pp.[ii].32. [300.]

T. I. YOUNG [TSUN-I YANG], Bibliography of chinese geology. National academy of Peiping: [Peiping] 1935. pp.[iii].xi.241. [2500.]

HSIAO-FANG LI, Bibliography of chinese geology. Bibliography of geology and geography of Sinkiang. National geological survey of China: Nanking 1947. pp.vii.216. [1750.]

Colombia

JESÚS EMILIO RAMÍREZ, Bibliografía de la biblioteca dos Instituto geofísica de los Andes colombianos sobre geología y geofísica de Colombia. Segunda edición. Archivo de la economía nacional (vol.20): Bogotá 1956. pp.521. [3847.]

Congo

J[ULES] CORNET, Bibliographie géologique du

bassin du Congo. Société géologique de Belgique: Annales (fasc. spéc., no.1): Liége 1916. pp.81. [865.]

THÉODORE [JULES MARIE JOSEPH] HEYSE, Bibliographie du Congo belge et du Ruanda-Urundi (1939–1945). Géologie et mines. Cahiers belges et congolais (no.4): Bruxelles 1946. pp.42. [371.]

BIBLIOGRAPHIE géologique du Congo belge et du Ruanda-Urundi. Musée royal du Congo belge: Tervuren.

i. 1818–1924. [By L. Cahen and others]. 1955. pp.175. [840.]
ii. 1925–1934. 1953. pp.188. [588.]
iii.
iv. 1945–1954. 1955. pp.xx.232. [885.]
v. 1955–1956. 1957. pp.xx.90. [313.]

in progress?

Corsica

[— NENTIEN], Résumé des travaux sur la géologie de la Corse. Société des sciences historiques et naturelles de la Corse: [Bulletin (vol.xxi, no.250):] Bastia 1900 [on cover: 1902]. pp.101. [75.]

Costa Rica

GABRIEL DENGO, Bibliografía de la geología de Costa Rica. Universidad de Costa Rica: Publicaciones: Serie ciencias naturales (no.3): [San José] 1954. pp.iii.27.

Cuba

PEDRO J[OAQUÍN] BERMUDEZ, Bibliografía geologica cubana. [Habana] 1938. pp.86. [605.]

Czechoslovakia

VLAD[IMIR] JOS. PROCHÁZKA, Repertorium literatury geologické a mineralogické království Českého, markrabství Moravského a vévodství Slezského od roku 1528 až do 1896. . . . Díl 1. Seznam auktorů. Praze 1897. pp.xxii.299. [5605.]
no more published?

MINERALOGICKO-geologická bibliografie ČSR. Štátny geologický ústav ČSR: Praha.
1930–1931. [By] Rudolf Schwarz, J. Sv. Procházka].
1932–1933. [By] R. Schwarz.
1936.
1937.
1938.
1939.
1940.
1941.
1942–1943.
1944.

1945–1946. [By] R. Schwarz, Milada Vavřínova. pp.44. [562.]

1947. [By] M. Vavřinová. pp.36. [392.]

1948. pp.47. [517.]

1949. [By] Václav Bartušek. pp.63. [789.]

1950. pp.124. [1245.]

1951. pp.158. [1227.]

1952. pp.116. [935.]

1953. pp.139. [947.]

1954. pp.955. [1122.]

1955. [By] Břetislav Tvrzník, Jan Havle. pp. 108. [1019.]

1956. pp.175. [1465.]

1957. pp.180. [1461.]

1958. [By] Ludvík Váňa [*and others*]. pp.133. [1262.]

1959. [By] B. Tvrzník [*and others*]. pp.203. [1648.]

1960. Vědecký redaktor Jan Kořan. pp.139. [1205.]

1961. pp.175. [1537.]

1962. pp.167. [1420.]

in progress.

Denmark

FORTEGNELSE over Danmarks geologiske undersøgelses skrifter. København 1923 &c.

in progress.

Geology

East Indies

OPGAVE van geschriften over geologie en mijnbouw van Nederlandsch oost-Indie. Geologisch mijnbouwkundig genootschap voor Nederland en koloniën: Verhandelingen (vols.i &c.): 's-Gravenhage.

i. Door R[ogier] D[iederik] M[arius] Verbeek. 1912–1915.

ii. 1913–1922.

[*continued as:*]

Geologisch mijnbouwkundige bibliographie van Nederlandsch-Indië.

iii. 1923–1932. Door R. D. M. Verbeek [parts 13–20: Nicolaas Wing Easton].

iv. 1933–1936.

Ecuador

CARLOS MANUEL LARREA, Bibliografía científica del Ecuador. II. Geología, petrografía, mineralogía, paleontología, vulcanología. Quíto 1952. pp.3–319. [2500.]

Egypt

LIST of publications of the Mines & quarries department, Geological survey of Egypt and Petroleum research board, to which is appended a list of Survey department papers. Cairo 1948. pp.10. [150.]

Geology

ELIAS HABIB KELDANI, A bibliography of geology and related sciences concerning Egypt. Department of survey and mines: Cairo 1941. pp.xxix.428. [2939.]

Estonia

D. KALJO, Eesti geoloogia bibliograafia 1840–1959. Eesti NSV teaduste akadeemia: Keskraamatukogu and Geoloogia instituut: Tallinn 1960. pp.176. [1819.]

Finland

AARNE LAITAKARI, Geologische bibliographie Finnlands, 1555–1933. Suomen geologinen toimikunta: Bulletin (no.108): Helsinki 1934. pp.224. [4000.]

LUETTELO Geologisen tutkimuslaitoksen julkaisuista. Helsinki 1946. pp.14. [200.]

MARJATTA OKKO and MARJETTA HANNIKAINEN, Guide to publications of the Geological survey of Finland, 1819–1960. Geologinen tutkimuslaitos: Helsinki 1960. pp.104.

France

a. *General*

[ALPHONSE LE TOUZÉ] DE LONGUEMAR, Revue des

études géologiques ayant pour objet le département de la Vienne publiées de 1830 à 1867. Poitiers 1867. pp.[ii].56. [20.]

E[RNST] W[ILHELM] BENECKE and H[ARRY] ROSENBUSCH, Einleitende bemerkungen über die geologische landes-aufnahme von Elsass-Lothringen. Verzeichniss der mineralogischen und geologischen litteratur. Abhandlungen zur geologischen specialkarte von Elsass-Lothringen (vol.i, no.1): Strassburg 1875. pp.xxvi.77. [750.]

— — Nachtrag. Geologische u. mineralogische literatur über Elsass-Lothringen. Von E. Schumacher. 1887. pp.vi.73. [700.]

BIBLIOGRAPHIE géologique de la Normandie: Société géologique de Normandie. Havre [1876]. pp.39. [384.]

[CHARLES CONSTANT] W[ILFRED] KILIAN and O[CTAVE] NICAUD, Répertoire de la bibliographie géologique du sud-est de la France (Alpes françaises et Provence). Société scientifique de l'Isère: Bulletin (vol.xliv): Grenoble 1923. pp.[iii].xiv. 413. [6733.]

b. *Colonies*

F. BLONDEL, Bibliographie géologique et minière de la France d'outre-mer. Édition définitive. Bureau d'études géologiques et minières coloniales: Publications (no.11): 1941. pp.x.422+[iii].423–1038. [12,500.]

Germany

a. *Bibliographies*

K[ONRAD] KEILHACK and E. ZIMMERMANN, Verzeichniss von auf Deutschland bezüglichen geologischen schriften und karten-verzeichnissen. . . . Ergänzt und zum druck vorbereitet durch R. Michael. Königlich preussische geologische landesanstalt: Abhandlungen (n.s., no.26): Berlin 1897. pp.ix.108. [992.]

b. *General*

ALFRED JENTZSCH, Die geologische und mineralogische literatur des königreiches Sachsen und der angrenzenden ländertheile von 1835 bis 1873. Geologische landesuntersuchung des königreiches Sachsen: Leipzig 1874. pp.xviii.132. [2438.]

C. [R. L.] CHELIUS, Chronologische uebersicht der geologischen und mineralogischen literatur über das grossherzogthum Hessen. Geologische landesanstalt zu Darmstadt: Abhandlungen (vol.i, no.1): Darmstadt 1884. pp.xvi.59. [750.]

H. VON DECHEN and HERMANN RAUFF, Geologische und mineralogische litteratur der Rheinprovinz und der provinz Westfalen sowie einiger angrenzenden gegenden. Naturhistorischer verein der preussischen Rheinlande, Westfalens und des reg.-bezirks Osnabrück: Verhandlungen (vol. xliv): Bonn 1887. pp.181–476. [3500.]

— — Sachregister . . . von H. & M. Rauff . . . (vol.lii, part 2): 1896. pp.xi.274.

HEINRICH ECK, Verzeichniss der mineralogischen, geognostischen, urgeschichtlichen und balneographischen literatur von Baden, Württemberg, Hohenzollern und einigen angrenzenden gegenden. Grossherzoglich badische geologische landesanstalt: Mitteilungen (vol.i): Heidelberg 1890[–1891]. pp.viii.1288. [12,500.]

— — Nachträge und . . . fortsetzung. . . . (vol.i, ergänzung).

i. 1893. pp.[iv].303. [2000.]
ii. 1898. pp.[iv].452. [3000.]
iii. 1901. pp.[iv].141. [1000.]

— — Nachträge und zusätze zu band 1 und zur 1ten bis 3ten fortsetzung sowie repertorium . . . zu dem verzeichnis und sämmtlichen fortsetzungen. Bearbeitet von Ewald Schütze 1909. pp.viii.292. [200.]

— — [continuation]. Verzeichnis der mineralogischen, geologischen, urgeschichtlichen und hydrologischen literatur von Württemberg, Hohenzollern, Baden und den angrenzenden gebieten. I. bd. (1901 bis 1905). Verein für vaterländische naturkunde in Württemberg: Jahreshefte (beilage: vols.lviii, lix, lx, lxii, lxiv): Stuttgart 1902–1908. pp.[v].251. [2500.]

J. H. KLOOS, Repertorium der auf die geologie, mineralogie und paläontologie des herzogthums Braunschweig und der angrenzenden landestheile bezüglichen litteratur. Braunschweig 1892. pp. xii.204. [2750.]

KONRAD KEILHACK, Zusammenstellung der geologischen schriften und karten über den ostelbischen theil des königreiches Preussen, mit anschluss der provinzen Schlesien und Schleswig-Holstein. Königlich preussische geologische landesanstalt: Abhandlungen (n.s., no.14): Berlin 1893. pp.xii.135. [1862.]

VERÖFFENTLICHUNGEN der Königlich preussischen geologischen landesanstalt und bergakademie zu Berlin. [1900]. pp.16. [500.]

— 34. verzeichnis der veröffentlichungen der Preussischen geologischen landesanstalt. 1930. pp. [ii].41. [1200.]

Geology

EWALD SCHÜTZE, Verzeichnis der mineralogischen, geologischen, urgeschichtlichen und hydrologischen literatur von Württemberg, Hohenzollern, Baden und den angrenzenden gebieten. I. bd. (1901 bis 1905). Stuttgart 1902–1908. pp. [v].251. [2000.]

ERICH KAISER, Die geologisch-mineralogische literatur des rheinischen Schiefergebirges und der angrenzenden gebiete für die jahre 1887–1900. Bonn 1903–1904. pp.iv.131+iv.182. [2000.]

also issued as parts of vols.lix–lx of the Verhandlungen des Naturhistorischen vereins der preussischen Rheinlande.

GEOLOGISCHE literatur Deutschlands. A. Jährlicher literaturbericht für bodenforschung: Berlin 1906 &c.

DANIEL HÄBERLE, Die geologische literatur der Rheinpfalz vor 1820 und nach 1880 bis zum jahre 1907 einschliesslich. Pollichia: Mitteilungen (vol.lxiv. no.23): Bad Dürkheim 1907. pp.161. [1060.]

this forms a supplement to A. Leppla, 'Die mineralogische und geologische literatur der Pfalz seit 1820', Jahresbericht der Pollichia (*1884*), *xl–xlii.12–53.*

P[AUL] KESSLER and H. WILLING, Die geologische, mineralogische und palaeontologische literatur des Saarbeckens. Saarbrücken 1911. pp.viii.139. [1175.]

ERWIN SCHULZE, Repertorium der geologischen litteratur über das Harzgebirge. Deutsche geologische landesanstalt: Geologische litteratur Deutschlands: B. Literatur über einzelne gebiete: Berlin 1912. pp.viii.602. [5000.]

R[ICHARD] MICHAEL and W[ILHELM] QUITZOW, Die geologische literatur der provinz Schlesien und der nachbargebiete. Deutsche geologische landesanstalten: Geologische literatur Deutschlands: B. Literatur über einzelne gebiete: Berlin 1914. pp.iv.253. [4000.]

KURT PIETZSCH, Die geologische literatur über den freistaat Sachsen aus der zeit 1870–1920. Leipzig 1922. pp.xii.232. [3000.]

KURT PIETZSCH, Die geologische literatur über Sachsen, 1921–1950. Geologie: Beiheft (no.5–6): Berlin 1953. pp.ix.320. [5000.]

MARTIN SCHWARZBACH, Geologische forschung in Schlesien. Eine bibliographie der jahre 1914–1945, gleichzeitig ein rückblick auf die arbeit deutscher Geologen. Osteuropa und der deutsche Osten (2nd ser., no.2): Köln-Braunsfeld 1955. pp.63. [1750.]

VERZEICHNIS verkäuflicher veröffentlichungen . . . des Hessischen landesamtes für bodenforschung. Wiesbaden 1957. pp.56. [1500.]

Gold Coast

W[ILLIAM] T[HOMAS] JAMES, A bibliography of Gold Coast geology, mining and archaeology. Gold Coast geological survey: Bulletin (no.9): Colchester [printed] 1937. pp.57. [508.]

Great Britain

A[LPHONSE] FAVRE, Notice sur les cartes géologiques de l'Angleterre. Genève 1847. pp.23. [20.]

CATALOGUE of the published maps, sections, memoirs, and other publications of the Geological survey of the United Kingdom. 1867. pp.42. [1000.]

— [another edition]. List of memoirs, maps, sections, &c., published by the Geological survey of Great Britain and the Museum of practical geology. 1937. pp.vi.125. [2500.]

Geology

W. WHITAKER, List of works on the geology, &c., of Cornwall. Truro 1875. pp.[ii].61–110. [651.]

CATALOGUE of maps and plans and other publications of the Ordnance survey of England and Wales, and Isle of Man. 1881. pp.216. [1000.]

— [another edition]. Catalogue of maps and other publications of the Ordnance survey 1924. pp.(ff.)[ii].78. [5000.]

numerous supplements and extracts have been issued.

CATALOGUE of the publications of the Geological survey of the United Kingdom. Ireland. 1884. pp.24. [350.]

ANNALS of british geology. . . . A critical digest of the publications . . . during the year. . . . By J[ohn] F[rederick] Blake.

1890. 1891. pp.xii.352. [746.]
1891. 1892. pp.xii.404. [652.]
1892. 1893. pp.[x].xliv.310. [555.]
1893. 1895. pp.[x].xxiv.365. [730.]

no more published; not wholly limited to british geology.

J. G. HAMLING, An index to the geological papers contained in the Reports and Transactions of the Devonshire association for the advancement of science, literature, and art . . . 1862, to . . . 1902. 1903. pp.20. [400.]

T[HOMAS] SHEPPARD, Bibliography of Yorkshire geology. Yorkshire geological society: Proceedings (vol.xviii): 1915. pp.xxxvi.630. [8000.]

A[RTHUR] MORLEY DAVIES, Local geology. A guide to sources of information on the geology of the British Isles. 1923. pp.16. [100.]

— — Second edition. 1927. pp.16. [100.]

Hawaii

NORAH D[OWELL] STEARNS, Annotated bibliography and index of geology and water supply of the island of Oahu, Hawaii. Division of hydrography: Bulletin (no.3): Hawaii 1935. pp.[iv].74. [436.]

GORDON ANDREW MACDONALD, Bibliography of the geology and water resources of the island of Hawaii. Division of hydrography: Bulletin (no. 10): Honolulu 1947. pp.191. [1000.]

Hungary

GYULA HALAVÁTS, A magyar pontusi emelet átalános és őslénytani irodalma. A Magyar kir.

földtani intézet kiadványai: Budapest 1904. pp. 134. [1250.]

VERZEICHNIS der deutschen publikationen der Königl. ungarischen geologischen reichsanstalt. Budapest 1918. pp.8. [100.]

India

R[ICHARD] D[IXON] OLDHAM, A bibliography of indian geology, being a list of books and papers relating to the geology of british India and adjoining countries. Calcutta 1888. pp.xiii.145. [2500.]

T[HOMAS] H[ENRY] D[IGGES] LA TOUCHE, A bibliography of indian geology and physical geography, with an annotated index of minerals of economic value. Geological survey of India: Calcutta 1917–1926. pp.[iii].xxviii.572+[iii].ii.490+[ii].xiii.143+[iii].348+[v].414. [5500.]

LIST of publications. Geological survey of India: Calcutta 1947. pp.40. [600.]

CATALOGUE of publications of the Geological survey of India and index of geological maps. Geological survey of India: Memoirs (vol.lxxvii): Calcutta &c. 1947 [*on cover:* 1948]. pp.[iii].129. [2000.]

Geology

Indochina

PUBLICATIONS du service géologique de l'Indochine. Catalogue et prix-courant. Hanoï 1931. pp.16. [100.]

— [another printed]. 1937. pp.27. [200.]

Ireland

R[OBERT] LLOYD PRAEGER, A bibliography of irish glacial and post-glacial geology. Belfast naturalists' field club: Systematic lists (vol.ii, no.6): [Belfast 1896]. pp.237–316. [767.]

Italy

A[NTONIO] D'ACHIARDI, Bibliografia mineralogica, geologica e paleontologica della Toscana. Roma 1875. pp.[ii].57. [1000.]

P. ZEZI, Indice bibliografico delle pubblicazioni italiane e straniere riguardanti la mineralogia, la geologia e la paleontologia della provincia di Roma ... con una appendice per le acque potabili, termali e minerali. Roma 1878. pp.19. [350.]

[ALESSANDRO PORTIS], Bibliographie géologique et paléontologique de l'Italie. Par les soins du Comité d'organisation du 2me Congrès géologique international. Bologne 1881. pp.[iv].viii.630. [6566.]

[A.] SILVESTRI, Bibliografia generale riguardante la vulcanologia, mineralogia, geologia, paleontologia e paleoetnologia della provincia di Catania e delle isole vulcaniche adiacenti alla Sicilia. Bologna 1881. pp.vii.64. [950.]

R[AFFAELE] V[ITTORIO] MATTEUCCI, Bibliografia scientifica delle provincie di Ancona, Pesaro-Urbino e limitrofe. Geologia e scienze affini. Napoli 1894. pp.52. [500.]

MICHELE GORTANI, Bibliografia geologica italiana (dalle origini al 1930). Parte prima. Opere di carattere generale. Bologna 1931. pp.100. [1570.]

GIAN FRANCESCO GAMURRINI, Bibliografia dell'Italia antica. . . . Preistoria, paleoetnologia italiana, il paese (geologia). Roma 1933. pp.iii–lxxv. 471. [10,000.]

BIBLIOGRAFIA geologica italiana. . . . Compilata a cura del R. ufficio geologico. Roma.

1915–1933. 1934–1938. pp.268. [750.]

1934. 1936. pp.175. [500.]

no more published; the 1915–1933 extends only to Castiglione.

BIBLIOGRAFIA geologica d'Italia. Consiglio nazionale delle ricerche: Comitato per la geografia, geologia e mineralogia: Napoli.

i. Lazio. A cura di Carmelo Maxia. 1956. pp.xii.174. [2500.]

ii. Lombardia. A cura di Ardito Desio. 1957. pp.[ii].vii.300. [4500.]

iii. Campania. A cura di G[eremia] d'Erasmo e M. L. Benassai Sgadari. 1958. pp.[iii].xix.555. [6500.]

iv. Calabria. A cura di Felice Ippolito. 1959. pp.[iii].viii.121. [1500.]

v. Puglia. A cura di G. d'Erasmo. 1959. pp. xxvi.252. [3000.]

vi. Friuli. A cura di Michele Gortani. 1960. pp.xxi.228. [2500.]

vii. Umbria. A cura di Cesare Lippi Boncambi. 1960. pp.xv.102. [1000.]

Japan

LIST of publications of the Imperial geological survey of Japan (1880–1932). Tokyo 1932. pp.[ii]. 2.2.58. [750.]

— (1880–1952). 1952. pp.[ii].124. [1000.]

Mexico

RAFAEL AGUILAR Y SANTILLÁN, Bibliografía geológica y minera de la republica Mexicana. Instituto geológico de México: Boletín (no.10): México 1898. pp.xi.159. [1953.]

— — [second edition]. . . . (no.17): 1908. pp. xiii.331. [4252.]

— — — [supplements].

1905–1918. Departamento de minas: Boletín minero: 1918. pp.99. [1662.]

1919–1930. Secretaría de la economía nacional: 1936. pp.iv.83. [1700.]

LISTA de publicaciones del Instituto de geología. Universidad nacional de México: México 1931. pp.[ii].xviii. [250.]

Netherlands

H[AGEN] G[ARRELD] JONKER, Lijst van geschriften welke handelen over of van belang zijn voor de geologie van Nederland (1734–1906). Commissie voor het geologisch onderzoek: Mededeelingen omtrent de geologie van Nederland (no. 36 = Koninklijke akademie van wetenschappen te Amsterdam: Verhandelingen, 2nd sect.xiii.2): Amsterdam 1907. pp.vii.154. [1413.]

— — Vervolg. Mededeelingen van's Rijks geologischen dienst (ser. A, no.1 &c.): Leiden ['s Gravenhage] 1922 &c.

1907–1920. Samengesteld door [Jakob] F[rederik] Steenhuis . . . (ser. A, no.1): 1922. pp. viii.134. [1373.]

Geology

1552–1920 . . . (ser. A, no.4): 1934. pp.109. [1000.]

1921–1925 . . . (ser. A, no.6): 1939. pp.65. [694.]

1926–1930 . . . (ser. A, no.7): 1940. pp.65. [703.]

1931–1935 . . . (ser. A, no.8): 1941. pp.66. [660.]

1936–1944, alsmede supplement 1864–1944. . . . (new ser., no.2): 1947. pp.99. [868.]

Newfoundland

RACHEL M[ARY] BETTS, Bibliography of the geology of Newfoundland, 1818–1936. Department of natural resources: Geological section: Bulletin (no.5: Princeton university contribution Jo the geology of Newfoundland, no.14): St. tohn's 1936. pp.v.35. [350.]*

New Zealand

G. LESLIE ADKIN, Bibliographic index of New Zealand stratigraphic names. Geological survey: Memoir (vol.ix = Lexicon international de stratigraphie: Océanie, no.i): [Wellington] 1954. pp.121. [4500.]*

Geology

Norway

[HANS H.] REUSCH, Geologisk literatur vedkommende Norge 1890–1895. Norges geologiske undersøgelse (no.21 = Aarbog for 1894 og 95): Kristiania 1896. pp.vi.148. [500.]

ANDERS K[RISTIAN] ORVIN, Bibliography of literature about the geology, physical geography, useful minerals, and mining of Svalbard. Kongelige industri-, handverk- og skipsfarts-departement: Norges Svalbard- og Ishavsundersøkelser skrifter (no.89): Oslo 1947. pp.[v].121. [1750.]

Pacific islands

CATALOGUE of translations of japanese geological literature of the Pacific islands. United States army forces, far east: Geological surveys branch: [*s.l.*] 1954. pp.[v].34. [350.]*

— [another edition]. Catalogue of translations of japanese geologic, soils and allied literature [&c.]. [Compiled by Roy C. Kepferle]. 1959. ff.iii.96. [1000.]*

ANNOTATED bibliography of geologic and soils literature of western north Pacific islands. United States army forces far east: Office of the engineer: [*s.l.*] 1956. pp.[iii].884. [7500.]*

Geology

Peru

L[EONIDAS] CASTRO B[ASTOS], Bibliografía geológica del Perú. Lima 1960. pp.[ii].317. [2868.]*

Philippines

JUAN S. TEVES, Bibliography of philippine geology, mining and mineral resources. Bureau of mines: Bibliography series (no.1): Manila 1953. pp.v.155. [2846.]

Poland

BIBLJOGRAFJA geologiczna Polski. [Warsaw].

i. 1914–1920. Zebrała: R[egina] Danysz-Fleszarowa. pp.24. [418.]
ii. 1921–1923. pp.24. [315.]
iii. 1924. pp.11. [145.]
iv. 1925. pp.11. [129.]
v. 1926. pp.12. [149.]
vi. 1927. pp.15. [203.]
vii. 1928. pp.20. [205.]
viii. 1929. pp.16. [216.]
ix. 1930. pp.17. [211.]
x. 1931. pp.16. [210.]
xi. 1932. pp.16. [200.]
xii. 1933. pp.19. [223.]
xiii. Uzupełnienia i skorowidz nazwisk autorów do nr.1–13, lata 1914–1933. pp.24. [29.]

xiv. 1934. pp.30. [216.]
xv. 1935. pp.28. [180.]
xvi. 1936. pp.59. [225.]
xvii. 1937. pp.51. [187.]
xviii. 1938–1939. Opracowała: R. Danysz-Fleszarowa. pp.25. [127.]
xix. 1940–1944. Pod redakcja E. Rühlego. pp.10. [127.]
xx. 1945–1947. Opracowała: C. Wardęska. pp.24.vii. [183.]
xxi. 1948.
xxii. 1949. pp.27. [228.]
xxiii. 1950.
xxiv. 1951. pp.16. [121.]
xxv. 1952. pp.32. [345.]
xxvi. 1953. pp.24. [234.]
xxvii. 1954. [By] C. Z. Bukowski, M. W. Bukowska, pp.119. [1250.]
xxviii. 1955. pp.164. [1500.]
xxix. 1956. pp.239. [2000.]
xxx. 1957. pp.267. [2500.]
xxxi. 1958. 1962. pp.255. [2000.]

in progress

RETROSPEKTYWNA bibliografia geologiczna Polski oraz prac polaków z zakresu nauk o ziemi. Instytut geologiczny: Muzeum ziemi: Warszawa.

i. 1900–1950. Zebrała . . . Regina [Danysz-]

Fleszarowa. 1957–1960. pp.xciv.478+624 +xix.778. [9734.]
in progress.

KATALOG wydawnictw Instytutu geologicznego. Warszawa 1956. pp.75. [750.]

Portugal

CATÁLOGO por ordem cronológica das publicações dos serviços geológicos de Portugal..., 1865–1948. Lisboa 1949. pp.36. [200.]

LUIZ DE MENEZES [CORRÊA] ACCIAIUOLI, Geologia de Portugal. Ensaio bibliográfico. Direcção-geral de minas e serviços geológicos: Lisboa [1957–1958]. pp.iii–xlviii.365+[vii].279[*sic*]–675. [6366.]

Romania

DAVID ROMAN and ALEXANDRU CODARCEA, Bibliografia geologică a României. Institutul geologic al României: Bucureşti 1926. pp.vii.155. [3000.]

— — 1-iul supliment. De Erich Jekelius. 1929. pp.iii–viii.103. [2000.]

Russia

ERNST VON BERG, Repertorium der literatur über die mineralogie, geologie, palaeontologie, berg- und hüttenkunde Russlands bis zum schlusse des XVIII. jahrhunderts. St. Petersburg 1862. pp. xx.228. [1000.]

РУССКАЯ геологическая библіотека. Геологическій комитетъ: С.-Петербургъ.

i. 1885. Издаваемая подъ редакціею С. Никитина [Sergyei Nikolaevich Nikitin]. pp.vii.126. [256.]
[ii]. 1886. pp.viii.200. [429.]
iii. 1887. pp.vi.178. [405.]
[iv]. 1888. pp.vi.201. [390.]
[v]. 1889. pp.vii.187. [410.]
[vi]. 1890. pp.vii.225. [467.]
vii. 1891. pp.vii.234. [452.]
[viii]. 1892. pp.vi.245. [474.]
[ix]. 1893. pp.[v].64.184. [498.]
x. 1894. pp.viii.202. [495.]
[xi]. 1895. pp.[iv].223. [517.]
[xii]. 1896. pp.[iv].244. [577.]
[xiii]. 1897.
[xiv]. 1898. Издаваемая Геологическимъ комитетомъ. 1914. pp.[vi].332. [920.]
[xv].
[xvi]. 1900. 1920. pp.[iv].252. [1026.]

К. К. FILIPOVICH, Указатель главной литературы по геологии и каменноугольной промышленности Подмосковного бассейна. [Moscow] 1920. pp.35. [372.]

B[ORIS] K[ONSTANTINOVICH] LIKHAREV, Обзор литературы по верхнепермским отложениям европейской России за истекшее десятилетие (1910–1919 гг.). Геологический комитет: Петербург 1920. pp.[ii].48. [30.]

КАТАЛОГ изданий Геологического комитита. Ленинград 1926. pp.27. [500.]
— [another edition]. 1931. pp.28. [600.]

УКАЗАТЕЛЬ изданий Центрального научно-исследовательского геолого - разведочого института. Ленинград &c. 1936. pp.76. [2000.]
— [another edition]. No.1. 1938. pp.20. [500.]
no more published.

ГЕОЛОГИЧЕСКАЯ литература СССР. Центральная [Всесоюзная геологическая библиотека: Ленинград. [Москва].
1934. 1937. pp.200. [3252.]
1937. 1939. pp.3–228. [4000.]
1951. 1956. pp.148. [2500.]
1952–1953. 1955. pp.248. [5352.]
1954. 1957. pp.187. [3000.]
1955. 1959. pp.335. [6000.]
1956. pp.282. [6318.]

1957. pp.[238]. [5311.]
1958. pp.[209]. [4179.]

in progress; the issues for 1935, 1936, 1938–1950 do not appear to have been published.

ГЕОЛОГИЯ в изданиях Академии наук. Академия наук СССР: Библиотека: Москва &c.

i. 1728–1928. Составили О. К. Смирнова, Э. П. Файдель, К. И. Шафрановский. 1938. pp.xv.471. [3607.]

in progress?

КАТАЛОГ книг, геологических карт и учебных плакатов. Госгеолтехиздат: [Moscow] 1956. pp.18. [100.]

Scotland

DESCRIPTIVE catalogue of the maps, sections, and memoirs (published and in preparation) of the Geological survey of Scotland. Edinburgh 1877. pp.31. [200.]

— [another edition]. Catalogue of the publications of the Geological survey of the United Kingdom. Scotland. 1884. pp.14. [150.]

Slovakia

JAN KOLIHA, Bibliografie Slovenska. [Geologie;

mineralogie, petrografie, hornictví, hutnictví, geomorfologie a podl]. Knihovna České společnosti zeměvědné v Praze (no.9): Praze 1919. pp.65. [2120.]

Spain

CATÁLOGO de las publicaciones del Instituto geológico y minero de España. Madrid 1929. pp.251. [1250.]

— [another edition]. Las publicaciones [&c.]. 1956. pp.326. [1500.]

BIBLIOGRAPHIE géologique de l'Espagne. Géologie de la Méditerranée occidentale (vol.i): Liége &c. 1930. pp.18.15. [250.]

Sweden

LISTE systématique des publications de l'Institut royal géologique de Suède, 1862–1890. Stockholm 1890. pp.15. [300.]

— [another edition]. Förteckning över Sveriges geologiska undersöknings publikationer. 1937. pp.54. [700.]

Switzerland

BIBLIOGRAPHIE géologique de la Suisse. Société helvétique des sciences naturelles: Matériaux pour la carte géologique de la Suisse (vol.xxix &c.): Berne.

1770–1900. Par Louis Rollier . . . (vol.xxxi): 1907–1909. pp.xlviii.541+[iii].541–1025. [17,500.]

1900–1910. Par Em. Gogarten et Walt[er] Hauswirth . . . (n.s. vol.xl): 1913. pp.x.412. [6000.]

1910–1920. Par Alphonse Jeannet . . . (n.s. vol.lvi): 1927. pp.xxxviii.462+[ii].463–855. [5000.]

1921–1930 . . . (n.s. vol.lxxiii): 1936. pp.xii. 496. [4135.]

in progress.

VERÖFFENTLICHUNGEN der Geologischen kommission der Schweizerischen naturforschenden gesellschaft. [Berne] 1928. pp.20. [300.]

— [another edition]. 1946. pp.36. [500.]

Tibet

T. C. TSENG, Bibliography of chinese geology. Bibliography of geology and allied sciences of Tibet and regions to the west of the Chinshachiang. National geological survey of China: Nanking 1946. pp.viii.116. [1000.]

Tunisia

G[ILBERT] CASTANY, Bibliographie géologique

de la Tunisie. Direction des travaux publics: Service des mines, de l'industrie et de l'énergie. Service géologique: Tunis.

i. [1950–1952]. 1954. pp.15. [110.]

no more published?

CATALOGUE des publications. Direction des travaux publics: Service géologique: Tunis [printed] 1954. pp.12. [40.]

Ukraine

OLEKSANDER BIRULYA, Геологія і гідрогеологія Подилля. Бібліографичний покажчик. Бібліографія Поділля (vol.i): [Kiev] 1930. pp.82. [419.]

United States

a. *General*

CATALOGUE of the publications of the United States geological survey of the territories. Washington 1874. pp.20. [50.]

— [another edition]. Publications of the Geological survey. 1953. pp.vi.354. [6000.]

there are numerous intermediate editions, with varying titles.

FREDERICK PRIME, A catalogue of the official reports upon geological surveys of the United States and territories, and of british north America.

Philadelphia 1879. pp.71. [750.]

— — Supplement I. [1880]. pp.13. [125.]

— — Supplement II. [1881]. pp.12. [125.]

PHILIP CREVELING WARMAN, Bibliography and index of the publications of the United States geological survey. Geological survey: Bulletin (no.100): Washington 1893. pp.495. [20,000.]

— — Catalogue and index . . . 1880 to 1901 . . . (no.177): 1901. pp.858. [75,000.]

— — 1901 to 1903 . . . (no.245): 1903. pp.234. [20,000.]

L[AURENCE] F[REDERICK] SCHMECKEBIER, Catalogue and index of the publications of the Hayden, King, Powell, and Wheeler surveys. Geological survey: Bulletin (no.222): Washington 1904. pp.208. [1000.]

TOPOGRAPHIC maps and folios and geological folios published by the United States Geological survey. Washington 1909. pp.102. [2000.]

— [another edition]. 1914. pp.131. [2500.]

RUSSEL C[LARKE] FLEMING, Source book. A directory of public agencies in the United States engaged in the publication of literature on mining and geology. American institute of mining and metallurgical engineers: New York 1933. pp.128. [1250.]

JAMES T[ATE] RUBY, A list of references on the United States geological survey and its works [Geological survey: Library: Bibliographical lit. (no.1): Washington] 1934. ff.[ii].21. [200.]

WILLARD ROUSE JILLSON, The geological map of the United States. A narrative outline and annotated bibliography (1752–1946). Frankfort, Ky. 1950. pp.23. [50.]

W. B. BEATTY and LEE LANGAN, Sources of information on geology and mining in the western states. Stanford university: School of mineral sciences: [Stanford 1957]. pp.[v].52. [300.]*

— — [another edition]. Mineral resource data in the western states. [1962]. pp.42. [300.]*

GEORGE H[OWARD] ALBERTSON, Geologic index to the publications of United States geological survey and Hayden, King. Powell, and Wheeler surveys. Denver [1963]. pp.vi.112. [20,000.]

b. *Individual states*

Alabama

LIST of publications of the Geological survey of

Geology

Alabama. [University 1917]. pp.[4]. [55.]

— [another edition]. Complete list [&c.]. [1943]. pp.11. [150.]

ELLEN HILDRETH and PAULINE MCCANDLESS FOSTER, Know Alabama! The geology and mineral resources of Alabama. A bibliography for schools. Geological survey of Alabama: Bulletin (no.51): University 1942. pp.56. [150.]

EARL L. HASTINGS, Bibliography of Alabama geology, 1935–1958. Geological survey of Alabama: Bulletin (no.67): University 1960. pp.138. [1235.]

Arizona

ESTELLE LUTRELL, Bibliography of Arizona mining, metallurgy and geology. University of Arizona: Bulletin (no.23): Tucson 1915–1916. pp.49. [900.]

OLAF P[ITT] JENKINS and ELDRED D[EWEY] WILSON, List of United States geological survey publications relating to Arizona. University of Arizona: Bulletin (no.104 = Geological series, no.1): Tucson [1920]. pp.[v].40. [150.]

ELDRED D[EWEY] WILSON, Bibliography of the geology and mineral resources of Arizona. Arizona bureau of mines: Bulletin (no.146 =

Geology

University of Arizona, Bulletin, vol.x, no.2): Tucson 1939. pp.164. [2000.]

— — 1939–1952. By John W. Anthony, E. D. Wilson and Richard T. Moore. . . . (no.161 = . . . vol.xxiv, no.1): 1953. pp.62. [1000.]

Arkansas

PUBLICATIONS. Geological survey of Arkansas: Little Rock [1942]. pp.14. [125.]

California

ANTHONY W[AYNE] VOGDES], A bibliography relating to the geology, palæontology, and mineral resources of California. California state mining bureau: Bulletin (no.10): San Francisco 1896. pp.viii.121. [1000.]

[—] — [second edition]. . . . (no.30): 1904. pp.290. [2000.]

SOLON SHEDD, Bibliography of the geology and mineral resources of California to December 31, 1930. Department of natural resources: Division of mines: Bulletin (no.104): Sacramento 1933. pp.xii.376. [7500.]

— — 1931–1936. . . . (no.115): 1937. pp.ix.125. xiv. [2500.]

Geology

[PHILIP J. LORENS and RAYMOND C. RICHTER]. Revised bibliography for geologic map index of unpublished mapping and reports in California, January 1937 to January 1954. Department of public works: Division of water resources: [Sacramento] 1954. ff.ii.51. [500.]*

Colorado

J[ESSE] HARLAN JOHNSON, Bibliography of the geology of north central Colorado. Colorado school of mines: Quarterly (xxii, no.4): Golden 1927. pp.38. [565.]

J[ESSE] HARLAN JOHNSON, Bibliography of the geology of the Denver basin of Colorado. Colorado school of mines: Circular of information: [Golden 1925]. pp.[7]. [173.]

J[ESSE] HARLAN JOHNSON, Bibliography of the geology of south central Colorado. Colorado school of mines: Quarterly (vol.xxiii, no.4): Golden 1928. pp.3–30. [268.]

J[ESSE] HARLAN JOHNSON, Bibliography of the geology of northeastern Colorado. Colorado school of mines: Golden 1925. pp.[12]. [305.]

J[ESSE] HARLAN JOHNSON, Bibliography of the

geology and related subjects of northwestern Colorado. Colorado school of mines: Quarterly (vol.xxi, no.3): Golden 1926. pp.52. [654.]

J[ESSE] HARLAN JOHNSON, Bibliography of the geology of southeastern Colorado. Colorado school of mines: Circular of information: Golden 1925. pp.[11]. [258.]

J[ESSE] HARLAN JOHNSON, Bibliography of the geology of southwestern Colorado. Colorado school of mines: Circular of information: Golden 1924. pp.[7]. [148.]

J[ESSE] HARLAN JOHNSON, Bibliography of Colorado maps. Colorado school of mines: Quarterly (vol.xx, no.4): Golden 1925. pp.40. [567.]
limited to geological maps.

HARRY M. CRAIN, Publications on work done by U.S. G[eological] S[urvey] in cooperation with state of Colorado. Colorado school of mines: Quarterly (vol.xl, no.4): Golden 1945. pp.28. [188.]

OLIVE M[ARY] JONES, Bibliography of Colorado geology and mining. State geological survey: Bulletin (no.7): Denver 1914. pp.493. [3500.]

Geology

Connecticut

HERBERT ERNEST GREGORY, Bibliography of the geology of Connecticut. State geological and natural history survey: Bulletin (no.8): [Hartford] 1907. pp.124. [600.]

Florida

PUBLICATIONS issued up to October 1, 1931. Florida state geological survey: Tallahassee [1931]. pp.4. [41.]

Georgia

BULLETINS of the Geological survey of Georgia. [Atlanta 1927]. pp.[4]. [42.]

Illinois

LIST of publications. State geological survey: [Urbana 1923]. pp.16. [100.]

— [another edition]. Publications on the geology, mineral resources, and mineral industries of Illinois. 1940. pp.116. [2000.]

Indiana

PUBLICATIONS of the Indiana geological survey. Indianapolis 1931. pp.4. [100.]

— [another edition]. [1939]. pp.[6]. [150.]*

GERALD S. WOODARD, List of geologic publications and maps of Indiana. Geological survey: Bloomington 1954. pp.66. [1000.]

Iowa

CHARLES [ROLLIN] KEYES, Annotated bibliography of Iowa geology and mining. Iowa geological survey (vol.xxii): Des Moines 1913. pp. 908. [4000.]

Kentucky

LIST of maps and publications issued by the Department of geology and forestry of Kentucky. Frankfort 1919. pp.10. [150.]

[W. R. JILLSON], A bibliography of the several books, reports, papers and maps principally relating to geology written and prepared by Willard Rouse Jillson. Kentucky 1920–1926.

details of this work are entered under Jillson, below.

WILLARD ROUSE JILLSON, The geologic map of Kentucky. Louisville, Ky. 1945. pp.3–12. [6.]

WILLARD ROUSE JILLSON, A bibliography of early printed books, pamphlets, articles and maps pertaining to the geology, paleontology and seismology of Kentucky, 1744–1854. Frankfort, Ky. 1950. pp.53. [150.]

WILLARD ROUSE JILLSON, A bibliography on the

geology of oil and gas in Kentucky . . . 1829–1956. Frankfort 1956. pp.35. [277.]

WILLARD ROUSE JILLSON, A bibliography of county geological reports in Kentucky during the past century, 1856–1956. Frankfort 1957. pp.18. [74.]

Louisiana

COMPLETE list of publications of the Louisiana geological surveys and partial list of publications of the State department of conservation. [New Orleans] 1938. pp.13. [125.]

— [another edition]. Publications. 1958. pp. v.23. [250.]*

Maine

CYRUS C[ATES] BABB, Bibliography of Maine geology. Waterville 1913. pp.68. [751.]

[JOSEPH C. TWINEM, WILLIAM T. FORSYTH and DABNEY W. CALDWELL], Bibliography on Maine geology 1836–1957. Department of economic development: Augusta 1958. pp.143. [1500.]*

Michigan

LIST of publications. [Geological survey: Lansing 1903]. pp.4. [50.]

— [another edition]. Catalog and table of contents of the publications of the Geological and biological survey division. [1922]. pp.36. [300.]

[HELEN MANDEVILLE MARTIN and MURIEL TARA STRAIGHT], An index of the geology of Michigan 1823–1955. Geological survey division: Publication (no.50): Lansing 1956. pp.ix.461. [5000.]

— — [continuation]. Edward A. Kirkby, Bibliography of Michigan geology, 1956–60. 1961. pp.ii.66. [650.]*

EDWARD A. KIRKBY, Geological map index of Michigan 1843–1962. Geological survey division: [Lansing] 1963. pp.vi.57. [500.]*

Minnesota

WINIFRED GREGORY, Bibliography of Minnesota mining and geology. University of Minnesota: Minnesota school of mines: Experiment station: Bulletin (no.4): Minneapolis 1915. pp.157. [991.]

— — Supplement . . . (no.8): 1920. pp.43. [233.]

MARGARET BRIGGS, Mesabi iron range. University of Minnesota: Division of library instruction: Bibliographical projects (no.4): [Minneapolis 1937]. ff.[iv].37. [500.]*

THEODORA G. MELONE and LEONARD W. WEIS, Bibliography of Minnesota geology. Minnesota geological survey: Bulletin (no.34): Minneapolis 1951. pp.xii.124. [2500.]

Mississippi

PUBLICATIONS of the Mississippi geological survey. [University 1922]. pp.[3]. [35.]
— [another edition]. [1926]. pp.[3]. [40.]

Missouri

DARLING K. GREGER, Bibliography of the geology of Missouri. Missouri geological survey and water resources: [Reports] (2nd ser., vol.xxxi): Rolla 1945. pp.294. [2750.]

F[RANCIS] A[SBURY] SAMPSON, A bibliography of the geology of Missouri. Geological survey of Missouri: Jefferson City 1890. pp.158.xix. [810.]

Montana

BIBLIOGRAPHY of the geology and mineral resources of Montana. Compiled by Works projects administration, Mineral resources survey. State of Montana: Bureau of mines and geology: Memoir (no.21): Butte 1942. pp.v.356. [1980.]*

J. M. CHELINI, Index of unpublished geologic studies in Montana. Preliminary compilation.

Geology

Bureau of mines and geology: Special publication (no.29): Butte 1963. pp.81.

Nevada

VINCENT P[AUL] GIANELLA, Bibliography of geologic literature of Nevada and bibliography of geological maps of Nevada areas, by Robert W. Prince. University of Nevada: Bulletin (vol. xxxix, no.64 = Geology and mining series, no.43): [Reno] 1945. pp.205. [4000.]

New Jersey

GEORGE F[RASER] BLACK, List of works relating to the geology, mineralogy, and palaeontology of New Jersey. Public library: New York 1916. pp.36. [450.]

New York

PLEISTOCENE geology of New York state. Published writings of Herman L[e Roy] Fairchild. [Rochester 1914]. pp.4. [75.]

North Carolina

FRANCIS BAKER LANEY and KATHARINE HILL WOOD, Bibliography of North Carolina geology, mineralogy and geography, with a list of maps. North Carolina geologic and economic survey:

Bulletin (no.18): Raleigh 1909. pp.[iii].428. [784.]

North Dakota

CHRISTIE E. BUDGE, Bibliography of the geology and natural resources of North Dakota, 1814–1944. North Dakota research foundation: Bulletin (no.1): Bismarck 1946. pp.214. [995.]

— — [another edition]. . . . (nos.1, 4, 5): 1959. pp.214.90.61.

Ohio

ALICE GREENWOOD DERBY and MARY WILSON PROSSER, A bibliography of Ohio geology. Geological survey of Ohio: Columbus 1906. pp.332. [6000.]

REPORTS and maps on the geology and mineral resources of Ohio. Geological survey of Ohio: Columbus 1947. pp.16. [100.]

LIST of publications and maps on the geology and mineral resources of Ohio. Division of geological survey: Columbus 1950. pp.16. [400.]

DOROTHY G. WATKINS, Bibliography of Ohio geology, 1819–1950. Department of natural resources: Division of geological survey: Bulletin (no.52): Columbus 1953. pp.xii.103. [2000.]

Geology

Oklahoma

L[AURENCE] E[MORY] TROUT and GEO[RGE] H. MYERS, Bibliography of Oklahoma geology. Oklahoma geological survey: Bulletin (no.25): Norman 1915. pp.105. [639.]

LIST of available publications. Oklahoma geological survey: Norman 1955. pp.[8]. [150.]

Oregon

CHA[RLE]S W. HENDERSON, J. B. WINSTANLEY and GRAHAM J. MICHAEL, Bibliography of the geology, paleontology, mineralogy, petrology, and mineral resources of Oregon. University of Oregon: Bulletin (n.s. x, no.4): Eugene 1912. pp.3–49. [350.]

—— [second edition]. Dorothy Elizabeth Dixon, Bibliography of the geology of Oregon. University of Oregon: Publication: Geology series (vol.i, no.1): 1926. pp.125. [1001.]

——— Supplement . . . by John Eliot Allen. State of Oregon: Department of geology and mineral industries: Bulletin (no.33): [Portland] 1947. pp.[iii].iii.110. [1500.]*

——— Second supplement . . . by Margaret L. Steere. . . . (no.44): 1953. pp.61. [1000.]*

WILLARD ROUSE JILLSON, A bibliography of the

geology and palaeontology of the John Day region, Oregon. Frankfort, Ky. 1923. pp.13. [102.]

RAY C[LARENCE] TREASHER and EDWIN T[HOMAS] HODGE, Bibliography of the geology and mineral resources of Oregon. Oregon state planning board: Portland 1936. pp.[vi].224. [2155.]*

Pennsylvania

WILLIAM A[RMSTRONG] INGHAM, A list of and brief guide to the publications of the Pennsylvania geological survey, 1874–1891 [*sic*, 1895]. [Harrisburg] 1896. pp.xxx. [100.]

PUBLICATIONS of the United States geological survey relating to Pennsylvania. Topographic and geologic survey of Pennsylvania: Harrisburg 1914. pp.10. [200.]

Rhode Island

REPORT on the geology of Rhode Island. Providence Franklin society: Providence 1887. pp. [iv].130. [300.]

ALONZO [WALLACE] QUINN and DAVID H[ENRY] SWANN, Bibliography of the geology of Rhode Island. Rhode Island industrial commission:

Mineral resources committee: [Providence 1944]. pp.23. [200.]

— — Second edition. 1950. pp.26. [225.]

South Dakota

CLEOPHAS C[ISNEY] O'HARRA, A bibliography of the geology and mining interests of the Black hills region. South Dakota school of mines: Bulletin (no.11): Rapid City 1917. pp.216.7. [1187.]

Tennessee

ELIZABETH COCKRILL, Bibliography of Tennessee geology, soils, drainage, forestry, &c. State geological survey: Nashville 1911. pp.119. [1000.]

LIST of publications. Division of geology: Nashville 1926. pp.3–11. [100.]

— [another edition]. 1939. pp.[ii].24. [150.]

CHARLES WILLIAM WILSON, Annotated bibliography of the geology of Tennessee through December 1950. Department of conservation: Division of geology: Bulletin (no.59): Nashville 1953. pp.v.308. [1850.]

Texas

ROSELLE M. GIRARD, Bibliography and index of Texas geology, 1933–1950. University of Texas:

Publication (no.5910): Austin 1959. pp.[iii].238. [3000.]

Utah

BRONSON F[ERRIN] STRINGHAM, Bibliography of the geology and mineral resources of Utah, to December 31, 1942. University of Utah: Bulletin (vol.xxxiv, no.15 = Geological series, vol.i, no.1): Salt Lake City 1944. pp.99. [1100.]

Virginia

THOMAS L. WATSON, A bibliography of the geological, mineralogical and paleontological literature of the state of Virginia. Bulletins of american paleontology (vol.ii, no.7): Ithaca N.Y. 1897. pp.111. [1000.]

JOSEPH K[ENT] ROBERTS and ROBERT O[LIVER] BLOOMER, Catalogue of topographic and geologic maps of Virginia. Richmond, Va. 1939. pp.[ii]. xix.246. [970.]

JOSEPH K[ENT] ROBERTS, Annotated geological bibliography of the Virginias. University of Virginia: Bibliographical series (no.2): Charlottesville 1942. pp.[iv].xi.726. [2563.]

ARTHUR [CHARLES] BEVAN, Publications on the geology and mineral resources of Virginia. Virginia geological survey: Circular (no.2): University 1942. pp.vii.57. [300.]

ARTHUR [CHARLES] BEVAN, Publications on the geology and mineral resources of Virginia. Virginia geological survey: Circular (no.2): University 1942. pp.viii.57. [235.]

LIST of publications and maps. Virginia division of mineral resources: Charlottesville 1959. pp.27. [200.]

Washington

GRETCHEN O'DONNELL, Bibliography of Washington geology and geography. Washington geological survey: Bulletin (no.12): Olympia 1913. pp.65. [420.]

PUBLICATIONS of the Washington geological survey, 1901–1922. Department of conservation and development: [Olympia 1922]. pp.[5]. [60.]

W[ILLIAM] A[LFRED] G[LENN] BENNETT, Bibliography and index of geology and mineral resources of Washington, 1814–1936. Division of geology: Bulletin (no.35): Olympia 1939. pp.140. [1250.]

[—] — 1937–1956. By William H[enry] Reichert. Division of mines and geology: Bulletin (no.46): 1960. pp.v.721. [3500.]*

West Virginia

LIST of publications of the West Virginia geological survey. Morgantown 1932. pp.11. [75.]
— [another edition]. 1951. pp.8. [100.]

JOHN B[ECKER] LUCKE, Bibliography and index of West Virginia geology and natural resources. West Virginia geological survey: Bulletin (no.4): [Morgantown 1937]. pp.[iv].84. [1000.]

Wisconsin

PUBLICATIONS. [Geological and natural history survey: Madison 1916]. pp.13. [100.]

Wyoming

GLADYS G. BOVEE, Bibliography and index of Wyoming geology, 1823–1916. Geologist's office: Cheyenne 1918. pp.[iii].319–446. [721.]

Urals

B. A. BEREZIN, Геолого-минералогическая литература по Ильменскому горному массиву (Южный Урал). Главное управление по заповедникам при СНК РСФСР: Москва 1940. pp.85. [750.]

Uruguay

ROGER LAMBERT, Bibliographie géologique de

la république orientale de l'Uruguay. Instituto geológico del Uruguay: Boletín (no.26): Montevideo 1939. pp.81. [292.]

Venezuela

H[OLLIS] D[OW] HEDBERG and F. HEDBERG. Bibliografía e índice de la geología de Venezuela. Carácas 1945. pp.81. [750.]

Wales

D[OUGLAS] A[NTHONY] BASSETT, Bibliography and index of geology and allied sciences for Wales and the welsh borders 1897–1958. Amgueddfa genedleathol Cymru: Cardiff 1961. pp.[ii].376. [3500.]*

West Indies

L[OUIS] M[ARTIN] R[OBERT] RUTTEN, Bibliography of west indian geology. Rijksuniversiteit: Geographisch en mineralogisch instituut: Geographische en geologische mededeelingen: Physiographisch-geologische reeks (no.16): Utrecht 1938. pp.vii.103. [2434.]

J[AKOB] F[REDERIK] STEENHUIS, De geologische literatuur over of van belang voor nederlandsch Guyana (Suriname) en de nederlandsche westindische eilanden. 's-Gravenhage 1934. pp.80.

Geology

Yugoslavia

SRETEN P. MILOJEVIĆ, *ed.* Geološka bibliografija Jugoslavije od XIV veka do 1944 godine. Bibliografski institut FNRJ: Beograd [1952]. pp.479. [6000.]

7. *Miscellaneous*

[PIERRE ARMAND DUFRÉNOY], Notice sur les travaux minéralogiques et géologiques de m. Dufrénoy. [1835]. pp.6. [30.]

— — [another edition]. [1838]. pp.8. [35.]

[VISCOUNT ÉTIENNE JULES ADOLPHE DEXMIER DE SAINT-SIMON D'ARCHIAC], Notice sur les travaux géologiques de m. le vicomte d'Archiac. [1847]. pp.31. [30.]

— — [another edition]. [1856]. pp.7. [45.]

[VISCOUNT É. J. A. DEXMIER DE SAINT-SIMON D'ARCHIAC], Liste bibliographique . . . des travaux géologiques de m. d'Archiac. 1856. pp.[i].7. [45.]

[ALEXANDRE F. G. A. LEYMERIE], Aperçu analytique des principaux travaux géologiques . . . par A. Leymerie. 1869. pp.48. [123.]

— — [another edition]. Catalogue des travaux [&c.]. 1869. pp.54. [127.]

CONTRIBUTIONS to geology and mineralogy by W. Lauder Lindsay. [*s.l.* 1875]. pp.19. [36.]

Geology

TITLES of the more important geological publications of Charles Henry Hitchcock. [*s.l.* 1890]. [150.]

SELECT bibliography on geological climates: Science library: Bibliographical series (no.6). [1931]. ff.[i].8. [178.]*

BIBLIOGRAPHY of military geology and geography. Geological survey of America: [New York] 1941. pp.18. [250.]

SEVERINE [HANSENNE] BRITT, Selected abstracts on engineering geology and related subjects. Geological survey: Circular (no.259): Washington 1954. pp.27.

Hydrography.

1. *Manuscripts*

[MARCEL DESTOMBES], Catalogue des cartes nautiques manuscrites sur parchemin, 1300–1700. . . . Cartes hollandaises: la cartographie de la Compagnie des Indes orientales. 1593–1743. Saïgon 1941. pp.[vi].99. [272.]

250 copies printed.

A MAGYARORSZAGI keziratos vizrajzi térképek katalógusa. Budapesti műszaki egyetem: Budapest 1954– . pp.64+156+220+ .

in progress.

2. *General*

CATALOGUE des cartes hydrographiques qui ont été dressées au dépôt des cartes, plans et journaux de la marine, pour le service des vaisseaux du roy . . . depuis 1737 jusqu'en 1768, par le s^{r} [Jacques Nicolas] Bellin. [1775]. pp.4. [80.]

— [another edition]. Catalogue des cartes, plans,

vues de côtes, mémoires, instructions nautiques, etc. qui composent l'hydrographie française. 1847. pp.ix.220. [1750.]

— [another edition]. Catalogue chronologique des cartes [&c.] 1860. pp.[iii].198. [2500.]

— [another edition]. Catalogue par ordre géographique des cartes [&c.] 1860. pp.viii.259. [2500.]

— [another edition]. Catalogue chronologique [&c.]. Dépôt des cartes et plans de la marine (no.515): 1873. pp.viii.286. [3000.]

— [another edition]. Catalogue par ordre géographique [&c.]. Ministère de la marine et des colonies (no.514): pp.viii.296. [3000.]

— [another edition]. Catalogue par ordre géographique [&c.]. [By Philippe Eugène Hatt]. Service hydrographique de la marine (no.810): 1900. pp.xi.431. [5000.]

— [another edition]. Catalogue des cartes [&c.]. . . . (no.986): 1915. pp.xv.427. [4000.]

— [another edition]. Inventaire des cartes et ouvrages. 1951. pp.48. [3000.]

only a selection of the editions is set out.

LIST of classes into which are distributed the plans of ports, &c. . . . by A. Dalrymple. [Nautical memoirs and journals (vol.v):] 1786. pp.[iv].52. [500.]

— [another edition]. List of charts, plans . . . by A. Dalrymple. 1789. pp.11.

CATALOGUE of pilots, charts, and hydrographic with other nautical works, printed for . . . Laurie and Whittle. 1809. pp.[20]. [275.]

A CATALOGUE of charts, plans, & views, printed at the Admiralty office for the use of his majesty's navy. 1829. pp.[iv].56. [1000.]

revised at frequent intervals with several charges of title, the current issue being:

Catalogue of Admiralty charts and other hydrographic publications. 1963. pp.ii.136. [5000.]

[JOSEPH LARTIGUE], Notice sur les travaux et les ouvrages d'hydrographie . . . de m. Lartigue. [1842]. pp.7. [7.]

CATALOGUE of maps of the United States and territories, coast survey, and northern and northwestern lakes. Washington 1862. ff.[74]. [500.]

— [another edition]. Catalog of nautical charts and related publications. Coast and geodetic survey: 1946. pp.viii.48. [1000.]

КАТАЛОГЪ атласовъ, картъ, плановъ, видовъ, рисунковъ, флаговъ и руководствъ для плаванія. Гидрографическій департемент Морскаго министерства: Санктпетербургъ 1868. pp.[ii].iv.89. [400.]

— [another edition]. Каталог карт и книг. Гидрографическое управление: Ленинград 1928. pp.[iii].81. [1000.]

CATALOGUE of charts and books published by the United States Hydrographic office. Bureau of navigation: Hydrographic office: Washington 1871. pp.85. [200.]

revised at intervals with minor changes of title, the current issue being:

General catalog of marines' and aviators' charts and books. 1935. pp.xxii.258. [5000.]

CATALOGUS der verzameling van kaarten van het Ministerie van marine. 's-Gravenhage 1872. pp.[iv].xii.1284. [8500.]

LIST of charts for the european station furnished from the Hydrographic office. Bureau of navigation: Washington 1872. pp.77. [1000.]

— [another edition]. Catalogue of charts, plans, and sailing directions issued to vessels on the european station. By J. R. Selfridge. Hydrographic office: 1884. pp.120. [1000.]

LIST of charts for the asiatic station, furnished from the Hydrographic office. Bureau of navigation: Washington 1872. pp.67. [750.]

— [another edition]. Catalogue of charts, plans and sailing directions issued to vessels on the asiatic station. Hydrographic office: 1885. pp.158. [1250.]

— [another edition]. Catalogue of charts and plans issued to vessels of the United States on the asiatic station. 1909. pp.[ii].178. [1500.]

CATALOGUE of books in the library of the United States Hydraulic office. Washington 1875. pp.182. [3000.]

printed on one side of the leaf.

FORTEGNELSE over det Kongelige søkaartarchiv's forhandlingsartikler. Kjøbenhavn 1890. pp.[11]. [150.]

— [another edition]. 1931.

CATÁLOGO de las cartas, planos, vistas, libros, etc. existentes en el Depósito hidrográfico. Madrid 1893. pp.viii.129. [1750.]

SØKAART-ARCHIVETS antike kaartsamling. Kjøbenhavn 1895. pp.[v].78. [1000.]

CATALOGUS van nederlandse zeekaarten en boekwerken. 1898.

— [another edition]. Ministerie van marine: [The Hague] 1950. pp.xiv.43. [500.]

A CONSECUTIVE list of the Admiralty charts.

Hydrographic office: 1900. pp.130. [3000.]
— [another edition]. 1912. pp.142. [3700.]
these are the first and last editions.

MARIO LONGHENA, Atlanti e carte nautiche del [*sic*, dal] secolo XIV al XVII conservati nella biblioteca e nell'archivo de Parma. R. deputazione di storia patria per le provincie Parmensi: [Parma] 1907. pp.46. [15.]

KATALOG over sjøkartverket. Norges geografiske opmaaling: Kristiania 1910. pp.15. [400.]
— [another edition]. Oslo 1932. pp.10. [300.]

CATALOGO alfabetico della biblioteca. Magistrato alle acque: Ufficio idrografico. Venezia 1916.
a–g only; no more published.

VERZEICHNIS der deutschen admiralitätskarten und nautischer bücher. Marineleitung: Berlin 1923. pp.viii.80. [750.]
— [another edition]. Verzeichnis der nautischen karten und bücher. Deutsches hydrographisches institut: Hamburg 1959. pp.viii.96. [1500.]

KATALOG sjökort och seglingsbeskrivningar m.m. utgivna af Kungl. sjökarteverket. Stockholm 1925. pp.18. [200.]
— [another edition]. 1937. pp.31. [250.]

CATALOGUE of official canadian government publications of use to mariners. [?1925].

— [another edition]. Catalogue of nautical charts, sailing directions, tidal information [&c.]. Hydrographic and map service: Ottawa 1948. pp.53. [1000.]

there are numerous intermediate editions.

CATÁLOGO de cartas y libros para navegación. Servicio hidrográfico: Buenos Aires 1926. ff.[i].15. [100.]

— 11ª edición. Catálogo de cartas, libros y otras publicaciones de ayuda a la navegación. Servicio de hidrografía naval: 1959. pp.[iv].85. [250.]

ʽΟΔΗΓΙΑΙ χρησεως ν. χαρτων και εὑρετηρια. ʽΥδρογραφικη ὑπηρεσια: ʼΑθηναις 1926. pp.37. [500.]

— [another edition]. 1932. pp.38. [500.]

FORTEGNELSE over det Kongelige søkortarkives forhandlingsartikler. København 1934. pp.22. [75.]

CATALOGUS van kaarten enz. uitgegeven door den [de] Topografischen dienst. [The Hague] 1935. ff.63. [1000.]*

— [another edition]. Delft 1956. pp.77. [1000.]

there are numerous intermediate editions.

CARTE nautiche ed altre pubblicazioni dell'Istituto idrografico della r. marina. Catalogo. [Genoa] 1937. pp.38. [500.]

— [another edition]. Catalogo generale dei documenti nautici in vendita. 1958. pp.ii.44. [500.]

ÍNDICE de las cartas en uso en la armada nacional. Departamento de navegación e hidrografía. Valparaiso 1941. pp.132. [1000.]

— [another edition]. Catalogo de cartas nauticas y otras publicaciones editadas por el Departamento de navegación e hidrografía de la armada de Chile. 1951. pp.73. [1000.]

CATÁLOGO de cartas y publicaciones. Servicio hidrográfico: [Montevideo] 1944. pp.(ff.)44. [200.]

— [another edition]. 1956. pp.121. [250.]

CATALOGUS van nederlandsche zeekarten en boekwerken uitgegeven door de afdeeling hydrografie van het Ministerie van marine. 's-Gravenhage 1946. pp.48.

ATLAS des index des cartes, plans et ouvrages qui composent l'hydrographie française. Service hydrographique: [1948].

consists of key-maps.

— [another edition]. Catalogue des cartes et ouvrages qui composent l'hydrographie française. Service hydrographique de la marine (no.7): 1960. pp.xiii.242. [2500.]

CATALOGUE of charts and publications of the hydrographic service. Bangkok 1948. pp.[iii].10. [41.]

MARINE atlases in the Dartmouth college library. A descriptive list. Hanover, N.H. 1950. pp.ix.29. [28.]

KATALOG pomorskih karata i navigaijskih publikacija. Hidrografski institut jugoslovenske: Split 1950. pp.8. [250.]

— [another edition]. 1962. pp.51. [1000.]

VERZEICHNIS der nautischen karten und bücher. Deutsches hydrographisches institut: Hamburg 1951. pp.vii.72. [1500.]

CATÁLOGO de cartas nauticas y otras publicaciones editadas por el Instituto hidrográfico de la marina. Cadiz 1951. pp.v.67. [500.]

KATALOG over sjøkarter . . . og farvannsbeskrivelser, vedkommende den norske kyst. Sjøkartverk: Oslo 1952. pp.14. [500.]

WALTER W. WEINSTEIN, Preliminary inventory

of the records of the Hydrographic office. National archives: [Preliminary inventories (no.39):] Washington 1952. pp.v.17. [large number.]*

ELENCO delle pubblicazioni del Servizio idrografico italiano. Roma 1953. pp.68. [400.]

KATALOG over sjøkarter . . . og farvannsbeskrivelser vedkommende den Norske kyst . . . samt arktiske sjøkarter og farvannsbeskrivelser utgitt av. Norsk polarinstitutt. Norges sjøkartverk: Oslo 1956. pp.14. [400.]

KATALOG over søkort og sejlhändbøger. Kongelige danske søkortarkiv: [Copenhagen] 1960. pp.34. [250.]

Mathematics.

Mathematics

1. *Bibliographies*

A[NNA] M[OISEEVNA] LUKOMSKAYA, Основные иностранные библиографические источники по математике и механике (1931–1957). Академия наук СССР: Библиотека: Москва &c. 1960. pp.184. [1005.]

A[NNA] M[OISEEVNA] LUKOMSKAYA, Библиография отечественной литературы по математике и физике. Обзор библиографических источников. Академия наук СССР: Библиотека: Москва &c. 1961. pp.159. [289.]

A[NNA] M[OISEEVNA] LUKOMSKAYA, Библиографические источники по математике и механике, изданные в СССР за 1953–1960 гг. Академия наук Союза ССР: Библиотека: Москва &c. 1963. pp.248. [1219.]

2. *Periodicals*

ERNST WÖLFFING, Verzeichniss der zeitschriften für die gebiete der mathematik, der physik . . .

und der verwandten wissenschaften welche auf württembergischen bibliotheken vorhanden sind. Stuttgart 1899. pp.18. [71.]

[W. J. GREENSTREET], Catalogue of current mathematical journals, etc. with the names of the libraries in which they may be found. Mathematical association: 1913. pp.40. [182.]

DAVID EUGENE SMITH and CAROLINE EUSTIS SEELY, Union list of mathematical periodicals. Bureau of education: Bulletin (1918, no.9): Washington 1918. pp.60. [81.]

VERZEICHNIS der naturwissenschaftlichen, mathematischen und medizinischen zeitschriften der universitätsbibliothek und der institute der universität. Freiburg i. Br. 1930. pp.113. [1333.]

S. R. RANGANATHAN [SĪGĀṚI RĀMĀMṚITA RAṄGĀNĀTHAN], P. N. KAULA and T. N. KORANNE, Union catalogue of periodical publications in the libraries of south Asia. Delhi 1950. pp.[71]. [300.]

a specimen only, limited to mathematics and physics.

PEDER KRISTIAN LYSTBOEK, Udenlandske tidsskrifter af matematisk eller fysisk indhold i danske biblioteker. Universitetsbibliotek: København 1952. pp.viii.176. [1569.]

Mathematics

PAUL BELGODÈRE, Liste des titres des périodiques régulièrement reçus par la Société mathématique de France. Documentation mathématique (no.21): 1952. ff.8. [300.]*

PAUL BELGODÈRE, Liste des titres des périodiques régulièrement reçus par la bibliothèque des sciences mathématiques de l'École normale supérieure. Documentation mathématique (no.20): 1952. ff.4. [100.]*

— — 3e édition. État des périodiques figurant à la bibliothèque [&c.]. . . . (no.26): 1954. ff.11. [200.]*

PAUL BELGODÈRE, Liste des titres des périodiques régulièrement reçus par la bibliothèque mathématique de l'Institut Henri Poincaré. Liste provisoire. Documentation mathématique (no.19): 1952. ff.8. [250.]*

— — [another edition]. Par Madeleine Estève. . . . (no.38): 1957. ff.40. [600.]*

[MARTHE] SIESTRUNCK, État des périodiques de mathématiques et de physique figurant à la bibliothèque de la Sorbonne. Édition provisoire. Documentation mathématique (no.24–25): 1953. ff.43 +[ii].44–80. [1500.]*

ПЕРИОДИЧЕСКАЯ печать СССР, 1917–1949. Библиографический указатель. Журналы, труды и бюллетени по естественным наукам и математике. Всесоюзная книжная палата: Москва 1956. pp.220. [1600.]

JEAN PELTIER, Inventaire collectif des périodiques mathématiques . . . de 13 bibliothèques parisiennes. Documentation mathématique (no. 36–37): 1957. ff.53+159. [1500.]*

UNION list of periodicals on mathematics and allied subjects in London libraries. University of London: Library: 1958. pp.v.62. [627.]

H[ENRY] A[LAN] STEEVES, Russian journals of mathematics. A survey and checklist. Public library: New York 1961. pp.19. [250.]

3. *History*

ABRAHAM GOTTHELF KÄSTNER, Geschichte der mathematik. Geschichte der künste und wissenschaften: Göttingen 1796–1800. pp.[ii].x.708+[ii].xiv.759+[ii].xiv.484+[ii].xiv.516. [5000.]

BOLLETTINO di bibliografia e storia delle scienze matematiche. Pubblicato per cura di Gino Loria. Torino.

i. 1898. pp.iv.164. [600.]
ii. 1899. pp.12.148. [750.]
iii. 1900. pp.[ii].136. [750.]
iv. 1901. pp.viii.128. [600.]
v. 1902. pp.iv.132. [750.]
vi. 1903. pp.iv.132. [750.]
vii. 1904. pp.iv.132. [750.]
viii. 1905. pp.iv.132. [750.]
ix. 1906. pp.iv.132. [750.]
x. 1907. pp.iv.131. [750.]
xi. 1908–1909. pp.iv.132. [750.]
xii. 1910. pp.iv.132. [750.]
xiii. 1911. pp.iv.140. [750.]
xiv. 1912. pp.iv.132. [750.]
xv. 1913. pp.iv.132. [750.]
xvi. 1914. pp.iv.132. [750.]
xvii. 1915. pp.iv.148. [750.]
xviii. 1916. pp.iv.148. [750.]
xix. 1917. pp.iv.100. [750.]
xx. 1918. pp.x.128. [1000.]
xxi. (nos.1–2). 1919. pp.64. [200.]

amalgamated with the Bollettino di matematica, *a non-bibliographical journal.*

JOSÉ A[UGUSTO] SÁNCHEZ PÉREZ, Las matemáticas en la biblioteca del Escorial. Madrid 1929. pp.[iv].367. [1000.]

CECIL B[YRON] READ, Articles on the history of mathematics: a bibliograph [*sic*] of articles appearing in five periodicals. Municipal university of Wichita: Bulletin (vol.xxvii, no.4 = University studies, no.36): Wichita 1952. pp.32. [450.]*

ELISABETH SAUVENIER-GOFFIN, Les sciences mathématiques et physiques à travers le fonds ancien de la bibliothèque de l'université de Liège. Première partie. De l'antiquité à la fin du seizième siècle. Bibliotheca universitatis leodiensis: Publications (no.10): Louvain 1958. pp.172. [508.]

4. *Manuscripts*

JONAS GURLAND, Kurze beschreibung der mathematischen, astronomischen und astrologischen hebraischen handschriften der Firkowitsch'schen sammlung in der Kayserlichen öffentlichen bibliothek zu St. Petersburg. Neue denkmäler der jüdischen literatur (no.2): St. Petersburg 1866. pp.[iv].57. [100.]

M[ORITZ] STEINSCHNEIDER, Intorno ad otto manoscritti arabi di matematica posseduti dal ch. sig. Guglielmo Libri.... Lettera v. Roma 1867. pp.[ii].71–92. [21.]

A. DE POORTER and M. ALLIAUME, Catalogue des

manuscrits mathématiques et astronomiques de la bibliothèque de Bruges. Bruges 1922. pp.13–50. [100.]

5. General

CORNELIUS À BEUGHEM, Bibliographia mathematica et artificiosa novissima perpetuò continuanda, seu conspectus primus catalogi librorum mathematicorum . . . quotquot currente hoc semiseculo . . . in quavis lingua . . . typis prodierunt. Amstelodami 1688. pp.[xiv].526. [3000.]
no more published.

[JOHANN EPHRAIM SCHEIBEL], Einleitung zur mathematischen bücherkentnis [bücherbekenntniss, bücherbekenntnis]. Breslau 1769–1798. pp. 112 + [vi].119–224 + 225–352 + [ii].353–470 + [ii].471–561 + [ii].563–690.[xxxv] + 116 + [ii]. 115–256 + [ii].257–401 + 405–508 + 309 [*sic*, 509]–426[626] + [ii].427[627] + 560[760] + [xiv].155 + [vi].182 + 193[*sic*, 183]–254[244] + [iii].232+233–380. [10,000.]
the first part was reprinted in 1772, described as a Neue auflage.

[JOHANN SAMUEL ERSCH], Allgemeines repertorium der literatur für die jahre 1785 bis 1790. XII. Mathematische literatur. Jena 1793. pp.[26.] [581.]

[—] — 1791–1795. Weimar 1800. pp.[38]. [778.]

[—] — 1796–1800. Weimar 1807. pp.[38]. [658.]

no more published.

FR[IEDRICH] WILH[ELM] AUG[UST] MURHARD, Litteratur der mathematischen wissenschaften. Leipzig 1797–1805. pp.[ii].xvi.[vi].256+[ii].xii. [vi].436 + xii.[iv].360 + [viii].343 + [iii].vi.243. [10,000.]

has the additional title Bibliotheca mathematica.

E. F. WREDE, Enzyklopädisch-szientifische literatur. Drittes heft, die enzyklopädisch-mathematische literatur enthaltend. Leipzig &c. 1812. pp.[ii].xxii.363. [3500.]

JOHANN WOLFGANG MÜLLER, Auserlesene mathematische bibliothek. Nürnberg 1820. pp.22.266. [4000.]

JOHANN WOLFGANG MÜLLER, Repertorium der mathematischen literatur. Augsburg &c. [1822–1825]. pp.144+126+68. [4000.]

REGLEMENT voor de leden, en catalogus van de boeken, uitmakende de boekerij van het Wiskundig genootschap: Een onvermoeide arbeid komt alles to boven. Amsterdam 1831. pp.47. [500.]

— [another edition]. Catalogus der bibliotheek van [&c.]. [By A. J. van Pesch]. 1883. pp.[vi].112. [1500.]

— — Vervolg. [1891]. pp.15. [200.]

the society is the Genootschap der mathematische wetenschappen.

L[UDWIG] A[DOLPH] SOHNCKE, Bibliotheca mathematica. Verzeichniss der bücher über die gesammten zweige der mathematik . . . welche in Deutschland und dem auslande vom jahre 1830 bis mitte des jahres 1854 erschienen sind. Leipzig 1854. pp.xviii.388. [7000.]

also issued with an english titlepage; intended as a continuation of Rogg's Handbuch *(1830), of which, however, only the volume devoted to arithmetic and geometry was published.*

[OLRY] TERQUEM, Bulletin de bibliographie, d'histoire et de biographie mathématique.

i. 1855. pp.[iii].204. [40.]
ii. 1856. pp.[ii].208. [20.]
iii. 1857. pp.[iii].100. [20.]
iv. 1858. pp.[iii].95. [20.]
v. 1859. pp.[iii].96. [40.]
vi. 1860. pp.[ii].96. [15.]

vii. 1861. pp.[iii].96. [25.]

CATALOGUE of books in Queen's college library, Belfast. Department of mathematics and natural philosophy: Belfast 1856. pp.18. [350.]

BULLETTINO di bibliografia e di storia delle scienze matematiche e fisiche. Pubblicato da B[aldassare] Boncompagni. Roma.

i. 1868. pp.[iii].466. [2000.]
ii. 1869. pp.[iii].552. [3000.]
iii. 1870. pp.[iii].468. [2500.]
iv. 1871. pp.[iii].556. [2500.]
v. 1872. pp.[iii].596. [3000.]
vi. 1873. pp.[iii].600. [3000.]
vii. 1874. pp.[iii].666. [3500.]
viii. 1875. pp.[iii].729. [4000.]
ix. 1876. pp.[iii].785. [5000.]
x. 1877. pp.[iii].736. [4000.]
xi. 1878. pp.[iii].887. [6000.]
xii. 1879. pp.[iii].984. [7000.]
xiii. 1880. pp.[iii].904. [6000.]
xiv. 1881. pp.[iii].821. [5000.]
xv. 1882. pp.[iii].791. [5000.]
xvi. 1883. pp.[iii].814. [5000.]
xvii. 1884. pp.[iii].960. [7000.]
xviii. 1885. pp.[iii].753. [4000.]
xix. 1886. pp.[iii].735. [4000.]

xx. 1887. pp.[iii].749. [4000.]

no more published; vol.xx contains an index to vols.i–xx.

JAHRBUCH über die fortschritte der mathematik. [vol.li &c.: Preussische akademie der wissenschaften:] Berlin.

i. 1868. Herausgegeben von Carl Ohrtmann und [Hermann] Felix Müller. pp.xxxiv. 428. [900.]
ii. 1869–1870. Herausgegeben von C. Ohrtmann, Fr. Müller, Albert Wangerin. pp. lvii.908. [1750.]
iii. 1871. pp.xxxiv.588. [900.]
iv. 1872. pp.xxxiv.633. [1000.]
v. 1873. pp.xxxiv.640. [1000.]
vi. 1874. pp.xl.777. [1250.]
vii. 1875. pp.xliv.764. [1250.]
viii. 1876. pp.xliv.800. [1250.]
ix. 1877. pp.xlix.850. [1500.]
x. 1878. pp.lii.844. [1500.]
xi. 1879. Herausgegeben von C. Ohrtmann. pp.liii.858. [1500.]
xii. 1880. pp.lii.912. [1750.]
xiii. 1881. pp.l.903. [1500.]
xiv. 1882. pp.lii.974. [1500.]
xv. 1883. Herausgegeben von Max Henoch

und Emil Lampe. pp.lii.1041. [1750.]

xvi. 1884. pp.lxvi.1167. [2000.]

xvii. 1885. pp.lxvi.1211. [2250.]

xviii. 1886. pp.lxvi.1176. [2250.]

xix. 1887. pp.lxxiv.1297. [2500.]

xx. 1888. Herausgegeben von E. Lampe. pp.lxxvi.6.1352. [2750.]

xxi. 1889. pp.lxvi.1309. [2500.]

xxii. 1890. pp.lxviii.1314. [2250.]

xxiii. 1891. pp.lxvi.1314. [2500.]

xxiv. 1892. pp.lxiv.1223. [2500.]

xxv. 1893–1894. pp.xcii.852+[ii].853–1996. [3500.]

xxvi. 1895. pp.lxii.1174. [2250.]

xxvii. 1896. pp.lxvi.883. [2250.]

xxviii. 1897. pp.lxviii.935. [2750.]

xxix. 1898. Herausgegeben von E. Lampe und Georg Wallenberg. pp.lxvi.892. [2500.]

xxx. 1899. pp.lxv.918. [2500.]

xxxi. 1900. pp.lxvii.970. [2750.]

xxxii. 1901. pp.lxvii.1013. [2750.]

xxxiii. 1902. Herausgegeben von E. Lampe. pp.lxxv.1039. [3000.]

xxxiv. 1903. pp.lxxiii.1099. [3250.]

xxxv. 1904. pp.lxxii.1060. [3250.]

xxxvi. 1905. pp.lxxiii.1088. [3250.]

xxxvii. 1906. pp.lxxiv.1072. [3250.]
xxxviii. 1907. pp.lxxvi.1060. [3500.]
xxxix. 1908. pp.lxxii.1102. [3250.]
xl. 1909. pp.lxxviii.1110. [3500.]
xli. 1910. pp.lxxix.1137. [3750.]
xlii. 1911. pp.lxxix.1131. [4000.]
xliii. 1912. pp.lxxxiii.1208. [4000.]
xliv. 1913. 1918. pp.lxxxvi.120. [4000.]
xlv. 1914–1915. Herausgegeben von E. Lampe und L. Lichtenstein. 1922. pp.xv. iii.7.1509. [4000.]
xlvi. 1916–1918. Herausgegeben von Leon Lichtenstein. 1923–1924. pp.xviii.1639. [5000.]
xlvii. 1919–1920. 1924–1926. pp.xxvii.1103. [4000.]
xlviii. 1921–1922. 1925–1928. pp.xxiii.1569. [5000.]
xliv. 1923. 1927–1928. pp.xix.808. [2250.]
l. 1924. 1928–1929. pp.xviii.749. [2250.]
li. 1925. Schriftleiter: Georg Feigl. 1931–1932. pp.xviii.902. [5000.]
lii. 1926. 1935. pp.xvi.1167. [5500.]
liii. 1927. 1931. pp.xviii.1019. [5000.]
liv. 1928. 1932. pp.xvii.1181. [5500.]
lv. 1929. 1932–1935. pp.[iv].582+xvii.583–1376. [7000.]

lvi. 1930. 1933–1935. pp.[iv].796+xvii.797–1525. [7000.]

lvii. 1931. [Part 2: Schriftleitung, Helmut Grunsky]. 1938–1939. pp.iv.1284+xxiv.1285–1795. [8000.]

lviii. 1932. 1938–1939. pp.iv.984+xxiii.985–1564. [6500.]

lix. 1933. 1934–1937. pp.iv.824+xxiv.825–1783. [7000.]

lx. 1934. [1934–]1935–1938. pp.iv.813+xxiii.815–1618. [5500.]

lxi. 1935. [1935–]1936–1939. pp.v.937+xxv.939–1736. [6500.]

lxii. 1936. [1936–]1938–1940. pp.iv.1006+xxviii.1007–1796. [6500.]

lxiii. 1937. [1938–]1939– . pp.[iv].792+

lxiv. 1938. [1939]–1942. pp.[iv].899+xxxi.901–1654. [6000.]

lxv. 1939. [1941–]1943. pp.[viii].800+xl.801–1683. [6500.]

lxvi. 1940. [1942–]1944. pp.[ix].800+xl.801–1541. [5500.]

lxvii. 1941. [1943–]1945. pp.[ix].966+967–1132. [3500.]

lxviii. 1942. 1944. pp.[x].657. [2000.]

vol.lix forms a second part of the Revue semestrielle

des publications mathématiques; *no more published.*

BARON FRIEDRICH VON LÖFFELHOLZ-COLBERG, Forstliche chrestomathie . . . III. 2. abteilung. Die literatur der geometrie, stereometrie, trigonometrie und höheren mathematik überhaupt. Berlin 1873. pp.viii.859–1092. [2500.]

LEO KOENIGSBERGER and GUSTAV ZEUNER, Repertorium der literarischen arbeiten aus dem gebiete der reinen und angewandten mathematik. Leipzig 1877–1879. pp.[iii].444+[ii].452. [350.]
a collection of abstracts.

LIST of mathematical works recently added to the Bowditch collection. Public library: Boston 1878. pp.8. [300.]

WOLF's naturwissenschaftlich - mathematisches vademecum. [Leipzig 1881]. pp.[ii].250. [5000.]

MATHEMATICS. Cornell university: Library: Special lists (no.1): Ithaca 1883. pp.92. [2500.]

[BENTO VIEIRA FERRAZ DIARAÚJO], Catálogo de mathematica da bibliotheca publica municipal. Porto 1885. pp.115. [1500.]

V[IKTOR] V[IKTOROVICH] BOBUININ, Русская физико-математическая библиографія. Москва 1885–1900.

details of this work are entered under Physics, below.

KATALOG der in den hamburger bibliotheken vorhandenen literatur aus der reinen und angewandten mathematik und physik. Hamburg 1890.

— Nachtrag. 1893.

— 2. nachtrag. 1906. pp.[ii].vi.352. [9000.]

REVUE semestrielle des publications mathématiques. Rédigées sous les auspices de la Société mathématique d'Amsterdam. Amsterdam [Groningen].

i. 1893. pp.[iv].104+[iv].114. [2000.]
ii. 1894. pp.[iv].136+[iv].152. [2500.]
iii. 1895. pp.[iv].171+[iv].172. [3000.]
iv. 1896. pp.[iv].164+[iv].167. [3000.]
v. 1896–1897. pp.[iv].163+[iv].156. [3000.]
— Table des matières . . . 1893–1897. pp.[iv].84.
vi. 1897–1898. pp.[iv].172+[iv].184. [3000.]
vii. 1898–1899. pp.[iv].179+[iv].159. [3000.]
viii. 1899–1900. pp.[iv].188+[iv].188. [3500.]
ix. 1900–1901. pp.[iv].188+[iv].180. [3500.]
x. 1901–1902. pp.[iv].180+[iv].188. [3500.]
— Table des matières . . . 1898–1902. pp.[iv].156.
xi. 1902–1903. pp.[iv].200+[iv].190. [3500.]
xii. 1903–1904. pp.[iv].195+[iv].184. [3500.]
xiii. 1904–1905. pp.[iv].184+[iv].200. [3500.]

xiv. 1905–1906. pp.[iv].178+[iv].178. [3000.]
xv. 1906–1907. pp.[iv].151+[iv].168. [3000.]
— Table des matières . . . 1903–1907. pp.[iv]. 164.
xvi. 1907–1908. pp.[iv].175+[iv].178. [3000.]
xvii. 1908–1909. pp.[iv].155 + [iv].176. [3000.]
xviii. 1909–1910. pp.[iv].179 + [iv].200. [3500.]
xix. 1910–1911. pp.[iv].160+[iv].175. [3000.]
xx. 1911–1912. pp.[iv].175+[iv].171. [3000.]
xxi. 1912–1913. pp.[iv].167+[iv].180. [3000.]
— Table des matières . . . 1908–1913. pp.[iv]. 173.
xxii. 1913–1914. pp.[iv].176 + [iv].178. [3000.]
xxiii. 1914–1915. pp.[iv].131 + [iv].119. [2000.]
xxiv. 1915–1916. pp.[iv].136 + [iv].108. [2000.]
xxv. 1916–1917. pp.[iv].119+[iv].85. [1370.]
— Table des matières . . . 1913–1917. pp.[iv]. 160.
xvi. 1917–1918. pp.[iv].81+[iv].108. [1650.]
xxvii. 1918–1919. pp.[iv].85+[iv].83. [1400.]
xxviii. 1919–1920. pp.[iv].141 + [iv].133. [2200.]

xxix. 1920–1921. pp.[iv].150 + [iv].189. [2900.]

xxx. 1921–1923. pp.[iv].129 + [iv].133. [2300.]

— Table des matières . . . 1918–1923. pp.[iv]. 149.

xxxi. 1923–1924. pp.[iv].207 + [iv].234. [4400.]

xxxii. 1924–1925. pp.[iv].230 + [iv].166. [4000.]

xxxiii. 1925–1926. pp.[iv].256 + [iv].280. [5000.]

xxxiv. 1929–1930. pp.[iv].179 + [iv].248. [4300.]

xxxv–xxxvi. 1931. pp.[iv].182+[iv].187+[iv].313. [7750.]

— Table des matières . . . 1924–1931. pp.[iv]. 271.

xxxvii. 1932. pp.[ii].198. [5750.]

xxxviii. 1933. pp.[ii].196. [5750.]

xxxix. 1934. pp.[ii].202. [6000.]

no more published; for 1932 there is a second part, comprising the Jahrbuch über die fortschritte der mathematik.

RÉPERTOIRE bibliographique des sciences mathématiques. Première[–vingtième] série: fiches 1 à

100[–2000]. 1894–1912. ff.2000. [20,000.]
consists of twenty packets of cards.

— — Analyse mathématique [*c.*1895]. pp.67.

— — — Deuxième édition. Index du répertoire [&c.]. 1898. pp.xiv.95.

— — — Nouvelle édition. 1908. pp.[iv].112.

— — — Troisième [*sic*] édition. 1916. pp.[v]. 115.

the index was also issued in a dutch edition.

VERZEICHNIS von abhandlungen aus der angewandten mathematik die im jahre . . . in technischen zeitschriften erschienen sind. Zusammengestellt von R[udolf] Mehmke. Zeitschrift für mathematik und physik: Beilage: Leipzig.

1897. pp.8. [350.]
1898. pp.10. [350.]
1899. pp.12. [400.]

subsequent issues form part of the text of the Zeitschrift.

MATHEMATICS. International catalogue of scientific literature (section A): Royal society.

i. 1902. pp.xv.201. [1506.]
ii. 1904. pp.viii.263. [1843.]
iii. 1904. pp.viii.228. [1610.]
iv. 1905. pp.viii.240. [1658.]
v. 1906. pp.viii.304. [2294.]

vi. 1907. pp.viii.195. [1297.]
vii. 1909. pp.viii.253. [2109.]
viii. 1910. pp.viii.240. [2266.]
ix. 1911. pp.viii.228. [2165.]
x. 1912. pp.viii.248. [2602.]
xi. 1913. pp.viii.196. [1956.]
xii. 1914. pp.viii.182. [1679.]
xiii. 1915. pp.viii.175. [1687.]
xiv. 1917. pp.xiii.163. [1530.]
no more published.

K. HIEMENZ, Katalog des mathematischen lesezimmers der universität Göttingen. Leipzig 1907. pp.xi.224. [2250.]

AANWINSTEN op het gebied der wis- en natuurkundige wetenschappen, gedeeltelijk behoorende tot . . . C. P. Burger. Universiteit: Bibliotheek: Amsterdam 1907. pp.[ii].32. [600.]

[HERMANN] FELIX MÜLLER, Führer durch die mathematische literatur, mit besonderer berücksichtigung der historisch wichtigen schriften. Abhandlungen zur geschichte der mathematischen wissenschaften (vol.xxvii): Leipzig 1909. pp.x.252. [6000.]

[PAUL HOHENEMSER], Katalog der mathematischen abteilung. Stadtbibliothek: Frankfurt a. M. 1909. pp.x.327. [6250.]

JIŘÍ KAVÁN, Katalog knihovny Jednoty českých mathematiků. Praze 1909. pp.[iv].xvi.211. [3000.]

CATALOGUE of the library. American mathematical society: New York 1910. pp.35. [1000.]

— [another edition. Edited by Arnold Dresden]. American mathematical society [: Bulletin (vol. lii, no.5, part II):] Menasha &c. 1946. pp.v.173. [6000.]

ST[ANISLAUS] GRÖNFELDT, Systematisk förteckning öfver [Magnus] G[ustaf] Mittag-Lefflers matematiska bibliotek. Stockholm &c. 1914 [1915]. pp.[iv].coll.xviii.692. [5500.]

БИБЛИОГРАФИЧЕСКИЙ справочник по математике. Математическая комиссия отдела реформы школы: Москва 1920. pp.52. [250.]

CATALOGUS van de boeken en tijdschriften geplaatst in de wiskundige leeszaal en in de laboratoria voor natuurkunde, dierkunde, plantkunde, plantenphysiologie en pharmacognosie. Universiteit: Bibliotheek: Amsterdam 1922. pp. 95. [1500.]

[E. H. NEVILLE], A first list of books & pamphlets

in the library of the Mathematical association. 1926. pp.47. [1200.]

[—] — A second list . . . books and pamphlets added during 1924 and 1925. 1929. pp.56. [1500.]

[—] — A third list . . . 1926 to 1929. 1930. pp.61. [1750.]

[—] — A fourth list . . . 1930 to 1935. 1936. pp.83. [1750.]

ZENTRALBLATT für mathematik und ihre grenzgebiete (reine und angewandte mathematik, theoretische physik, astrophysik, geophysik). Berlin.

i. Schriftleitung: O. Neugebauer. 1931. pp. v.466. [1000.]
ii. 1932. pp.v.472. [1000.]
iii. 1932. pp.[iii].458. [1000.]
iv. 1932. pp.[iii].458. [1000.]
v. 1933. pp.[iii].480. [1000.]
vi. 1933. pp.[iii].466. [1000.]
vii. 1934. pp.[iii].464. [1000.]
viii. 1934. pp.[iii].464. [1000.]
ix. 1934. pp.[iii].463. [1000.]
x. 1935. pp.[iii].486. [1000.]
xi. 1935. pp.[iii].466. [1000.]
xii. 1936. pp.[iii].467. [1000.]
xiii. 1936. pp.[iii].466. [1000.]
xiv. 1936. pp.[iii].466. [1000.]
xv. 1937. pp.[iii].490. [1000.]

xvi. 1937. pp.[iii].465. [1000.]
xvii. 1938. pp.[iii].465. [1000.]
xviii. 1938. pp.[iii].465. [1000.]

READERS' guide to books on mathematics, physics and chemistry. Library association: County libraries section [no.11]: 1938. pp.[iv].36. [900.]

FRIEDA M. HELLER, Mathematics — queen of the sciences. A bibliography of materials for . . . elementary and high schools. Reading for background (no.8): New York 1938. pp.16. [75.]

MATHEMATICS and physics. Library of Congress: Washington 1944. ff.2. [12.]*

PEDRO PI CALLEJA, Sobre orientación bibliográfica en matematica. Buenos Aires 1946. pp.17. [188.]

NATHAN GRIER PARKE, Guide to the literature of mathematics and physics including works on engineering science. New York &c. 1947. pp.xv. 205. [3000.]
— — Second . . . edition. [1958]. pp.xviii.436. [11,000.]

РЕФЕРАТИВНЫЙ журнал. Математика. Академия наук СССР: Институт научной информации: Москва.

1953–1954. coll.216.pp.v.+coll.236.pp.v+coll.230.pp.v + pp.iii.v + 122.vi + 66.iv + 98.iv + 96.iv + 96.iv + 87.iv + 94.iv + 77.ii + 87.iv + 126.vi + 116 + 148. [5919.]

1955. pp.136.v + 123.iv + 130.vi + 116.iv + 116.iv + 127.iv + 122.vi + 127.iv + 136.v+135.vii+140.vi+112. [6307.]

156. pp.181.iv + 176.iv + 159.iv + 152.iv + 132 + 152 + 121 + 136 + 136 + 151 + 121+148+124+231. [9228.]

1957. pp.169+157+148+165+162+148+153 + 148 + 148 + 154 + 153 + 120 + 188. [9035.]

1958. pp.168+153+157+165+160+172+192+204+ +200+204+151. [10,443.]

1959. pp.212+226+228+188+176+184+224+192+192+192+216+175. [11,725.]

1960. pp.196+251+184+179+248+220+250 + 272 + 227 + 216 + 187 + 216 + 192*+226. [14,640.]

1961. pp.80.88.72 + 72.80.48 + 80.104.64 + 72.104.56 + 88.88.48 + 88.88.56 + 88.104.64 + 84.96.56 + 96.104.72 + 80.104.

64 + 64.80.64 + 96.96.64 + 224 + 252.* [17,500.]

1962. pp.80.88.76 + 84.116.96 + 72.100.84 + 76.92.68 + 84.124.104 + 80.124.92 + 84.92.92 + 84.88.80 + 68.76.76 + 68.84.84 + 68.92.92 + 60.92.88 + 260 + 328. [20,000.]

1963. pp.91.80.76 + 68.88.100 + 64.100.110 + 60.100.96 + 92.108.108 + 80.104.84 + 80.96.100 + 76.112.92 + 72.104.88 + 72.92.79 + 68.120.112 + 96.120.120 + + . [20,000.]

in progress.

REGINA BUFFARA and JAYME MACHADO CARDOSO, Matemática. Bibliografia seletiva. Universidade do Paraná: Serie bibliografia e documentação (no.1): Curitiba 1959. pp.[vii].90. [1476.]

WILLIAM L[EONARD] SCHAAF, The high school mathematics library. National council of teachers of mathematics: Washington [1960]. pp.36. [500.]

— — Revised edition. [1963]. pp.55. [700.]

JOHN E. PEMBERTON, How to find out in mathematics. A guide to sources of mathematical information. Commonwealth and international library of science: Libraries and technical information division (vol.2): Oxford &c. [1963]. pp.x.158. [200.]

6. *Countries &c.*

[see also section 3, above]

America

MATHEMATICA americana. A guide to an exhibition in the William L. Clements library. University of Michigan: William L. Clements library: Bulletin (no.xxiii): Ann Arbor 1935. pp.7. [30.]

LOUIS C[HARLES] KARPINSKI, Bibliography of mathematical works printed in America through 1850. University of Michigan studies and publications: Ann Arbor 1940. pp.[ii].xxvi.697. [2998.]

Brazil

BIBLIOGRAFIA brasileira de matemática e física. Instituto brasileiro de bibliografia e documentação: Rio de Janeiro.

1900–1957. [1959]. pp.246. [1870.]

in progress?

China

T'UNG-LI YÜAN, A bibliography of chinese mathematics 1918–1960. Washington 1963. pp. x.154. [2000.]*

Denmark

NIELS NIELSEN, Matematiken i Danmark, 1528–

1800. Bidrag til en bibliografisk-historisk oversigt. København &c. 1912–1910. pp.[ii].vi.336+viii.194. [7500.]

Egypt

RAYMOND CLARE ARCHIBALD, Bibliography of egyptian mathematics, with special reference to the Rhind mathematical papyrus. Mathematical association of America: Oberlin, O. 1927. pp.[ii].84. [250.]

— — Supplement. [1929]. pp.[ii].85–106. [50.]

the Library of Congress copy contains a duplicated leaf for corrections and additions.

France

[MARQUIS A. R. DE VOYER D'ARGENSON DE PAULMY and A. G. CONTANT D'ORVILLE], De la lecture des livres françois. Livres concernant les sciences mathématiques & les arts qui en dépendent, imprimés au seizième siècle. Mélanges tirés d'une grande bibliothèque (vol. Bb): 1782. pp.viii.381. [100.]

E[UGÈNE] ESTANAVE, Nomenclature des thèses de sciences mathématiques soutenues en France dans le courant du XIXe siècle devant les facultés des sciences de Paris et des départements. 1903. pp.viii.44. [292.]

Germany

JOHANN GABRIEL DOPPELMAYR, Historische nachricht von den nürnbergischen mathematicis und künstlern. Nürnberg 1730. pp.[xxviii].314.[xviii]. [250.]

JOHANN SAMUEL ERSCH, Handbuch der deutschen literatur seit der mitte des achtzehnten jahrhunderts. . . . Des zweyten bandes erste . . . abtheilung, die literatur der mathematik, natur- und gewerbskunde enthaltend. Amsterdam &c. 1813. pp.xii.coll.760. [mathematics: 600.]

— — Neue . . . ausgabe, von Franz Wilhelm Schweigger-Seidel. Leipzig 1828. pp.[x].coll. 1740. [1500.]

A[LBERT] ERLECKE, Bibliotheca mathematica. Systematisches verzeichniss der bis 1870 in Deutschland . . . erschienenen werke. . . . Erster band, die encyclopaedisch-mathematische literatur umfassend. Halle a. S. [1872–]1873. pp.[iv]. 307. [7500.]

sets out the contents of mathematical periodicals; no more published.

DR. BUTTNERS literaturführer. Abteilung I, enthält sämtliche literarische erscheinungen auf dem gebiete der mathematik aus den jahren 1884–1889. Leipzig [1890]. pp.43. [850.]

VERZEICHNISS der seit 1850 an den deutschen universitäten erschienenen doctor-dissertationen und habilitationsschriften aus der reinen und angewandten mathematik. Deutsche mathematik-vereinigung: München 1893. pp.[iii].35. [1500.]

Hungary

JÓZSEF SZINNYEI and JÓZSEF SZINNEYI, Magyarország természettudományi és mathematikai könyvészete, 1472–1875. Kir. magyar természettudományi társulat: Budapest 1878. pp.viii. coll.1008. [15,000.]

ILONA GÁSPÁR, A magyar mathematikai irodalom bibliografiája 1901–1925. A magyar tudományos irodalom bibliografiája 1901–1925 (vol. iv): Budapest 1930. pp.xi.86. [2000.]

Italy

PIETRO RICCARDI, Biblioteca matematica italiana dalla origine della stampa ai primi anni del secolo XIX. Modena 1870–1880. pp.xxx.coll.656 +480.54.pp.[iv].156+pp.xxiii.312. [15,000.]

the above collation is of the work as correctly bound; it was issued erratically in parts; a facsimile was published Milano 1952.

— — Correzioni ed aggiunte. R. accademia di

scienze, lettere ed arti: 1928. pp.iii–vi.coll.106. [400.]

BIBLIOGRAFIA matematica italiana. Unione matematica italiana: Roma.

i. 1950. [By Alfredo Perna]. pp.xi.151. [1250.]
ii. 1951. pp.xi.159. [1750.]
iii. 1952. pp.xi.115. [1250.]
iv. 1953. pp.xii.119. [1250.]
v. 1954. pp.xiii.116. [1250.]
vi. 1955. pp.xiii.132. [1500.]
vii. 1956. pp.xiii.125. [1250.]
viii. 1957. pp.iii–xvi.137. [1500.]
ix. 1958. pp.iii–xvi.131. [1250.]
— Indice dei volumi I a IX. 1960. pp.[iii].67.
x. 1959. pp.iii–xii.104. [1093.]
xi–xii. 1960–1961. 1963. pp.vi–xvi.352. [4741.]

in progress.

Jews

ANNA GOLDBERG, Die jüdischen mathematiker und die jüdischen anonymen mathematischen schriften. Frankfurt a. M. 1901. pp.12. [500.]

Netherlands

D[AVID] BIERENS DE HAAN, Bibliographie néerlandaise historique-scientifique des ouvrages importants dont les auteurs sont nés aux 16ᵉ, 17ᵉ et 18ᵉ siècles sur les sciences mathématiques et physiques. Rome 1883. pp.424. [6500.]

a facsimile was published Nieuwkoop 1960.

CATALOGUS van boeken in Noord-Nederland verschenen van de vroegsten tijd tot op heden. [IX. Wis- en natuurkunde]. Samengesteld door de Tentoonstelling-commissie der Nationale tentoonstelling van het boek . . . 1910. Vereeniging ter bevordering van de belangen des boekhandels: 's-Gravenhage 1911. pp.[ii].coll.96.[viii]. [1250.]

Poland

TEOFIL ŻEBRAWSKI, Bibliografija piśmiennictwa polskiego z działu matematyki i fizyki. Na obchód czterechsetletniéj rocznicy urodzin Kopernika. Krakowie 1873. pp.[v].620. [2640.]

— — Dodatki. 1886. pp.[vii].155. [605.]

MARJA LORIOWA, Materjały do bibljografji piśmiennictwa kobiet polskich (do roku 1929). Nauki matematyczno-przyrodnicze i nauki stosowane. Polskie stowarzyszenie kobiet z wyższem wykształceniem: Oddział lwowski: Lwów 1934. pp.[vii].275. [4500.]

[ROMAN SIKORSKI *and others*], Polska bibliografia analityczna. Matematyka. Polska akademia nauk:

Mathematics

Ośrodek bibliografii i dokumentacji naukowej: Warszawa.

i. 1955. pp.   . [160.]
ii. 1956. pp.84.31.28.31. [190.]
no more published.

Portugal

RODOLPHO GUIMARÃES, Les mathématiques en Portugal au XIXe siècle. Aperçu historique et bibliographique. Coïmbre 1900. pp.167. [1500.]

— — Deuxième édition. 1909. pp.659. [3000.]

— — — Appendice II. 1911. pp.107. [100.]
the first appendix forms part of the main work.

Russia

УКАЗАТЕЛЬ русской литературы по математикѣ, чистымъ и прикладнымъ естественнымъ наукамъ. Общества естествоиспытателей: Кіевъ 1872–1891.
details of this work are set out under Science, below.

РУССКАЯ библіографія по естествознанію и математикѣ. Императорская академія наукъ: С.-Петербургское бюро международной библіографіи: Санктпетербургъ 1908–1912.
details of this work are set out under Science, below.

Mathematics

БИБЛИОГРАФИЯ математической литературы (1/1 1930 г.–1/VI 1934 г.). Книги, вышедшие в СССР на русском языке. Научно-исследовательский критико-библиографический институт: Москва &c. 1934. pp.72. [1500.]

МАТЕМАТИКА [ii–iii: и механика] в изданиях Академии наук. Библиографический указатель. Москва.

[i]. 1728–1935. Составили О. В. Динзе и К. И. Шафрановский. 1936. pp.xx. 315. [2439.]

ii. 1936–1947. Составила В. П. Алексеева. 1955. pp.516. [5154.]

iii. 1948–1952. Составили В. Б. Португаль, Г. И. Натансон, В. П. Алексеева. 1957. pp.363. [4152.]

GEORGE E[LMER] FORSYTHE, Bibliography of russian mathematics books. New York 1956. pp. [ii].106. [500.]

A[NNA] M[OISEENNA] LUKOMANSKAYA, Библиографические источники по математике и механике, изданные в СССР за 1917–1952 гг. Академия наук СССР: Библиотека: Москва &c. 1957. pp.355. [1305.]

Mathematics

United States

HENRY C[LAY] BADGER, Mathematical theses of junior and senior classes, 1782–1839. Harvard university: Library: Bibliographical contributions (no.32): Cambridge, Mass. 1888. pp.14. [406.]

BIBLIOGRAPHIA hopkinsiensis, 1876–1893. Parts IV, V, and VI. Physics, astronomy, mathematics. Johns Hopkins university: Baltimore 1894. pp. vii.57. [1500.]

7. *Applied mathematics*

ALWIN WALTHER [*and others*], Applied mathematics. Office of military government for Germany: Field information agencies technical: Fiat review of german science, 1939–1946: Wiesbaden 1948.

i. [Graphical methods, nomography, statistics &c.]. pp.[viii].307. [1250.]
ii. Mechanics of rigid and elastic bodies. pp.[iv].161. [1000.]
iii. Mathematical foundations of fluid mechanics. pp.[iv].202. [750.]
iv. Geodesy. pp.[iv].116. [558.]
v. Kinematics, geometrical optics, cristallography, applications in electrotechnics, geophysics, astronomy, ballistics, gyro-

scope devices. pp.[iv].263. [650.]
the text is in german.

M. CHLEBÍKOVÁ, Aplikácia matematiky v ekonómii. Výber literatúry. Ekonomické aktuality: Bibliografický zpravodaj (1960, no.3): Bratislava 1960. ff.[ii].409. [409.]*

8. *Pure mathematics*

BASIL ANDERTON and R. T. RICHARDSON, Catalogue of the books and tracts on pure mathematics in the Central library. Public libraries: Newcastle-upon-Tyne 1901. pp.vi.50. [2000.]

ERNST WÖLFFING, Mathematischer bücherschatz. Systematisches verzeichnis der wichtigsten deutschen und ausländischen lehrbücher und monographien des 19. jahrhunderts auf dem gebiete der mathematischen wissenschaften. . . . 1. teil: reine mathematik. Abhandlungen zur geschichte der mathematischen wissenschaften (no.xvi): Leipzig 1903. pp.xxxvi.416. [15,000.]
no more published.

CATALOGUE of scientific papers, 1800–1900. Subject index. Volume I. Pure mathematics. [Edited by] Herbert McLeod. Royal society: Cambridge 1908. pp.lviii.667. [40,000.]

WILHELM SÜSS [*and others*], Pure mathematics. Office of military government for Germany: Field information agencies technical: Fiat review of german science, 1939–1946: Wiesbaden 1948. pp.[viii].316+[iv].265. [1250.]
the text is in german.

9. *Teaching of mathematics*

DAVID EUGENE SMITH and CHARLES GOLDZIHER, Bibliography of the teaching of mathematics, 1900–1912. Bureau of education: Bulletin (no. 503): Washington 1912. pp.95. [1849.]

A LIST of books suitable for school libraries. Mathematical association: 1926. pp.16. [100.]
— [another edition]. 1936. pp.24. [171.]

MAX BLACK, The teaching of mathematics. A bibliography. University of London: Institute of education: [1938]. pp.14. [85.]

A BIBLIOGRAPHY of materials for the enrichment of mathematics instruction in elementary schools. State education department: Albany 1939. pp.15. [125.]

WILLIAM L[EONARD] SCHAAF, A bibliography of mathematical education. A classified index of the periodical literature since 1920. Forest Hills, N.Y. [1941]. pp.ix.144. [4250.]*

ROBERT A. and LOUISE J. ROSENBAUM, Bibliography of mathematics for secondary school libraries. Wesleyan university: Middletown, Conn. 1956. pp.16. [100.]*

— — [third edition]. 1959. pp.23. [150.]

G[ASTON] MIALARET, L'enseignement des mathématiques. Unesco: Revue analytique de l'éducation (vol.xi, no.9): 1959. pp.21. [79.]*

WILLIAM L[EONARD] SCHAAF, The high school mathematics library. National council of teachers of mathematics: Washington [1960]. pp.36. [450.]

CLARENCE ETHEL HARDGROVE, The elementary and junior high school mathematics library. National council of teachers of mathematics: Washington [1960]. pp.32. [200.]

JACK N. SPARKS and KENNETH I. TAYLOR, A collection of supplementary books in mathematics for the secondary school. Revised edition. West Leyden high school: Center for instructional materials: Northlake, Ill. [1960]. ff.18. [150.]*

10. *Mathematical tables*

RAYMOND CLARE ARCHIBALD, Mathematical

table makers. Scripta mathematica studies (no.3): New York 1948. pp.[v].82. [1250.]

HAROLD T[HAYER] DAVIS and VERA FISHER, *edd.* A bibliography and index of mathematical tables. Evanston, Ill. [1949]. ff.[ii].xxiii.286. [3000.]*

—— Second edition. By Alan Fletcher [*and others*]. 1962.

KARL SCHÜTTE, Index mathematischer tafelwerke und tabellen aus allen gebieten der naturwissenschaften. München 1955. pp.143. [1000.]

A. V. LEBEDEV and R. M. FEDOROVA, A guide to mathematical tables. English edition prepared by D. G. Try. Oxford &c. 1960. pp.xlvi.586. [3000.]*

—— N. M. Burunova, Supplement no.1. pp. [ii].190. [500.]*

DIRECTORY of continuing numerical data projects. National research council: Publication (no. 837): Washington 1961. pp.xiii.66. [100.]*

11. *Miscellaneous*

A CATALOGUE of the mathematical works of the learned mr. Thomas Baker. [*c.*1680]. pp.7. [13.]

[MICHEL CHASLES], Notice des travaux mathématiques de m. Chasles. [1843]. pp.39. [44.]

[JOSEPH BERTRAND], Notice des travaux mathématiques de m. J. Bertrand. 1850. pp.[ii].13. [27.]

— — [another edition]. 1856. pp.24. [28.]

[O. BONNET], Notice sur les travaux mathématiques de m. Ossian Bonnet. [1851]. pp.6. [23.]

— — [another edition]. 1856. pp.14. [36.]

[CLAUDE BOUQUET], Notice sur les travaux mathématiques de m. Bouquet. 1870. pp.9. [13.]

[A. MANNHEIM], Notice sur les travaux mathématiques de m. Am. Mannheim. 1870. pp.30. [25.]

— — [another edition]. 1885. pp.64. [91.]

— — — Supplément. 1886. pp.4. [4.]

[EDMOND NICOLAS LAGUERRE], Notice sur les travaux mathématiques de m. Laguerre. 1875. pp.16. [46.]

[L. SALTEL], Notice sur les travaux mathématiques de m. Louis Saltel. Bordeaux [printed] 1881. pp.20. [100.]

[GEORGE HENRI HALPHEN], Liste des travaux mathématiques de G.-H. Halphen. [1883]. pp.8. [75.]

— — [another edition]. Notice sur les travaux [&c.]. 1885. pp.52. [90.]

ПЕРЕЧЕНЬ статей по математикѣ А. П. Старкова. Одесса 1884. pp.3.

[PHILIBERT MAURICE D'OCAGNE], Notice sur les travaux mathématiques de m. Maurice d'Ocagne. Rochefort 1885. pp.29.*

[E. LEBON], Notice sur les travaux mathématiques de m. Ernest Lebon. 1900. pp.32. [38.]

MAURICE LECAT, Bibliographie des séries trigonométriques. Avec un appendice sur le calcul des variations. Louvain &c. 1921. pp.viii.ff.168. [3012.]

— — Compléments. 1924. pp.16. [266.]

[N. M. KRUILOV], Annotated list of the mathematical works of . . . Nicolas Kryloff. Chartres [printed] [1925]. pp.13. [61.]

A FIRST list of references on mathematical models. Science library: Bibliographical series (no.1): [1931]. ff.[i].3. [50.]*

VARIATION of van der Waal's 'b'. Science library: Bibliographical series (no.87): [1933]. ff.[i].2. [22.]*

FERNANDO L. GASPAR, Crónica bibliográfica de las publicaciones de la Facultad de ciencias matemáticas de la Universidad nacional del Litoral. Rosario 1940. pp.10. [9.]

G[EORGES] BOULIGAND, Index d'analyse et de géométrie, suivi de suppléments sur la théorie de l'intégrale, sur la spatialisation, sur la transformation de Sophus Lie. 1940. pp.vi.88. [1500.]

JOHANN JAKOB BURCKHARDT, Der mathematische nachlass von Ludwig Schläfli (1814–1895) in der Schweizerischen landesbibliothek. Bern 1942. pp.22. [large number.]

WILLIAM L[EONARD] SCHAAF, Recreational mathematics. A guide to the literature. National council of teachers of mathematics: Washington [1955]. pp.vii.129. [2750.]

VERA RILEY and SAUL I. GASS, Linear programming and associated techniques. A comprehensive bibliography in linear, nonlinear, and dynamic programming. Johns Hopkins university: Operations research office: Bibliographic reference series (no.5): Baltimore 1958. pp.x.613.

REFERENCES on mathematical programming (excluding linear programming). Hertfordshire county council: Technical information service [Hatfield] 1964. pp.40. [300.]*

Meteorology.

Meteorology

1. *Periodicals*

H. DARNAJOUX, Catalogue des principaux périodiques et séries météorologiques, scientifiques et techniques reçus à la bibliothèque. Direction de la météorologie nationale: Bibliothèque: Catalogue (no.1): 1963. ff.6. [150.]

H. DARNAJOUX, Catalogue des revues climatologiques françaises. Direction de la météorologie nationale: Bibliothèque: Catalogue (no.4): 1963. pp.[i].13. [120.]

H. DARNAJOUX, Catalogue des principaux bulletins de prévision du temps. Direction de la météorologie nationale: Bibliothèque: Catalogue (no.2): 1963. pp.[i].6. [50.]

2. *General*

[JOSEPH LARTIGUE], Notice sur les travaux et les ouvrages d'hydrographie et de météorologie de m. Lartigue. [1842]. pp.7.

CATALOGUE of books in the library of the society. British meteorological society: 1862. pp.12. [195.]
— [fourth edition]. Catalogue of the library of the Royal meteorological society. Compiled by J. S. Harding. 1891. pp.viii.214. [4750.]

[ALBERT LANCASTER], Catalogue des ouvrages d'astronomie et de météorologie qui se trouvent dans les principales bibliothèques de la Belgique. Bruxelles 1878. pp.[iii].xxiii.645. [meteorology: 1516.]

ALEXANDER RAMSAY, The scientific roll. A bibliography, guide and index to climate. [1880–] 1884. pp.[iii].449. [3000.]
published as a quarterly.

OLIVER L[ANARD] FASSIG, *ed.* Bibliography of meteorology. A classed catalogue of the printed literature of meteorology from the origin of printing to the close of . . . 1887. . . . Prepared under the direction of . . . A. W. Greely. War department: Washington.*

i. Temperature. 1889. pp.[viii].385. [15,000.]
ii. Moisture. 1889. pp.[v].478. [20,000.]
iii. Winds. 1891. pp.[vii].210.ff.[v]. [10,000.]
iv. Storms. 1891. pp.[v].383. [15,000.]

[C. KOCH], Katalog der bibliothek der Deutschen seewarte zu Hamburg. Hamburg 1890. pp.x.619. [10,660.]

[—] — Nachtrag.
i. 1894. pp.vi.92. [1500.]
ii. 1899. pp.vi.142. [2500.]
iii. 1901. pp.viii.64. [1000.]

iv. 1903. pp.vi.62. [1000.]
v. 1904. pp.vi.36. [500.]
vi. 1905. pp.vi.53. [750.]
vii. 1907. pp.vi.86. [1500.]
viii. 1909. pp.vi.78. [1250.]
ix. 1911. pp.vi.82. [1500.]

JAHRBUCH der astronomie und geophysik. Enthaltend die wichtigsten fortschritte auf den gebieten der astrophysik, meteorologie und physikalischen erdkunde. Leipzig 1890–1913.

details of this work are set out under Astronomy, above.

J. S. HARDING, Catalogue of the . . . Royal meteorological society. 1891.

LÁSZLÓ SZALAY, A M. Kir. országos meteorologiai és földmágnességi intézet könyvtárának név- és tárgymutatója. Budapest 1902. pp.viii.423. [5500.]

[—] — [supplements].
[i]. 1902. pp.[ii].25. [500.]
ii. 1903. pp.[ii].34. [600.]
iii. 1904. pp.[ii].32. [600.]
iv. 1905. pp.[ii].41. [750.]
v. 1906.
vi. 1907.

vii. 1908. pp.52. [750.]
viii. 1909. pp.54. [750.]
ix. 1910. pp.44. [600.]
x. 1911. pp.53. [750.]
xi. 1912. pp.47. [650.]
xii. 1913. pp.87. [1250.]
xiii. 1914. pp.57. [750.]
xiv. 1915. pp.38. [600.]

METEOROLOGY, including terrestrial magnetism. International catalogue of scientific literature (section F): Royal society.

i. 1902. pp.xiv.184. [1218.]
ii. 1904. pp.viii.296. [1988.]
iii. 1905. pp.viii.235. [1305.]
iv. 1906. pp.viii.223. [1235.]
v. 1907. pp.viii.283. [1645.]
vi. 1908. pp.viii.257. [1527.]
vii. 1909. pp.viii.284. [1900.]
viii. 1910. pp.viii.293. [2159.]
ix. 1912. pp.viii.238. [1106.]
x. 1913. pp.viii.224. [1884.]
xi. 1914. pp.viii.245. [1816.]
xii. 1915. pp.viii.194. [1350.]
xiii. 1916. pp.viii.137. [692.]
xiv. 1919. pp.viii.181. [1124.]

no more published.

Meteorology

ANTAL RÉTHLY, Az Ó-Gyallai M. Kir. orsz. meteorologiai és főldmágnességi observatorium könyvtárának név- és tárgymutatója. Budapest 1903. pp.x.115. [2000.]

[—] — Az 1903 évi szaporulat jegyzéke. 1904. pp.[iii].43. [750.]

SUBJECT list of works on aerial navigation and meteorology, in the library of the Patent office. Patent office library series (no.17 = Bibliographical series, no.14): 1905. pp.64. [337.]

C. FITZHUGH TALMAN, Brief list of meteorological text-books and reference books. A selection of works suitable for general, scientific, and university libraries in the United States. Department of agriculture: Weather bureau: Washington 1909. pp.16. [150.]

— — Second edition. 1910. pp.18. [200.]

SELECT list of references on descriptive meteorology. Library of Congress: Washington 1911. ff.3. [18.]*

BIBLIOGRAPHY of meteorological literature. Prepared by the Royal meteorological society with the collaboration of the Meteorological office. Royal meteorological society.

[i. 1920–1925]. pp.367. [6000.]

[ii. 1926–1931]. pp.462. [11,000.]
iii. [1932–1935]. pp.344. [8000.]*
iv. [1936–1939]. pp.504. [12,000.]*
v. [1940–1945]. pp.284. [7000.]*
vi. [1946–1949]. [7000.]*

in progress; previously published in the society's Quarterly journal.

BIBLIOGRAPHIE météorologique. Office national météorologique et Société météorologique de France.

i–iv. 1921–1924. pp.210. [3662.]
v. 1925. Publiée sous la direction générale du général Delcambre. pp.93. [1396.]
vi. 1926. pp.235. [1505.]
vii. 1927. pp.219. [1289.]
viii. 1928. pp.181. [1052.]
ix. 1929. pp.163. [1227.]
x. 1930. pp.151.xxix.9.6.33. [1051.]
xi. 1931.
xii. 1932. pp.291. [1545.]

[*continued as:*]

Bibliographie internationale de météorologie générale. Nouvelle série.

i. 1933. ff.256. [2162.]
ii. 1934. pp.(ff.)237. [1500.]
iii. 1935. pp.(ff.)262. [1000.]
iv. 1936. pp.(ff.)359. [1250.]

v. 1937. pp.(ff.)449. [2000.]
vi. 1938. pp.(ff.)415. [2000.]
vii. 1939. pp.(ff.)487. [2500.]
[*continued as:*]

Bibliographie météorologique internationale. Météorologie nationale [1952 &c.: Organisation météorologique mondiale]: Genève.

1948. pp.(ff.)140+147+174+244. [1850.]
1949. pp.(ff.)265+263+194+189. [2000.]
1950. pp.(ff.)260+214+196+207. [2500.]
1951. pp.(ff.)430+375+194+213. [3250.]
1952. pp.(ff.)213+329+225+269. [3000.]

in progress.

W. GEORGII, Wettervorhersage. Wissenschaftliche forschungsberichte: Naturwissenschaftliche reihe (vol.xi): Dresden &c. 1924. pp.viii.114. [250.]

G[USTAV] HELLMANN, Versuch einer geschichte der wettervorhersage im XVI. jahrhunderte. Preussische akademie der wissenschaften: Physikalisch - mathematische klasse: Abhandlungen (1924, no.1): Berlin 1924. pp.54. [500.]

THE BIBLIOGRAPHY of Ford Ashman Carpenter. Meteorology and aeronautics. Chamber of commerce: Library: Los Angeles 1929. ff.[ii].24. [228.]*

BIBLIOGRAPHIE der aerologischen literatur. Internationale kommission für die erforschung der freien atmosphäre: Berlin.*

1929–1934. Von K[arl] Keil. ff.[138]. [1000.]
1934–1935. ff.[85]. [684.]
1935–1936. ff.[67]. [520.]
1936–1937. ff.[59]. [440.]
1937–1938. ff.[73]. [566.]
1938–1939. ff.[59]. [456.]

no more published.

BIBLIOGRAPHY of aeronautics. Part 2. Meteorology.... Compiled from the Index of aeronautics of the Institute of the aeronautical sciences. Works progress administration: [New York] 1937. pp. [vii].641. [8500.]*

A BIBLIOGRAPHY of scientific papers on climatic variations. International geographical union: Lwów 1938. ff.[ii].iv.254. [4153.]*

ZENTRALBLATT für geophysik, meteorologie ünd geodesie. Berlin.

i. Redaktion: Karl Jung. 1938. pp.iv.490. [2000.]
ii. 1938. pp.iv.499. [2000.]

CATÁLOGO de publicaciones periódicas existentes en la biblioteca de la Dirección de meteorología, geofísica e hidrología. Ministerio de agricultura: Buenos Aires 1938–1939. pp.124. [1000.]

GRACE HADLEY FULLER, Weather in relation to war. A list of references. Compiled . . . under the direction of Florence S[elma] Hellman. Library of Congress: Division of bibliography: [Washington] 1943. pp.[ii].24. [350.]*

SELECTED bibliography on meteorology and related subjects. Weather bureau: Library: Washington 1943. pp.31. [350.]*

RATJE MÜGGE [*and others*], Meteorology and physics of the atmosphere. Office of military government for Germany: Field information agencies technical: Fiat review of german science, 1939–1946: Wiesbaden 1948. pp.[viii].291. [1000.]
the text is in german.

BIBLIOGRAPHY of meteorological instruments. Harvard university: Blue Hill meteorological observatory: [*s.l.*] 1950. ff.[165]. [500.]*

BIBLIOGRAFIE van meteorologiese streekliteratur. Weather bureau: Pretoria.

i. 1486–1948. Compiled by R. J. Venter. 1950. pp.[iv].412. [2911.]

ZUGÄNGE der Wetterdienstbibliothek. Deutscher wetterdienst in der US-zone: Bad Kissingen 1949 &c.*

in progress?

H[ANS] CORDES, Bibliographie "klima und mensch" 1945–1949. Bibliographie Deutscher wetterdienst in der US-zone: Bad Kissingen 1950. pp.[ii].ii.96. [445.]*

[E. H. MUELLER], Meteorology and climate and related subjects. Linda hall library: Bibliographic bulletin (no.7): Kansas City 1954. ff.[ii].48. [600.]*

EFFECTS of sunspots on weather. Science library: Bibliographical series (no.725): 1954. ff.4. [67.]*

HANS CORDES and GERHARD ZIMMERMAN, Bibliographie "klima und mensch" 1950–1951. Deutscher wetterdienst: Bibliographien (no.2): Bad Kissingen 1955. pp.x.59. [762.]

G[ALINA] P[ETROVNA] BOGATOVA, Погода и ее научное предвидение. Краткий рекомендательный указатель. Государственная... библиотека СССР имени В. И. Ленина: В помощь лектору: Москва 1955. pp.19. [40.]

CONTROLLED atmospheric environment in relation to people, animal and plants. A list of publications by staff members of the university of Illinois and the State natural history survey division. [Urbana] 1957. pp.43. [600.]

[GEORGE H. SIEHL], An annotated meteorological bibliography for secondary schools. Weather bureau: Washington 1960. pp.16. [100.]

[—] — Second edition. [By Bette M. Mock]. 1963. pp.31. [250.]

A. S. CHAPLUIGINA, Аннотированная библиография работ по проблеме численных методов краткосрочного прогноза. Академия наук СССР: Междуведомственный геофизический комитет: Москва 1962. pp.336+337–698 +701–1064. [2000.]*

RUDOLF ANIOL and MAX SCHLEGEL, Klimaatlanten und neuere klimakarten. Deutscher wetterdienst: Bibliographien (no.14): Offenbach a. M. 1963. pp.31. [434.]*

METEOROLOGIE, klimatologie. Literatuur overzicht. Technisch documentatie [en informatie] centrum voor de krijgsmacht: 's-Gravenhage.*

[i]. 1955.
[ii]. 1956.
[iii]. 1957. pp.179. [300.]
iv. 1958. pp.156. [257.]
v. 1959. pp.77. [228.]

vi. 1960. pp.120. [323.]
vii. 1961. pp.126. [300.]
in progress.

HAROLD L. CRUTCHER and GEORGIA L. WHITING, Ocean station vessel meteorological records survey, Atlantic and Pacific. Weather bureau: Key to meteorological records documentation (no.2-3): Washington 1956. pp.vi.63.

JUDITH HAYES, Investigating the lunar atmosphere and planetary atmosphere. California institute of technology: Jet propulsion laboratory: Astronautics information literature search (no. 196): Pasadena 1960. pp.iii.16.*

METEOROLOGICAL & geoastrophysical titles. American meteorological society: [Washington] 1961 &c.*
in progress.

CUMULATIVE list of foreign literature relating to the atmospheric sciences translated by American meteorological society 1 January 1952–31 December 1961. Boston [1962]. pp.128.

ELEMER KISS, Bibliography on meteorological satellites, 1950–1962. Weather bureau: Washington 1963. pp.xx.380. [988.]*

Meteorology

JOHN R. SWARTZ, Selective guide to published climatic data sources. Weather bureau: Key to meteorological records documentation (no.4.11): Washington 1963. pp.xv.84.

CATALOGUE of meteorological satellite data Tiros II television photography. Weather bureau: Key to meteorological records documentation (no.5.32): Washington 1963. pp.iii.19.*

MICROMETEOROLOGY, aerosols, and air pollution. Bibliography. Library of Congress: Aerospace information division: AID report (no.B-63-50): [Washington] 1963. ff.[ii].13. [108.]*

HISTORY and catalogue of upper air data for the period 1946–1960. Weather bureau: Key to meteorological records documentation (no.5.21): Washington 1964. pp.v.352.

3. *Countries &c.*

Africa

A[MILCARE] FÀNTOLI, Bibliografia meteorologica e geofisica dell'impero Etiopico-Eritrea e Somalia. Ministero dell'Africa italiane: Ufficio studi e propaganda: Roma 1937. pp.11. [2000.]

Meteorology

America, south

MARGARET M[ORRIS] WELCH, Bibliography on the climate of South America. Weather bureau: Monthly weather review: Supplement (no.18): Washington 1921. pp.42. [1550.]

Argentina

ENRIQUE SPARN, Bibliografía meteorologica y climatologica de la republica Argentina. Con un apéndice: "Literatura sobre meteorología y climatología de las regiones antártica y subantártica americanas". Academia nacional de ciencias: Miscelanea (no.7): Córdoba 1923. pp.72. [850.]

— — 1924–1931. 1934. pp.47. [500.]

SELECCIÓN de reseñas bibliográficas meteorológicas y geofísicas. Servicio meteorológico nacional: Serie G: Buenos Aires.

1952. pp.[iii].107. [465.]
1953. pp.[iii].89. [377.]
1954. pp.[ii].83. [343.]

no more published?

Brazil

TANCREDO [ESTRADA] DE BARROS PAIVA, Bibliographia do clima brasílico. Ministerio da agricultura, industria e commercio: Serviço de informações: Rio de Janeiro 1928. pp.[ii].34. [600.]

Meteorology

Canada

MORLEY K. THOMAS, A bibliography of canadian climate 1763–1957. National research council [&c.]: Ottawa 1961. pp.114. [1250.]

Caucasus

NINA A[LEKSEEVNA] STEPANOVA, Annotated bibliography of the climate of the Caucasus. Weather bureau: Washington 1960. pp.[v].43. [134.]*

Germany

GUSTAV HELLMANN, Repertorium der deutschen meteorologie. Leistungen der deutschen schriften, erfindungen und beobachtungen. Leipzig 1883. pp.xxiv.coll.996. [10,000.]

G[USTAV] HELLMANN, Die meteorologie in den deutschen flugschriften und flugblättern des XVI. jahrhunderts. Preussische akademie der wissenschaften: Physikalisch-mathematische klasse: Abhandlungen (1921, no.3): Berlin 1921. pp.96. [200.]

India

PUBLICATIONS of the India meteorological department. Complete list. [Delhi 1944]. pp.12. [300.]

Meteorology

Italy

GIUSEPPE BOFFITO, Per la storia della meteorologia in Italia. Real collegio Carlo Alberto in Moncalieri: Osservatorio centrale: Torino 1898. pp. 111. [250.]

Jordan

DARTHULA M. CARRAWAY, Annotated bibliography of climatic maps of Jordan. Weather bureau: Washington 1961. pp.v.10. [33.]*

Mexico

RAFAEL AGUILAR Y SANTILLAN, Bibliografía meteorológica mexicana. Sociedad científica "Antonio Alzate": México 1890. pp.47. [223.]
supplements by the author appear in the Sociedad's Memorias *(1890–1897), iv–x.*

Netherlands

LIJST van uitgaven. Liste des publications. 1850–51 mei 1904. Koninklijk nederlandsch meteorologisch instituut: Utrecht 1904. pp.12. [90.]

LIJST van uitgaven. Meteorologisch instituut (uitgave 93a): Utrecht 1913. pp.27. [250.]

Meteorology

Peru

ANGEL J. INDACOCHEA G., Bibliografía climatológica del Peru. Dirección de minas y petróleo: Boletín (no.4): Lima 1946. pp.81. [700.]

Poland

BIBLJOGRAFJA prac opublikowanych przez pracowników i dawnych współpracowników Państwowego instytutu meteorologicznego. Warszawa 1935. pp.113. [800.]

WYKAZ prac naukowych ogłoszonych w ciągu 38 lat: 1898–1935. Władislaw Gorczyński. Państwowy instytut meteorologiczny: Warszawa 1935. pp.viii.32. [300.]

POLISH meteorology. Bibliography. [Library of Congress:] Aerospace information division: AID report (no.62–3): [Washington] 1962. ff.[i].vi.31. [122.]*

Russia

BIBLIOGRAPHY of russian texts and translations. [Department of the air force:] Air weather service: [Washington] 1947. ff.[ii].68. [278.]*

R. M. AKCHURINA and YU[LIYA] O[SIPOVNA] BIBILO, Климат и воды суши Юго-Востока европейской части СССР. Библиографический указатель. Университет: Научная библиотека: Библиография Саратовской области (vol.5): [Saratov] 1961. pp.268. [3065.]

Switzerland

R. BILLWILL, Klimatologie. Centralkommission für schweizerische landeskunde: Bibliographie der schweizerischen landeskunde (section IV.4): Bern 1927. pp.xv.140.9. [3000.]

United States

WEATHER bureau bulletins, circulars, maps, and monthly weather review. Superintendent of documents: Price list (no.48): Washington [1909]. pp.12. [150.]

frequently re-edited under various titles.

[LEWIS JEFFERSON DARTER], List of climatological records in the National archives. National archives: Special list (no.1): Washington 1942. pp.lxii.138. [very large number.]*

HAROLD T. PINKETT, HELEN T. FINNERAN and KATHERINE H. DAVIDSON, Preliminary inventory of the climatological and hydrological records of the Weather bureau. National archives: [Preliminary inventories (no.38):] Washington 1952. pp.v.76. [very large number.]*

Meteorology

UNITED STATES navy weather charts. Office of naval operations: [Washington] 1956. pp.19.

consists of key maps.

JAMES M. HAVENS, An annotated bibliography of meteorological observations in the United States 1715–1818. Weather bureau: Key to meteorological records documentation (no.5–11): Washington 1958. pp.14.

Physics.

1. *Bibliographies*

KARL K[ELCHNER] DARROW, Classified list of published bibliographies in physics, 1910–1922. National academy of sciences: National research council: Bulletin (vol.viii, part 5 = no.47): Washington 1924. pp.v.102. [2500.]

PEDER KRISTIAN LYSTBOEK, Udenlandske tidsskrifter af matematisk eller fysisk indhold i danske biblioteker. Universitets bibliotek: København 1952. pp.viii.176. [1569.]

A[NNA] M[OISEEVNA] LUKOMSKAYA, Основные иностранные библиографические источники по физике (1931–1955). Академия наук Союза CCP: Библиотека: Москва &c. 1959. pp.115. [415.]

A. M. LUKOMSKAYA, Библиография отечественной литературы по математике и физике. Обзор библиографических источников. Академия наук СССР: Библиотека: Москва &c. 1961. pp.159. [289.]

2. *Periodicals*

ERNST WÖLFFING, Verzeichniss der zeitschriften für die gebiete der mathematik, der physik . . . und der verwandten wissenschaften welche auf württembergischen bibliotheken vorhanden sind. Stuttgart 1899. pp.18. [71.]

S. R. RANGANATHAN, P. N. KAULA and T. N. KORANNE, Union catalogue of periodical publications in the libraries of south Asia. Delhi 1950. pp.[71]. [300.]

a specimen only, limited to mathematics and physics.

ROBERT S. BRAY, List of periodicals of physics interest. Department of commerce: Office of technical services: [Washington] 1950. pp.[iv].37. [900.]*

REAL Sociedad española de fisica y química. Índice de las revistas existentes en su biblioteca. [Madrid 1951]. pp.14. [300.]

*PEDER KRISTIAN LYSTBŒK, Udenlandske tidsskrifter af matematisk eller fysisk indhold i danske biblioteker. Universitetsbiblioteket: København 1952. pp.viii.176. [1569.]

[MARTHE SIESTRUNCK], État des périodiques de mathématiques et de physique figurant à la bibliothèque de la Sorbonne. Édition provisoire. Documentation mathématique (no.24–25): 1953. ff.43 +[ii].44–80. [1500.]*

3. *General*

THE PHYSIOLOGICAL library begun by mr. Steuart, and some of the students of natural philosophy in the university of Edinburgh. [?Edinburgh] 1725. pp.64. [410.]

J[EREMIAS] D[AVID] REUSS, Repertorium commentationum a societatibus litterariis editarum. ... Tom. IV. Physica. Gottingae 1805. pp.viii.416. [5000.]

JOHANN WILHELM HEINRICH CONRADI, Catalogus bibliothecae medico-physicae Ern. Godofr. Baldingeri. Tom. II. Bibliotheca physica. Marburgi 1805. pp.iv.729–1168. [5055.]

there are two sets of pages numbered 729–738 in the whole work.

[JÖNS] JACOB BERZELIUS, Jahres-bericht über die fortschritte der physischen wissenschaften. Tübingen 1822–1850.

details of this work are set out under Chemistry: General, above.

GUSTAV THEODOR FECHNER, Repertorium der experimentalphysik, enthaltend eine vollständige zusammenstellung der neuern fortschritte dieser wissenschaft. Leipzig 1832. pp.xx.492+xii.498+vi.258. [2500.]

DIE FORTSCHRITTE der physik im jahre .... Dargestellt von der Physikalischen gesellschaft zu Berlin [*afterwards:* Deutschen physikalischen gesellschaft]. Berlin [vols.xliv–lxxiv: Braunschweig].

i. 1845. Redigirt von G. Karsten. 1847. pp. xxxii.622. [1000.]
ii. 1846. 1848. pp.xlviii.639. [1000.]
iii. 1847. 1850. pp.xliv.703. [1000.]
iv. 1848.
v. 1849. Redigirt von W. Beetz und G. Karsten. 1853. pp.xxiv.500. [250.]
vi. 1850.
vii. 1851.
viii. 1852. Redigirt von A. Krönig. 1855. pp.lxxii.794. [1000.]
ix. 1853. 1856. pp.xliv.753. [1000.]
x. 1854. 1857. pp.l.814. [1000.]
xi. 1855. 1858. pp.lxxvi.834. [1000.]

xii. 1856. 1859. pp.lxx.826. [1000.]

xiii. 1857. Redigirt von A. Krönig und O. Hagen. 1859. pp.xliv.628. [1000.]

xiv. 1858. Redigirt von O. Hagen. 1860. pp.xlviii.735. [1000.]

xv. 1859. Redigirt von E[mil] Jochmann. 1861. pp.lxxviii.806. [1250.]

xvi. 1860. 1862. pp.lx.928. [1250.]

xvii. 1861. 1863. pp.lx.810. [1250.]

xviii. 1862. 1864. pp.lvi.856. [1250.]

xix. 1863. 1865. pp.lvi.747. [1250.]

xx. 1864. 1867. pp.lxiii.950. [1250.]

— Namen- und sach-register zu . . . band 1. bis xx. Bearbeitet von W. Barentin. 1872. pp.[ii].500.

xxi. 1865. Redigirt von H. Kiessling und A. Kundt. 1868. pp.lii.764. [1250.]

xxii. 1866. Redigirt von G. Quincke, B. Schwalbe und [Albert] Wangerin. 1869. pp.lvi. [1000.]

xxiii. 1867. Redigirt von G. Quincke und B. Schwalbe. 1870. pp.lii.656. [1500.]

xxiv. 1868. Redigirt von B. Schwalbe. 1872. pp.lxviii.792. [1500.]

xxv. 1869. 1873. pp.lxxvii.1084. [2500.]

xxvi. 1870. 1875. pp.lxxiv.1021. [3000.]

xxvii. 1871. 1876. pp.lxxxvi.1146. [3500.]
xxviii. 1872. 1877. pp.xcviii.1296. [3500.]
xxix. 1873. 1878. pp.c.1289. [3500.]
xxx. 1874. Redigirt von B. Schwalbe und [F.] Neesen. 1879. pp.cxvi.1524. [4500.]
xxxi. 1875. Redigirt von [F.] Neesen. 1880. pp.civ.1204. [3000.]
xxxii. 1876. Redigirt von B. Schwalbe. 1881. pp.cxvi.1636. [5000.]
xxxiii. 1877. 1882. pp.cxxviii.1675. [5500.]
xxxiv. 1878. Redigirt von F. Neesen. 1884. pp.cxl.1391. [3000.]
xxxv. 1879. 1886. pp.cxx.1504. [4000.]
xxxvi. 1880. Redigirt von F. Neesen und B. Schwalbe. 1883. pp.cli.1250.634. [5500.]
xxxvii. 1881. 1887. pp.xcviii.1237+xlvii.788. [5500.]
xxxviii. 1882. Redigirt von F. Neesen und E. Rosochatius. 1887–1888. pp.lx.400+xlvii.761+xlvii.792. [5500.]
xxxix. 1883. Redigirt von E. Rosochatius. 1889–1890. pp.lxiv.460+lviii.1055+xlix.850. [6000.]
xl. 1884. Redigirt von E. Budde [part 3: B. Schwalbe]. 1890. pp.lxvii.538+lvi.1058+lv.920. [8000.]

xli. 1885. 1891. pp.lxvi.534+l.896+lxiii. 1113. [8000.]

xlii. 1886. 1892. pp.lxx.602+xlviii.820+ lxxvii.1177. [8000.]

xliii. 1887. [part 3: Redigirt von (Richard) Assmann]. 1893–1894. pp.xviii.586+liii. 936[*sic*, 944]+lviii.815. [8000.]

— Namenregister nebst einem sach-ergänzungsregister zu . . . bd.XXI (1865) bis XLIII (1887) unter berücksichtigung der in den bänden I–XX enthaltenen autoren-namen. Bearbeitet von B. Schwalbe. 1897[–1898]. pp.vii.1094.

xliv. 1888. Redigirt von Richard Börnstein [part 3: R. Assmann]. 1894. pp.lxii.479+ xlv.809+l.771. [6500.]

xlv. 1889. 1895. pp.lxvii.580+xlix.821+ lvii.793. [7500.]

xlvi. 1890. 1896–1897. pp.lxvii.523+xliv.781 +liv.780. [7500.]

xlvii. 1891. 1897. pp.lxiv.418+xlii.752+ xlvii.821. [7000.]

xlviii. 1892. 1897–1898. pp.lxvi.449+xliii. 778+l.597. [7500.]

xlix. 1893. 1895. pp.lxx.562+liv.900+xliv. 727. [7000.]

l. 1894. 1895. pp.lxxiii.600+xlv.853+li.716. [7500.]

li. 1895. 1896. pp.lxxii.510+xlvii.843+liv. 686. [7500.]

lii. 1896. 1897. pp.lxx.476+xlix.820+xlv. 531. [8000.]

liii. 1897. 1898. pp.lxxiii.573+lii.912+xlv. 566. [8000.]

— Namenregister nebst einem sach-ergänzungsregister zu . . . band XLIV (1888) bis LIII (1897), unter mitwirkung von E. Schwalbe . . . bearbeitet von G. Schwalbe. 1903. pp.xix.1044.

liv. 1898. 1899–1900. pp.lxxvi.694+liv.984 +xliii.586. [8000.]

lv. 1899. Redigirt von R. Börnstein und Karl Scheel [part 3: R. Assmann]. 1900. pp. lxxviii.693+liii.935+xliv.544. [8000.]

lvi. 1900. Redigirt von K. Scheel [part 3: R. Assmann]. 1901. pp.xxxvi.357+lii.794 +xlviii.472. [6500.]

lvii. 1901. 1902. pp.xxxix.421+lxiv.810+ lviii.610. [8500.]

lviii. 1902. 1903. pp.xl.496+liv.906+lxviii. 680. [9500.]

lix. 1903. 1904. pp.xlviii.691+xlvi.675+ lxiv.581. [9500.]

lx. 1904. 1905. pp.l.765+lii.810+lxv.638. [10,000.]

lxi. 1905. 1906. pp.xxxix.670+xl.714+ lxxiii.658. [10,000.]

lxii. 1906. 1907. pp.xxxix.714+xlii.727+ lxxii.641. [10,000.]

lxiii. 1907. 1908. pp.xxxvi.614+xliv.714+ lxxiv.604. [10,000.]

lxiv. 1908. 1909. pp.xxxii.573+xliii.678+ lxvii.590. [10,000.]

lxv. 1909. 1910. pp.xxx.507+xliii.654+ lxxi.584. [9000.]

lxvi. 1910. 1911. pp.xxviii.485+xlvi.758+ lxxiv.686. [10,000.]

lxvii. 1911. 1912. pp.xxxiv.544+xlviii.755+ lxx.650. [10,000.]

lxviii. 1912. 1913. pp.xxx.499+l.789+ lxxviii.717. [10,000.]

lxix. 1913. 1914. pp.xxxii.487+l.816+lviii. 582. [10,000.]

lxx. 1914. 1915. pp.xxviii.431+xliii.664+ xliv.430. [9000.]

lxxi. 1915. 1916. pp.xxviii.424+xxxv.544+ li.562. [8500.]

lxxii. 1916. 1917. pp.xxviii.432+xxxiv.497 +l.508. [8000.]

lxxiii. 1917. 1918–1919. pp.xiv.227+xix.270 +xxiv.276. [6000.]

lxxiv. 1918. Redigirt von K. Scheel [part 3: Reinhard Joachim Süring]. 1919. pp.xvi. 208+xx.288+xx.243. [5000.]

— Die fortschritte der physik. . . . Halbmonatliches litteraturverzeichniss. Deutsche physikalische gesellschaft: Braunschweig.

i. 1902. Redigirt von Karl Scheel, Richard Assmann. pp.[ii].xiv.452. [7500.]

[*continued as:*]

Halbmonatliches literaturverzeichnis der 'Fortschritte der physik'.

ii. 1903. pp.[ii].407. [7500.]
iii. 1904. pp.[ii].438. [7500.]
iv. 1905. pp.[ii].444. [7500.]
v. 1906. pp.[ii].446. [7500.]
vi. 1907. pp.[ii].456. [7500.]
vii. 1908. pp.[ii].454. [7500.]
viii. 1909. pp.[ii].450. [7500.]
ix. 1910. pp.[ii].470. [8000.]
x. 1911. pp.[ii].464. [8000.]
xi. 1912. pp.[ii].490. [8000.]
xii. 1913. pp.[ii].568. [8000.]
xiii. 1914. pp.[ii].414. [7500.]
xiv. 1915. pp.[ii].382. [7000.]
xv. 1916. pp.[ii].368. [7000.]

xv [*sic*, xvi]. 1917. pp.[ii].346. [6000.]
xvii. 1918. pp.[ii].302. [5000.]
xviii. 1919. Redigirt von K. Scheel, R[einhard Joachim] Süring. pp.[ii].300. [5000.]

amalgamated, in both its forms, with the Beiblätter zu den Annalen der physik, *to form the* Physikalische berichte, *which is entered below; in the* Fortschritte *the second part of vols.xxxvi–xxxvii is devoted to the physics of the earth; the three parts of the subsequent volumes deal respectively with the physics of matter (afterwards general physics, acoustics, physical chemistry), the physics of ether (afterwards electricity and magnetism, optics, heat), and the physics of the earth (afterwards cosmic physics); the second of the three parts contains an index to the first two.*

FRIEDRICH ZAMMER, Jahresbericht über die fortschritte der physik . . . für 1857. Giessen 1858. pp.xii.288. [400.]

forms the second part of the issue for 1857 of the Jahresbericht über die fortschritte der chemie, *which is entered under Chemistry above.*

[MARCELLIN BERTHELOT], Notice sur les travaux de physique de m. Berthelot. 1872. pp.33. [60.]

TEOFIL ŻEBRAWSKI, Bibliografija piśmiennictwa polskiego z działu matematyki i fizyki. Na

obchód czterechsetletnièj rocznicy urodzin Kopernika. Krakowie 1873. pp.[v].620. [2640.]
— — Dodatki. 1886. pp.[vii].155. [605.]

D[AVID] BIERENS DE HAAN, Bibliographie néerlandaise historique-scientifique des ouvrages importants dont les auteurs sont nés aux 16e, 17e et 18e siècles sur les sciences mathématiques et physiques. Rome 1883. pp.424. [5000.]
a facsimile was published Nieuwkoop 1960.

V[IKTOR] V[IKTOROVICH] BOBUININ, Русская физико-математическая библиографія. Москва.
i.I. [1587–1720]. 1885. pp.102. [99.]
[i.II]. [1726–1737]. 1886. pp.116. [250.]
i.III. [1746–1763]. 1890. pp.162. [1000.]
ii.I. 1764–1774. 1889. pp.[ii].109. [1000.]
ii.II. [1775–1786]. 1890. pp.[ii].160. [1500.]
ii.III. [1787–1791]. 1892. pp.[ii].162. [1500.]
ii.IV. [1792–1799]. 1893. pp.[ii].272. [2500.]
iii.I. [1800–1805]. 1895. pp.[ii].172. [1500.]
iii.II. [1806–1809]. 1895. pp.[ii].170. [1500.]
iii.III. [1810–1816]. 1900. pp.[ii].iv.248. [2000.]

KATALOG der in den hamburger bibliotheken vorhandenen literatur aus der reinen und ange-

wandten mathematik und physik. Hamburg 1890.
— Nachtrag. 1893.
— 2. nachtrag. 1906. pp.[ii].vi.352. [9000.]

BIBLIOGRAPHIA hopkinsiensis, 1876–1893. Parts IV, V, and VI. Physics, astronomy, mathematics. Johns Hopkins university: Baltimore 1894. pp. vii.57. [1500.]

ROBERT ALFRED LEHFELDT, A list of the chief Memoirs on the physics of matter. Physical society: 1894. pp.ix.41. [511.]
interleaved.

SEYDEL's führer durch die litteratur der theoretischen physik und elektrotechnik. . . . 6. . . . auflage. Berlin 1898. pp.[iv].91. [1400.]
limited to works in german.

ABSTRACTS of physical papers from foreign sources. [Edited by J. Swinburne]. Physical society.
i. 1895. pp.iii.621. [798.]
ii. 1896. pp.vi.499. [877.]
iii. 1897. pp.vi.469. [726.]

SCIENCE abstracts. Physics and electrical engineering. Issued under the direction of the Institution of electrical engineers, the Physical society of London.

i. 1898. pp.viii.834. [1423.]
ii. 1899. pp.xi.961. [2000.]
iii. 1900. pp.xii.1100. [2525.]
iv. 1901. pp.xii.1192. [2506.]
v. 1902. pp.xii.1056. [2362.]

[*continued as:*]

Science abstracts. Section A. Physics. Institution of electrical engineers.

vi. 1903. pp.xv.636. [1532.]
vii. 1904. pp.xvi.1148. [3339.]
viii. 1905. pp.xvi.820. [2308.]
ix. 1906. pp.xvi.751. [2171.]
x. 1907. pp.xvi.800. [2132.]
xi. 1908. pp.xvi.814. [2070.]
xii. 1909. pp.iv.812. [2161.]
xiii. 1910. pp.iv.695. [1810.]
xiv. 1911. pp.iv.663. [1749.]
xv. 1912. pp.xvi.695. [1916.]
xvi. 1913. pp.iv.755. [2022.]
xvii. 1914. pp.iv.811. [2041.]
xviii. 1915. pp.iv.771. [1789.]
xix. 1916. pp.iv.598. [1397.]
xx. 1917. pp.xi.644. [1385.]
xxi. 1918. pp.iv.576. [1283.]
xxii. 1919. pp.iv.663. [1583.]
xxiii. 1920. pp.vii.752. [1669.]
xxiv. 1921. pp.viii.907. [2012.]

xxv. 1922. pp.viii.996. [2477.]
xxvi. 1923. pp.viii.1091. [2542.]
xxvii. 1924. pp.viii.1240. [3100.]
xxviii. 1925. pp.viii.1046. [2808.]
xxix. 1926. pp.viii.1068. [2996.]
xxx. 1927. pp.xii.1100. [3238.]
xxxi. 1928. pp.xii.1164. [3355.]
xxxii. 1929. pp.xii.1206. [3687.]
xxxiii. 1930. pp.xii.1343. [4165.]
xxxiv. 1931. pp.xii.1339. [4365.]
xxxv. 1932. pp.xvi.1610. [5364.]
xxxvi. 1933. pp.xvi.1657. [5491.]
xxxvii. 1934. pp.xvi.1563. [5269.]
xxxviii. 1935. pp.xvi.1571. [5251.]
xxxix. 1936. pp.xviii.1675. [5716.]
xl. 1937. pp.xviii.1594. [5494.]
xli. 1938. pp.xviii.1457. [5081.]
xlii. 1939. pp.xviii.1414. [4705.]
xliii. 1940. pp.xvi.1003. [3230.]
xliv. 1941. pp.x.467. [2737.]
xlv. 1942. pp.vii.376. [3152.]
xlvi. 1943. pp.xii.312. [2968.]
xlvii. 1944. pp.xvi.313. [2688.]
xlviii. 1945. pp.339. [3148.]
xlix. 1946. pp.xvi.431. [3273.]
l. 1947. pp.xii.434. [3765.]
li. 1948. pp.xv.509. [4088.]

lii. 1949. pp.xx.1089. [7500.]
liii. 1950. pp.xii.1313. [9198.]
liv. 1951. pp.xii.1460. [9908.]
lv. 1952. pp.xii.1421. [9292.]
lvi. 1953. pp.xvi.1416. [8830.]
lvii. 1954. pp.xii.1869. [11,693.]
lviii. 1955. pp.xii.1640. [10,160.]
lix. 1956. pp.viii.1277. [9165.]*
lx. 1957. pp.viii.1183. [10,000.]*
lxi. 1958. pp.viii.1132. [9201.]*
lxii. 1959. pp.xxiv.1766. [14,016.]*
lxiii. 1960. pp.xxviii.2906. [21,387.]*
lxiv. 1961. pp.xxx.2918. [21,167.]*
lxv. 1962. pp.xxxii.3203. [24,236.]*
lxvi. 1963. pp.2474.22.496. [26,000.]*

in progress; section B is entered under Electricity, above.

EUGÈNE ESTANAVE, Nomenclature des mémoires de physique expérimentale et de physique mathématique présentés en France dans le courant du XIXe siècle devant les facultés des sciences en vue du doctorat. 1903. pp.20. [400.]

PHYSICS. International catalogue of scientific literature (section C): Royal society.

i. 1902–1903. pp.xiv.239+xii.207. [3208.]
ii. 1903. pp.viii.327. [2433.]

iii. 1904. pp.viii.395. [3136.]
iv. 1906. pp.viii.408. [3150.]
v. 1907. pp.viii.617. [4937.]
vi. 1908. pp.viii.568. [5180.]
vii. 1909. pp.viii.504. [4504.]
viii. 1910. pp.viii.451. [4066.]
ix. 1911. pp.viii.424. [4328.]
x. 1912. pp.viii.397. [3848.]
xi. 1913. pp.viii.406. [4421.]
xii. 1914. pp.viii.416. [4394.]
xiii. 1915. pp.viii.303. [3625.]
xiv. 1917. pp.viii.342. [3187.]

no more published.

CATALOGUE of scientific papers, 1800–1900. Subject index. Volume III. Physics. [Edited by] Herbert McLeod. Royal society [London]: Cambridge 1912–1914. pp.c.550.vii+xv.551–927.vii. [57,500.]

SUBJECT list of works on general physics (including measuring, calculating and mathematical instruments, and meteorology). Patent office library: Subject lists (n.s. FS–GF): 1914. pp.iv.192. [1750.]

LE JOURNAL de physique et le radium. . . . Revue bibliographique.

sixth series.

i. 1920. pp.xl.256. [1000.]
ii. 1921. pp.lxiv.384. [1250.]
iii. 1922. pp.lxxiv.480. [1500.]
iv. 1923. pp.ccxx.1173. [4000.]
v. 1924. pp.clii.560. [3000.]
vi. 1925. pp.cxc.704. [3500.]
vii. 1926. pp.clxxxviii.704. [4000.]
viii. 1927. pp.cclxxxiii.927. [5000.]
ix. 1928. pp.ccli.896. [5500.]
x. 1929. pp.cclii.888. [5500.]

seventh series.

i. 1930. pp.cccxxiv.992. [6500.]
ii. 1931. pp.ccxciv.848. [5500.]
iii. 1932. pp.cccvii.944. [6000.]
iv. 1933. pp.ccxxii.664. [4500.]
v. 1934. pp.ccxiv.344. [3500.]
vi. 1935. pp.cccxi.376. [4000.]
vii. 1936. pp.cclxx.384. [4000.]
viii. 1937. pp.cclx.344. [3500.]
ix. 1938. pp.ccxxxi.312. [3000.]
x. 1939. pp.   . 296. [3000.]

eighth series.

i. 1940.
ii. 1941. pp.xxxix.100. [1000.]

iii. 1942. pp.xliii.144. [1500.]
iv. 1943. pp.xxxi.120. [1250.]
v. 1944. pp.xix.80. [1000.]
vi. 1945. pp.xxxii.32. [500.]
vii. 1946. pp.60. [1000.]
viii. 1947. pp.160. [2000.]
ix. 1948.
x. 1949.
xi. 1950. pp.63. [1000.]

earlier series and later volumes contain no separately published Revue bibliographique.

PHYSIKALISCHE berichte. Als fortsetzung der 'Fortschritte der physik' und des 'Halbmonatlichen literaturverzeichnisses' sowie der 'Beiblätter zu den Annalen der physik' gemeinsam herausgegeben von der Deutschen physikalischen gesellschaft und der Deutschen gesellschaft für technische physik. Braunschweig.

i. 1920. Redaktion von Karl Scheel. pp.cxx. 784+[ii].785–1722. [5000.]
ii. 1921. pp.cxxviii.712 + [ii].713–1454. [5000.]
iii. 1922. pp.cxxv.592+[ii].593–1430. [6000.]
iv. 1923. pp.cxxxix.704 + [ii].705–1741. [7000.]
v. 1924. pp.clviii.864 + [ii].865–1912. [7500.]

vi. 1925. pp.clxiii.880 + [ii].881–1856. [7500.]

vii. 1926. pp.cc.976+[ii].977–2297. [9000.]

viii. 1927. pp.ccxxviii.1104 + [ii].1105–2500. [9000.]

ix. 1928. pp.ccxlviii.1136+[ii].1137–2523. [9000.]

x. 1929. pp.cclviii.1200 + [ii].1201–2582. [9000.]

xi. 1930. pp.cccx.1312 + [ii].1313–3064. [11,000.]

xii. 1931. pp.cclxxxix.1416+[ii].1417–3213. [12,000.]

xiii. 1932. pp.cclii.1192 + [ii].1193–2529. [9000.]

xiv. 1933. pp.cclxxii.968 + [ii].969–2389. [9000.]

xv. 1934. pp.cclxxii.1008+[ii].1009–2436. [11,000.]

xvi. 1935. pp.cclxxxi.1044+[ii].1045–2670. [12,000.]

xvii. 1936. pp.cccii.1240 + [ii].1241–2786. [12,000.]

xviii. 1937. pp.cccxxv.1176+[ii].1177–2808. [12,000.]

xix. 1938. Redaction von L. Dede. pp. cclxxv.1156+[ii].1157–2837. [12,000.]

xx. 1939. pp.cdl.1360 + [ii].1361–3099. [14,000.]
xxi. 1940. pp.2770. [12,000.]
xxii. 1941. pp.xxiv.1308 + 1309–3344. [14,000.]
xxiii. 1942. pp.xxix.2908. [12,000.]
xxiv. 1943. pp.1726. [7500.]
xxv.
xxvi. 1947.
xxvii. 1948. pp.955. [3000.]
xxviii. 1949. pp.636.xi. [3000.]
xxix. 1950. pp.1820. [8000.]
xxx. 1951. pp.1844. [8000.]
xxxi. 1952. pp.1924. [8500.]
xxxii. 1953. pp.2156. [10,000.]
xxxiii. 1954. pp.3432. [15,539.]
xxxiv. 1955. pp.2284. [10,274.]
xxxv. 1956. pp.2584. [12,881.]
xxxvi. 1957. pp.2176. [11,822.]
xxxvii. 1958. pp.2332. [13,177.]
xxxviii. 1959. pp.2444. [14,667.]
xxxix. 1960. pp.2492. [15,000.]
xl. 1961. pp.2404. [15,000.]
xli. 1962. pp.2724. [16,000.]
xlii. 1963. pp.xlviii.3396.851. [20,000.]

in progress.

PHYSIK, bearbeitet von Ludwig Schiller. Chemie, bearbeitet von Walter Becker. Jahresberichte des Literarischen zentralblattes über die wichtigsten wissenschaftlichen neuerscheinungen des gesamten deutschen sprachgebietes (1924, vol.xx): Leipzig 1925. pp.104. [physics: 500.]

THE PHYSICAL review supplement. American physical society: New York &c.

i. July–October 1929. pp.[iv].240. [400.]

[*continued as:*]

Reviews of modern physics. American physical society [July 1932 &c.: American institute of physics].

ii. 1930. pp.[iv].508. [1200.]
iii. 1931. pp.[iv].495. [600.]
iv. 1932. pp.[iv].766. [700.]
v. 1933. pp.iv.288. [800.]
vi. 1934. pp.vii.280. [700.]
vii. 1935. pp.iv.271. [900.]
viii. 1936. pp.iv.398. [900.]
ix. 1937. pp.iv.457. [1600.]
x. 1938. pp.iv.269. [800.]
xi. 1939. pp.iv.303. [800.]
xii. 1940. pp.iii.358. [900.]
xiii. 1941. pp.iii.327. [900.]
xiv. 1942. pp.iv.410. [1300.]
xv. 1943. pp.iii.264. [800.]

xvi. 1944. pp.iii.324. [900.]
xvii. 1945. pp.iv.490. [1500.]
xviii. 1946. pp.iii.544. [1700.]
xix. 1947. pp.iii.364. [900.]
xx. 1948. pp.iv.728. [1800.]
xxi. 1949. pp.iv.652. [1700.]
xxii. 1950. pp.iii.422. [1300.]
xxiii. 1951. pp.iii.411. [1300.]
xxiv. 1952. pp.iii.402. [1300.]
xxv. 1953. pp.v.860. [2500.]
xxvi. 1954. pp.iii.472. [1400.]
xxvii. 1955. pp.[ii].473. [1300.]
xxviii. 1956. pp.iii.542. [1600.]
xxix. 1957. pp.iv.836. [2300.]
xxx. 1958. pp.vi.1196. [3000.]
xxxi. 1959. pp.iv.1077. [2900.]
xxxii. 1960. pp.v.1033. [2900.]
xxxiii. 1961. pp.iii.625. [1700.]
xxxiv. 1962. pp.[ii].892. [2400.]

in progress; vol.xxvii includes an index to vols.i–xxvii.

[ALVARO VICENTE LEAL], Catálogo. Biblioteca nacional: Eposição de física: Lisboa [1930]. pp.95. [750.]

CATALOGUS van de laboratorium-bibliotheken. I. . . . van 't Hoff laboratorium, Organisch-chemisch lab., Pharmaceutisch laborat., Physisch laboratorium. Rijksuniversiteit: Bibliotheek: Utrecht 1933. pp.[iv].100. [1100.]

REPORTS on progress in physics. Physical society.
[i]. 1934. pp.iv.371. [1000.]
ii. General editor Allan Ferguson. 1936. pp. iv.371. [1000.]
iii. 1937. pp.iv.394. [1000.]
iv. 1938. pp.vi.389. [1000.]
v. 1939. pp.v.446. [1000.]
vi. General editor J. H. Awbery. 1940. pp. v.434. [1000.]
vii. 1940. 1941. pp.[iii].362. [1000.]
— Cumulative name index of volumes I–VII. pp.78.
viii. 1941. General editor W. B. Mann. 1942. pp.[iv].372. [1000.]
ix. 1942–1943. 1943. pp.[iv].353. [1000.]
x. 1944–1945. 1946. pp.[iv].442. [1300.]
xi. 1946–1947. pp.[iv].461. [1250.]
xii. 1948–1949. pp.[iv].382. [1000.]
xiii. 1950. pp.[iv].424. [1250.]
xiv. 1951. pp.[iv].512. [1500.]
xv. 1952. pp.[iv].338. [1000.]
xvi. 1953. pp.[iv].407. [1250.]
xvii. 1954. pp.[iv].280. [1000.]

xviii. 1955. pp.[iv].477. [1250.]
xix. 1956. pp.[iv].367. [1000.]
xx. 1957. pp.[iv].568. [1500.]
xxi. 1958. pp.[iv].383. [1000.]
xxii. 1959. pp.[iv].634. [1750.]
xxiii. 1960. Executive editor A. C. Stickland. pp.[iv].629. [1750.]
xxiv. 1961. pp.[iv].424. [1000.]
xxv. 1962. pp.[iv].529. [1250.]
xxvi. 1963. pp.[iv].472. [1000.]

in progress; in vols.i–vi only the year of publication is shown.

ABSTRACTS of papers published in the year. National physical laboratory.

1936. pp.iii.65. [140.]
1937. pp.iii.71. [141.]
1938. pp.iii.63. [119.]

no more published; limited to papers published from the National physical laboratory.

READERS' guide to books on mathematics, physics and chemistry. Library association: County libraries section [no.11]: 1938. pp.[iv].36. [900.]

MARY BATTELL QUAM, Basic requirements of a physical science collection in a branch library.

Public library: New York 1940. ff.[iii].65. [125.]*

NATHAN GRIER PARKE, Guide to the literature of mathematics and physics including related works on engineering science. New York &c. 1947. pp.xv.205. [3000.]

— — Second . . . edition. [1958]. pp.xviii.436. [11,000.]

КНИГИ по физике, 1947–1953. Аннотированный каталог. Москва 1954. pp.40. [111.]

РЕФЕРАТИВНЫЙ журнал. Физика. Академия наук СССР: Институт научной информации: Москва.

1954. pp.268.xii+250.x+278.xii+270.xii+232.ix + 236.xii + 276.xiii + 268.xii + 269.xii + 297.xiv + 358.xvi + 411 + 225. [14,200.]

1955. pp.372.xvi+365.xviii+387.xvii+389.xviii + 399.xx + 412.xx + 386.xviii + 410.xix + 395.xx + 425.xxii + 428.xx. 596. [26,550.]

1956. pp.437.xiv + 449.xiv + 497.xiv + 443.xiii + 433 + 425 + 368 + 392 + 396 + 384+252+408+540+604. [36,500.]

1957. pp.332+332+332+344+360+404+

364+388+388+345+345+321+ .
1958. pp.336+328+324+319+328+304+ 305 + + + 308 + 315 + 324 + . [28,970.]
1959. pp.312+305+332+328+308+336+ 324+336+324+328+320+332+ .
1960. pp.344+332+380+373+388+424+ 456 + 448 + 460 + 448 + 488 + 371 + 428+647. [34,450.]
1961. pp.451+431+464+439+432+420+ 380 + 356 + 304 + 344 + 344 + 500 + 708+456. [40,000.]
1962. pp.[509]+[476]+[512]+[496]+[480] + [488] + [480] + [456] + [504] + [484] +[522]+[658]+730+456. [40,000.]
1963. pp.[488]+[400]+[476]+[482]+[440] + [486] + [468] + [408] + [424] + [410] +[416]+[446]+ .

in progress; from 1957 the abstracts relating to geophysics were formed into a separate series.

ROBERT H. WHITFORD, Physics literature. A reference manual. Washington 1954. pp.228. [1200.]

[JAN WEYSSENHOFF *and others*], Polska bibliografia analityczna. Fizyka. Polska akademia nauk: Ośrodek bibliografii i dokumentacji naukowej: Warszawa.

i. 1955. pp. . [93.]
ii. 1956. pp.37. [65.]
iii. 1957. pp.36.45. [96.]
no more published.

CONSTANTIN G. BEDRERG, Bibliografia fizicii romîne. Societăs de ştiinte matematice şi fizice din R.P.R.: Biblioteca (vol.28): Buccureşti 1957. pp. 293.

S. I. EREMEEVA [*and others*], Выдающиеся физики мира. Рекомендательный указатель. Государственная ... библиотека СССР имени В. И. Ленина: Центральная политехническая библиотека: Москва 1958. pp.436. [2000.]

ELISABETH SAUVENIER-GOFFIN, Les sciences mathématiques et physiques à travers le fonds ancien de la bibliothèque de l'université de Liège. Première partie. De l'antiquité à la fin du seizième siècle. Bibliotheca universitatis leodiensis: Publications (no.10): Louvain 1958. pp.172. [508.]

BIBLIOGRAFIA brasileira de matemática e física. Instituto brasileiro de bibliografía e documentação: Rio de Janeiro.
1900–1957. [1959]. pp.246. [1870.]
in progress?

M. LOIS MARCKWORTH, Dissertations in physics. An indexed bibliography of all doctoral theses accepted by american universities, 1861–1959. Stanford 1961. pp.xii.803. [8418.]*

ANNUAL physics book list. American institute of physics: New York.
1962. pp.[iii].26. [600.]
in progress.

4. *Miscellaneous*

METHODS of analogy in physics and mechanics (exclusive of photoelasticity) 1930 to date. Science library: Bibliographical series (no.629): [1946]. ff.2. [25.]*

GEORG JOSS [*and others*], Physics of solids. Office of military government for Germany: Field information agencies technical: Fiat review of german science, 1939–1946: Wiesbaden 1947–1948. pp. [xv].228+[iv].236. [1500.]
the text is in german.

Statistics.

[*the libraries of statistical organizations usually have wide scope, and nearly all their catalogues have therefore been excluded.*]

1. *Periodicals*

LÉON LEBON, Recherches bibliographiques sur les annuaires statistiques existant dans les différents pays. 1880. pp.56. [100.]

PERIODICALS received in the library: statistics. International labour office: Bibliographical reference list (no.63): Geneva 1953. pp.9. [200.]*

RÉPERTOIRE des périodiques statistiques enregistrés à la bibliothèque. Communauté européenne du charbon et de l'acier: Haute autorité: Service de la documentation: Bibliothèque: [Luxembourg] 1959. pp.iii.106. [900.]

BIBLIOGRAPHIE des périodiques statistiques catalogués à la bibliothèque. 4ᵉ édition. Organisation européenne de coopération économique: Service de presse et de documentation: 1954. pp.111. [700.]

— 7ᵉ édition. 1957. pp.159. [1500.]

2. *General*

CATALOGUE of the library of the [Royal] Statistical society of London. 1859. pp.iv.142. [3750.]

— [another edition]. 1908. pp.viii.276. [10,000.]

KATALOG der bibliothek des Königl. statistischen bureaus. Berlin 1874–1879. pp.[iv].xii.632. lxxi+[ii].xxxii.983. [60,000.]

ÍNDICE de los trabajos históricos y estadísticos de Manuel Landaeta Rosales. Carácas 1909. pp.31. [401.]

HANDLIST of statistical publications and state papers of the countries of continental Europe (including certain of their colonies) preserved in the library of the British museum. 1910. pp. [ii].coll.44. [1250.]

nine single sheets of accessions were issued at intervals.

HAND- en leerboeken voor de theorie der statistiek, internationale statistische werken, jaarboeken der verschillende landen en staten. Centraal bureau voor de statistiek: Bibliotheek: Lijst (no.1): [The Hague 1919]. pp.v.17. [200.]

a copy in the Library of Congress contains additions in manuscript.

BOEKWERKEN in de leeszaal geplaatst ten gebruike van ambtenaren en bezoekers. Centraal bureau voor de statistiek: Bibliotheek: Lijst (no.2): [The Hague 1923]. pp.vii.104. [1100.]

LIST of references on statisticians with a glance at statistical methods. Library of Congress: Washington 1930. ff.5. [48.]*

GERLOF VERWEY and D[IRK] C[ORNELIS] RENOOIJ, The central economist's handbook. A manual of statistical sources. Amsterdam 1934. pp.ix.460. [4000.]

— — Supplement. 1937. pp.[vii].79. [350.]

JEAN CAUMARTIN, Les principales sources de documentation statistique. 1935. pp.40. [300.]

RÉPERTOIRE des données publiées régulièrement dans les journaux officiels. Société des nations: Bibliothèque: Listes bibliographiques (no.4): Genève 1935. pp.[ii].24. [300.]*

HELEN F[IELD] CONOVER, List of recent references on statistical methods and their application. Compiled under the direction of Florence S[elma] Hellman. Library of Congress: Division of bibliography: [Washington] 1938. pp.23. [204.]*

— — [supplement]. 1941. pp.[ii].18. [169.]*

[EMILIO ALANIS PATIÑO, GILBERTO LOYO and ENRIQUE SANTAMARÍA], Bibliografía del método estadístico y sus aplicaciones. Secretaría de la economía nacional: Dirección general da estadística: México 1938. pp.169. [1811.]
covers the period from 1920, exclusive of mexican writers.

RESEARCH and statistical methodology. Books and reviews. New Brunswick, N.J.
1933–1938. By Oscar [Krisen] Buros. pp.100.
1938–1941. pp.383.
[*continued as:*]
Statistical methodology reviews.
1941–1950. pp.457. [342.]

G. M. MORANT and B. L. WELCH, A bibliography of the statistical and other writings of Karl Pearson. University college: Biometrika: 1939. pp.viii.119. [670.]

PUBLICATIONS on foreign countries. An annotated [An] accession list. Bureau of the census: Washington 1947–1958.*

CATÁLOGO de la biblioteca. Instituto nacional de estadística: Madrid 1947–1952. pp.197+566. [7500.]
— Suplemento. 1951. pp.153. [1500.]

STATISTICS. A guide to current material. Public libraries: Westminster 1951. pp.16. [150.]

[WILLIAM R. BUCKLAND], Bibliography of basic texts and monographs on statistical methods. International institute of statistics: The Hague 1951. pp.84. [100.]

— — Second edition. By W. R. Buckland and Ronald A[rthur] Fox. International statistical institute: Edinburgh 1963. pp.vii.297. [500.]

[G. DARMOIS and E. MORICE], Bibliographie sur la méthode statistique et ses applications. Liste des principaux ouvrages en langue française. Institut international de statistique [&c.]: 1952. pp.49. [100.]*

PHYLLIS G[WENDOLIN] CARTER, Statistical yearbooks. An annotated bibliography of the general statistical yearbooks of major political subdivisions of the world. Library of Congress: Washington 1953. pp.vii.123. [247.]*

TERVGAZDASÁGI, statisztikai és számviteli bibliográfia. Budapest.

[1945–1953. By Eleméméné Hajdú, Béla Hámori and Gyula Haraszthy]. pp.715. [6078.]

1954. pp.168. [1222.]

Statistics

PHYLLIS G[WENDOLIN] CARTER, Statistical bulletins. An annotated bibliography of the general statistic bulletins of major political subdivisions of the world. Library of Congress [&c.]: Washington 1954. pp.[ii].x.93. [200.]*

ANNOTATED list of international standards for statistics. United nations [E./CN.3/181): [New York] 1954. pp.31. [56.]*

[PHYLLIS GWENDOLIN CARTER], International statistical standards: an annotated bibliography of the statistical recommendations of international conferences, meetings, and agency. Preliminary edition. Pan american union [&c.]: Washington 1954. pp.vi.121. [201.]*

[—] — Revised edition. 1955 [1956]. pp.ii.144.*

FOLYÓIRATCIKK-bibliográfia. Központi statisztikai hivatal könyvtaranak kiadvanyai (no.2): [Budapest] 1955 &c.

in progress?

[CORRADO GINI], Bibliografia sui metodi statistici e le loro applicazioni. Elenco dei principali scritti di autori italiani. Società italiana di statistica: Roma 1957. pp.[iii].56. [391.]

ALMA C. DICKMAN, Accounting and statistical

services. Pacific air forces: PACAF basic bibliographies: [*s.l.*] 1957. pp.[iii].ii.11. [50.]*

INTERNATIONAL journal of abstracts. Statistical theory and method. International statistical institute.*

i. 1959–1960. [Edited by William R. Buckland]. pp.xii.ff.[376]. [751.]
ii. 1961. pp.x.xxxii.ff.[390]. [780.]
iii. 1962. pp.x.xxxi.ff.[337]. [674.]
iv. 1963. pp.x.xxxii.ff.[397]. [794.]

in progress.

MAURICE G[EORGE] KENDALL and ALISON G[RANT] DOIG, Bibliography of statistical literature 1950–1958. Edinburgh &c. 1962. pp.xii.297. [9802.]

3. *Countries*

Africa

DRAFT bibliography of african statistical publications. United nations: Economic commission for Africa: [New York] 1961. pp.[ii].xv.190. [1000.]*

America

BIBLIOGRAPHY of selected statistical sources of the american nations. . . . Preliminary edition.

Statistics

Inter american statistical institute: Was6in3ton 1945. pp.xxviii.677. [2500.]*

Austria

VERZEICHNIS der veröffentlichungen des Österreichischen zentralamtes, 1945–1959. Wien 1960. ff.[i].ii.18. [100.]*

Bulgaria

L[YUBA] DZHEROVA and N. TOTEVA, Библиография на българската статистическа литература, 1878–1960. Централно статистическо управление: София 1961. pp.106. [1726.]*

Costa Rica

MANUEL GARCÍA VALVERDE, Bibliografía estadística de Costa Rica. Dirección general de estadística: San José 1950. ff.[i].9. [150.]*

Europe

[JAN GRAY ROSS *and others*], Guide to sources of statistics in common market countries. British nylon Spinners ltd.: [1962]. pp.[v].17. [150.]*

France

BULLETIN bibliographique. Ministère de la France d'outre-mer: Service colonial de statistique: Service des statistiques [1961– : Ministère de la coopération].*

1–5. 1948. pp.126. [600.]
6–17. 1949. pp.310. [1500.]
18–29. 1950. pp.311–662. [1600.]
30–41. 1951. pp.663–1152. [2500.]
42–50. 1952. pp.357. [1500.]
50–56. 1953. pp.394. [1500.]
57–62. 1954. pp.64.59.56.66.63.80. [1200.]
63–67. 1955. pp.87.76.80.94.73. [1500.]
68–72. 1956. pp.77.71.90.84.79. [1300.]
73–77. 1957. pp.86.62.93.92.66. [1300.]
78–82. 1958. pp.97.113.95.85.84. [1400.]
83–87. 1959. pp.110.130.113.99.115. [2000.]
88–92. 1960. pp.101.109.155.137.133. [1200.]
93–97. 1961. pp.119.122.138.144.131. [1200.]
98–102. 1962. pp.117.135.146.153.135. [1300.]

in progress.

BIBLIOGRAPHIE sur la méthode statistique et ses applications. Liste des principaux ouvrages en langue française. Institut international de statistique *and* Institut national de la statistique et des études économiques: 1952. pp.49. [100.]

Germany

XAVIER HEUSCHLING, Bibliographie historique

de la statistique en Allemagne. Bruxelles 1845. pp.112. [600.]

VERZEICHNISS der officiellen publicationen des Königl. preussischen statistischen bureaus in Berlin. 1868. pp.18. [125.]

— [another edition]. Verzeichnis der veröffentlichungen [&c.]. 1904. pp.40. [500.]

VERZEICHNIS der veröffentlichungen des Statistischen bundesamtes. Wiesbaden.

1951. pp.19. [400.]
1952. pp.23. [500.]
1953. pp.16+15. [800.]
1954. pp.18+20. [900.]
1955. pp.23+ .

in progress.

Great Britain

GUIDE to current official statistics [of the United Kingdom]. Permanent consultative committee on official statistics.

i. 1922. pp.162. [500.]
ii. 1923. pp.306. [1500.]
iii. 1924. pp.252. [600.]
iv. 1925. pp.262. [600.]
v. 1926. pp.274. [750.]
vi. 1927. pp.288. [750.]
vii. 1928. pp.312. [750.]

viii. 1929. pp.316. [750.]
ix. 1930. pp.320. [750.]
x. 1931. pp.358. [750.]
xi. 1932. pp.344. [750.]
xii. 1933. pp.346. [750.]
xiii. 1934. pp.350. [750.]
xiv. 1935. pp.366. [750.]
xv. 1936. pp.404. [750.]
xvi. 1937. pp.406. [750.]
xvii. 1938. pp.404. [750.]

vol.ii contains an 'Appendix relating to selected statistical publications prior to 1923'; no more published.

WILLIAM H. PATCH and RUTH SANDERSON, A finding list of statistical abstracts in the british sessional papers. University of Kentucky: Library: Occasional contribution (no.106): [Lexington] 1959. pp.12. [500.]*

Hungary

JOSEPH KŐRÖSI, Catalogue raisonné de l'exposition du Bureau de statistique de la ville de Budapest. Exposition nationale de 1885 à Budapest: Berlin 1885. pp.54. [50.]

A MAGYAR királyi központi statisztikai hivatal Nyilvános könyvtárának és térképgyüjteményének czímjegyzéke. Budapest 1898. pp.xvi. coll.1198.

Statistics

India

GEORGE BUIST, Index to books and papers on the physical geography, antiquities, and statistics of India. Bombay 1852. pp.5–103. [4000.]

Italy

LUIGI BODIO, Sui documenti statistici del regno d'Italia. Cenni bibliografici. Firenze 1867. pp. xxxix.152. [200.]

[—] — Annali di statistica. Saggio di bibliografia statistica italiana. Terza edizione. Ministero di agricoltura, industria e commercio: Direzione generale della statistica: Roma 1890. pp.xix.213. [2000.]

PIETRO MAESTRI, Le pubblicazioni della Direzione di statistica. Firenze 1869. pp.vii.71. [40.]

CATALOGO delle pubblicazioni disponibili per la vendita. Istituto centrale di statistica: Roma 1951 [*on cover:* 1950]. pp.32. [750.]

BIBLIOGRAFIE con brevi cenni biografici. Biblioteca di statistica: [Bologna 1959]. pp.3–551. [15,000.]

Mexico

BIBLIOGRAFÍA mexicana de estadística. Secretaría de la economía nacional: Dirección mexicana de estadística: México [1941–1942]. pp.696+773. [13,642.]

Netherlands

LIBRAIRIE MARTINUS NIJHOFF, Aperçu de la littérature statistique des Pays-Bas. La Haye 1911. pp.16. [110.]

PUBLICATIES van het Centraal bureau voor de statistiek. 's-Gravenhage 1938. pp.40.

— [another edition]. 1942. pp.84.

OVERZICHT van de na-oorlogsche publicaties. Centraal bureau voor de statistiek: 's-Gravenhage 1946. pp.10.*

New Zealand

STATISTICAL publications, 1840–1960, mainly those produced by the Registrar general 1853–1910 and the government statistician 1911–1960. Department of statistics: Wellington 1961. pp.66.

Norway

FORTEGNELSE over Norges offisielle statistikk 1828–1950. Statistisk sentralbyrå: Norges offisielle statistikk (11th ser., no.63): Oslo 1951. pp.97. [2000.]

there were several earlier editions.

Russia

V. PFAF, Систематическій каталогъ книгъ

на русскомъ языкѣ Императорской публичной библіотеки по части правовѣдѣнія, политическихъ и статистическихъ наукъ. С.-Петербургъ 1863. pp.445. [statistics: 486.]

there is another titlepage, on which the author's name does not appear, but instead that of prince [Nikolai Borisovich] Yusupov as editor.

V[LADIMIR] I[ZMAILOVICH] MEZHOV, Труды центральнаго и губернскихъ статистическихъ комитетовъ. Библіографическій указатель книгъ и заключающихся въ нихъ статей, обнимающій дѣятельность статистическихъ комитетовъ съ самаго начала ихъ учрежденій вплоть до 1873 г. Библіографическія монографіи (vol.i, part 1): С.-Петербургъ 1873. pp.[ii].iv.128. [522.]

APERÇU bibliographique des publications statistiques de la Russie. Центральный статистический комитет: Санктпетербургъ 1895–1897. pp.216+212+209. [5000.]

КАТАЛОГЪ библіотеки Архангельскаго губернскаго статистическаго комитета. Архангельскъ 1896. pp.[vi].129. [2446.]

БИБЛИОГРАФИЧЕСКИЙ обзор изданий Центрального статистического управления Союза ССР... (1918–1928). Москва 1928. pp.72. [287.]

БИБЛИОГРАФИЧЕСКИЙ указатель по статистике. Серия А. Литература на языках народов СССР. Том I [&c.]–1930 [&c.] год. Московский институт народнохозяйственного учема: Москва 1934 &c.

Switzerland

W. SPEISER, Bankwesen. — Traugott Geering, Handelsstatistik. Central kommission für schweizerische landeskunde: Bibliographie der schweizerischen landeskunde (section v.9gε.[i]): Bern 1893. [statistics: 200.]

VERÖFFENTLICHUNGEN des Eidgenössischen statistischen amtes seit 1930. [Bern 1936]. pp.[3].
— [another edition]. [1939]. pp.[4].

SCHWEIZERISCHE bibliographie für statistik und volkswirtschaft. . . . Bearbeitet vom Eidgenössischen statistischen amt. Schweizerische gesellschaft für statistik und volkswirtschaft [: Zeitschrift für schweizerische statistik und volkswirtschaft (beilage)]: Bern.

i. 1937. [Edited by Leo M. Kern]. pp.
ii. 1938. pp.xv.107. [2000.]
iii. 1939–1940. pp.xvi.130. [2500.]

iv. 1940–1941. pp.xx.113. [2000.]
v. 1941–1942. pp.xx.128. [2500.]
vi. 1942–1943. pp.xx.112. [2000.]
— Verfasserverzeichnis 1–6. [1943]. pp.36.
vii. 1943–1944. pp.xx.128.9. [2250.]
viii. 1944–1945. pp.xx.134.11. [2500.]
ix. 1945–1946. pp.xx.256.18. [5000.]
x. 1946–1947. pp.xx.180.14. [3500.]
xi. 1947–1948. pp.xx.178.14. [3500.]
xii. 1948–1949. pp.xx.180.15. [3500.]
xiii. 1949–1950. pp.xx.174.14. [3500.]

Turkey

İSTATISTIK umum müdürlüğü neşriyatı 1927–43. Ankara 1943. pp.12.

Ukraine

PUBLICATIONS de la Direction centrale de la statistique de l'Ukraine. Kharkoff 1923. pp.10. [100.]

United States

INDEX of state sources of statistical data. Bureau of statistics: [Harrisburg 1955]. ff.22.*

JOHN L. ANDRIOT, U. S. government statistics. . . . Pilot edition. Arlington, Va. 1956 &c.*

GAISELLE KERNER, Preliminary inventory of the

records of the Central bureau of planning and statistics. National archives: Preliminary inventories (no.98): Washington 1957. pp.iii.15. [50,000.]*

BIBLIOGRAPHY of statistical text books and other teaching material. Second edition. Pan american union: Washington 1960. pp.xii.3–120. [2500.]*

PAUL WASSERMAN [*and others*], Statistics sources. Detroit [1962]. pp.288.[viii]. [6000.]

4. *Miscellaneous*

BRIEF list of references on index numbers. Library of Congress: [Washington] 1919. ff.3. [37.]*

— [another edition]. 1929. ff.15. [166.]*

there are intermediate editions.

NELLIE G[ENEVA] LARSON, The sampling method in social and economic research. A partial list of references. Compiled . . . under the direction of Mary G. Lacy. Department of agriculture: Bureau of agricultural economics: Agricultural economics bibliography (no.90): Washington 1941. pp.vi.155. [426.]*

— — [another edition]. . . . (no.94): 1942. pp.iii–vi.314. [742.]*

Statistics

GRANT I[LLION] BUTTERBAUGH, A bibliography of statistical quality control. [University of Washington:] College of economics and business: Bureau of business research: Seattle 1946. pp.viii. 114. [550.]

— — Supplement. 1951. pp.141. [850.]

A LIST of books on statistical quality control. Linda hall library: Bibliographical bulletin (no.2): Kansas City 1951. pp.[ii].6. [25.]*

— [another edition]. A list of books on statistical quality control available in Linda hall library. . . . (no.18): 1954. ff.12. [37.]*

INTERNATIONAL journal of abstracts on statistical methods in industry. International institute of statistics: [The Hague].*

i. 1954. General editor: Grant I. Butterbaugh. pp.viii.ff.120.pp.16. [357.]
ii. 1955. pp.vi.ff.117.pp.16. [351.]
iii. 1956. pp.vi.ff.112.pp.19. [336.]
iv. 1957. pp.vii.ff.103.pp.16. [309.]
v. 1958. pp.vi.ff.103.pp.15. [309.]
vi. 1959. pp.viii.ff.111.pp.20. [322.]
vii. 1960. pp.ix.ff.109.pp.18. [309.]
viii. 1961. [Edited by William R. Buckland]. pp.vii.ff.[95].pp.xvi. [284.]
ix. 1962. pp.viii.ff.[104].pp.xvi. [310.]

x. 1963. pp.xvi.ff.[86]. [255.]
in progress.

BIBLIOGRAPHY of the Monte Carlo method and associated material, including abstracts. University of Florida: Statistical laboratory: [Gainesville 1955]. ff.73.*

BIBLIOGRAPHIE sur la méthode statistique appliquée à la médecine et aux sciences anthropologiques. Liste d'ouvrages et d'articles en langue française. Centre d'études statistiques de la chaire d'hygiène et de médecine préventive and Institut national de la statistique et des études économiques: 1955. pp.34. [200.]*

VERA RILEY and THEODORE CUTLER, Bibliography on work sampling and ratio delay. Johns Hopkins university: Operation research office: BRS [Bibliography reference series] (no.6): Chevy Chace, Md. 1956. pp.22. [125.]*

RUTH KRAUSE, Bibliographie zur statistischen qualitätskontrolle. Auswahl deutschsprachiger veröffentlichungen der jahre 1955–1959. Deutsche staatsbibliothek: Bibliographische mitteilungen (no.18): Berlin 1961. pp.59. [356.]

I. RICHARD SAVAGE, Bibliography of nonparametric statistics. [Cambridge, Mass. 1961]. ff.[iii]. xxxvi.373.12. [2000.]*

— — [another edition]. 1962. pp.[ix]. 284. [3250.]

Special Subjects

Aceto-acetic ester.

PAUL H. SEYMOUR, Bibliography of aceto acetic ester and its derivatives. Smithsonian institution: Publication (no.970 = Smithsonian miscellaneous collections, vol.xxxviii, no.2): Washington 1894. pp.[x].148. [250.]

Acetylacetones.

SOME references to substituted acetylacetones, 1938–1955. Science library Bibliographical series (no.759): [1958]. pp.3. [60.]*

Acides.

WALTER SIDNEY SINGLETON, The physical properties of fatty acides, glycerides and derivatives. The technical literature and patents of the Southern regional research laboratory, U.S. Department of agriculture. U.S. Bureau of agricultural and industrial chemistry (no.361): [*s. l.*] 1953. pp.33.

Adsorption.

VICTOR R[ICHARD] DEITZ, Bibliography of solid adsorbents. An annotative bibliographical survey of the scientific literature on bone char, activated carbons, and other technical solid adsorbents, for . . . 1900 to 1942 inclusive. Washington 1944. pp.lxxxi.877. [6002.]

— — 1943 to 1953. National bureau of standards: Circular (no.566): 1956. pp.vi. 1528. [13,763.]*

Airglow.

AIRGLOW, gegenschein, zodiacal light. A review of soviet literature for the period 1959–1962. (Preliminary). Library of Congress: Aerospace information division: AID report (no.62–162: [Washington] 1962. ff.iv.49. [116.]*

Alchemy.

PIERRE BOREL, Bibliotheca chimica. Sev catalogvs librorvm philosophicorvm hermeticorvm. Parisiis 1654. pp.[xii].276. [4000.]

a copy in the Bodleian library contains voluminous ms. notes and additions by Elias Ashmole.

— — [another edition]. Heidelbergæ 1656. pp.[x].254. [4000.]

[NICOLAS LENGLET DU FRESNOY], Histoire de la philosophie hermétique. Accompagnée d'un catalogue raisonné des écrivains de cette science. . . . Tome III. 1742. pp.[xxiv].432. [1500.]

this volume consists of the Catalogue; *reissued in the same year with the author's name, Paris, and again La Haye, and in Paris, 1744; a copy of the 1742 La Haye issue in the Bibliothèque nationale contains ms. notes by the author.*

[ALFRED] H[UMBERT JACQUIER] DE TERRE BASSE, Recherches bibliographiques. Voiron [1887]. pp.4. [3.]

on alchemical works printed in Lyons in 1557 by Macé Bonhomme.

J[AMES] CAMPBELL BROWN and ALEXANDER BROWN, Catalogue of books illustrating the history of alchemy and early chemistry exhibited at the associated soirée. Liverpool 1890 pp.[100.]

[HENRY CARRINGTON BOLTON], Catalogue of works on alchemy and chemistry exhibited at the Grolier club. New York 1891. pp.32. [100.]

a catalogue of works from the author's collection.

JOHN FERGUSON, Bibliotheca chemica: a catalogue of the alchemical, chemical and pharmaceutical books in the collection of the late James Young. Glasgow 1906. pp.xxiii.488+[iii].598. [2500.]

privately printed; a facsimile was issued London 1954.

JOHN FERGUSON, Some english alchemical books. 1913. pp.16. [125.]

CATALOGUE des manuscrits alchimiques grecs. Union académique internationale: Bruxelles.

i. Henry Lebègue, Les Parisini. 1924. pp.x. 320. [600.]
ii. C. O. Zuretti, Les manuscrits italiens. 1927. pp.vi.369. [750.]
iii. Dorothea Waley Singer, Les manuscrits des Îles Britanniques. 1924. pp.7.84. [50.]
iv. G[ünther] Goldschmidt, Manuscrits d'Allemagne, d'Autriche, de Danemark, de Hollande et de Suisse. 1932. pp.xxxvi. 448. [1000.]
v. C. O. Zuretti, Les manuscrits d'Espagne. . . . Les manuscrits d'Athènes, décrits par A[lb.] Severyns. 1928. pp.v.175. [250.]

vols. vi–viii contain texts.

DOROTHEA WALEY SINGER, Catalogue of latin

and vernacular alchemical manuscripts in Great Britain and Ireland dating from before the XVI century. Union académique internationale: Brussels 1928–1931. pp.xxv.326+ix.331–755+[vii].759–1179. [1116.]

[WILLIAM JEROME WILSON], Catalogue of latin and vernacular alchemical manuscripts in the United States and Canada. Osiris (vol.vi): Brugis 1939. pp.ix.844. [2500.]

CATALOGUE des manuscrits alchimiques latins. Union académique internationale: Bruxelles.

i. James [Arthur] Corbett, Manuscrits des bibliothèques publiques de Paris. 1939. pp.367. [1250.]

ii. — Manuscrits des bibliothèques publiques des départements français antérieurs au XVIIe siècle. 1951. pp.200. [750.]

[KATHERINE R. THOMSON and MARY MARGARET SERVICE], Catalogue of the [John] Ferguson collection of books, mainly relating to alchemy, chemistry, witchcraft and gipsies, in the library of the university of Glasgow. [Edited by William Ross Cunningham]. Glasgow university publications: [Glasgow] 1943. pp.xv.384+[ii].385–820. [6000.]

K[URT] K[ARL] DOBERER, A bibliography of books on alchemy in the British museum. 1946. ff.[8]. [250.]

the British museum press-marks are added in MS. to the copy in that library.

BIBLIOTHECA alchemica et chemica. An annotated catalogue of printed books on alchemy, chemistry and cognate subjects in the library of Denis I[an] Duveen. 1949. pp.vii.669. [2000.]

ALFRED SIGGEL, Katalog der arabischen alchemistischen handschriften Deutschlands. . . . Im auftrage der Deutschen akademie der wissenschaften zu Berlin bearbeitet. Union académique internationale: Paris 1949–1956. pp.144+118+60. [1000.]

A HUNDRED alchemical books. Science museum: Book exhibitions: Catalogue (no.1): [1952]. pp. 32. [100.]

Algebra.

LAO GENEVRA SIMONS, Bibliography of early american textbooks on algebra . . . through 1850. Yeshiva college: Scripta mathematica (no.1): New York 1936. pp.[ii].68. [100.]

A. G. KUROSH, Алгебраический реферативный сборник за 1941–1946 гг. Москва 1948. pp.248+199+168. [883.]

Alginic acid.

ALGINIC acid and alginates. A list of references, mainly 1945 to 1947. Science library: Bibliographical series (no.655): 1948. ff.5. [80.]*

Alkaloids.

PAUL I[NGOLD] MURRILL, Alkaloidal estimation: a bibliographical index of chemical research. Committee of revision and publication of the pharmacopoeia of the United States of America: Ann Arbor 1898. pp.[ii].iii.58. [400.]

Americium.

BIBLIOGRAPHY on americium. Atomic energy research establishment: Harwell 1952. pp.6. [83.]*

Amidines.

AMIDINES in general. Science library: Bibliographical series (no.148): 1934. ff.[i].5. [99.]*

ANNOTATED bibliography. Diamidines. Merck & co.: Rahway, N.J. 1943. ff.[i].25. [50.]*

— Supplement. [1944]. pp.[i].20. [50.]*

Arithmetic.

I[GNAZ] ROGG, Handbuch der mathematischen literatur vom anfange der buchdruckerkunst. . . . Erste abtheilung, welche die arithmetischen und geometrischen wissenschaften enthält. Tübingen 1830. pp.[ii].vi.579. [7500.]
no more published.

AUGUSTUS DE MORGAN, Arithmetical books from the invention of printing to the present time. 1847. pp.[iii].xxviii.124. [500.]

ELENCO cronologico delle opere di computisteria e ragioneria venute alla luce in Italia dal 1202. Quarta edizione. Ministero del tesoro: Roma 1889. pp.xi.280. [1578.]

DAVID EUGENE SMITH, Rara arithmetica. A catalogue of the arithmetics written before the year MDCI, with a description of those in the library of George Arthur Plimpton of New York. Boston &c. 1908. pp.xvi.508. [250.]
— — Addenda. 1939. pp.x.52. [100.]
the collection now forms part of the library of Columbia university.

ALPHONS JOHANNES ÉMILE MARIE SMEUR, De zestiende-eeuwse nederlandse rekenboeken. Proefschrift. 's-Gravenhage 1960. pp.xii.167. [112.]

Arsenic.

LIST of references on asenic. Library of Congress: [Washington] 1921. ff.8. [81.]*

R[OBERT] A[MBROSE] OSBORN and L[INNIE] A[RAH] PEASE, Selected references on the toxicity of arsenic. [Washington 1934]. pp.45. [450.]*

BIBLIOGRAPHY on the influence of arsenic and tin on the properties of iron and steel . . . 1864 to 1936. Iron and steel institute: Bibliographical series (no.4): 1937. pp.24. [100.]

TOXICITY to plants of arsenic in soil. Science library: Bibliographical series (no.753): 1957. ff.2. [50.]*

Aspergillus niger.

CITRIC acid production by the mould aspergillus niger. Public library of South Australia: Research service: [Adelaide] 1955. ff.4. [33.]*

Astrology.

JONAS GURLAND, Kurze beschreibung de mathematischen, astronomischen und astrologischen hebräischen handschriften der Firkowitsch'schen sammlung in der Kayserlichen öffentlichen bibliothek zu St. Petersburg. Neue denkmäler der jüdischen literatur (no.2): St. Petersburg 1866. pp.[iv].57. [100.]

CATALOGUS codicum astrologorum graecorum. Bruxellis.

[i]. Alessandro Olivieri, Codices florentinos. 1898. pp.vii.182. [750.]

ii. G. [Wilhelm] Kroll and A. Olivieri, Codices venetos. 1900. pp.viii.224. [750.]

iii. E[midio] Martini and D[omenico] Bassi, Codices mediolanenses. 1901. pp.[v].60. [200.]

iv. D. Bassi [*and others*], Codices italicos praeter florentinos, venetos, mediolanenses, romanos. 1903. pp.viii.192. [750.]

v. Franz Cumont and Franz Ball [part II: Wilhelm Kroll; III: Josef Heeg], Codices Romanorum. 1904–1910. pp.viii.256+[v].163+viii.160. [1750.]

vi. Wilhelm Kroll, Codices vindobonenses. 1903. pp.viii.122. [500.]

vii. Franz Ball, Codices germanicos. 1908. pp.viii.268. [800.]

viii. Franz Cumont [part II: Charles Émile Ruelle; III–IV: Pierre Boudreaux], Codices Parisinorum. 1911–1929. pp.vi.292+viii.

195+viii.222+vii.283. [3500.]
part 1 was published last.

x. Armand Delatte, Codices athenienses. 1924. pp.viii.291. [500.]

xi. C[arlo] O[reste] Zuretti, Codices hispanienses. 1932–1934. pp.vii.288+vii.217. [1000.]

xii. Mstislav Antonin F. Šangin. Codices rossicos. 1936. pp.viii.268. [750.]

in progress.

HṚISHÍKEŚA ŚÁSTRÍ and ŚIVA CHANDRU GUI, A descriptive catalogue of sanskrit manuscripts in the library of the Calcutta sanskrit college. Volume IX. Astrology & astronomy manuscripts. Calcutta [1904–] 1906. pp.[ii].180.ii.[193.]

F. LEIGH GARDNER, A catalogue raisonné of works on the occult sciences . . . Vol.II. Astrological books. 1911. pp.xx.164. [1340.]
privately printed.

MARC HENRIOUD, Les astrologues de Combremont-le-Petit et leurs almanachs (1696–1839). Lausanne 1913. pp.53. [30.]

FRITZ SAXL, Verzeichnis astrologischer und mythologischer illustrierter handschriften des lateinischen mittelalters. Heidelberger akademie der wissenschaften: Sitzungsberichte: Philosophisch-historische klasse (1915 &c.).

[i]. In römischen bibliotheken... (1915, no.6–7). pp.xviii.143. [500.]

ii. Die handschriften der National-bibliothek in Wien. . . . (1925–1926, no.2). pp.254. [1000.]

iii. Handschriften in englischen bibliotheken. Von F. Saxl und Hans Meier. Herausgegeben von Harry Bober. Warburg institute: 1953. pp.lxxv.447+65. [3000.]

CATERINA SANTORO, I codici di medicina e astrologia della biblioteca Trivulziana. [*s.l.*] 1945. pp.8. [9.]

FRANCIS J. CARMODY, Arabic astronomical and astrological sciences in latin translation. A critical bibliography. Berkeley &c. 1956. pp.vi.193. [2000.]

Atmospheric scattering.

PAPERS on atmospheric scattering 1936 to date. Science library: Bibliographical series (no.569): 1942. ff.2. [40.]*

Audition.

[GEORGE A. MILLER *and others*], A bibliography in audition. Harvard university: Psychoacoustic laboratory: Cambridge, Mass. [1950]. ff.[225]+[200]. [4000.]*

— — [another edition]. Bibliography on hearing. J. G. C. Loring, compiler. 1955. pp.[vii].599. [10,000.]*

ARNO BLAU, Bibliographie, praktische audiometrie und hörerziehung 1945–1957. Straubing/Niederbayern 1957. pp.46. [1250.]

Aurora borealis.

AMI BOUÉ, Chronologischer katalog der nordlichter bis zum Jahre 1856, sammt einer bibliographie über diese erscheinung. Wien 1856. pp.74. [2000.]

THE AURORA-borealis. 1. Books; 2. General articles; 3. Photographs; 4. Physics; 5. Height and geographical distribution; 6. Periodicity. Science library: Bibliographical series (no.491): 1939. ff.46. [875.]*

Balancing.

PRACTICAL dynamic balancing, especially of radially symmetrical parts. Science library: Bibliographical series (no.507): 1939. ff.3. [45.]*

Barbiturics.

SOME references to phenylethylbarbituric acid, 1916–1934. Science library: Bibliographical series (no.175): 1935. ff.[i].5. [94.]
reproduced from typewriting.

Barium.

BIBLIOGRAPHY of literature relating to barium titanate. Admiralty centre for scientific information and liaison: Library: Bibliography (no.4): [1951]. pp.22.

SIMONE B[UISSET] SCHWIND, Barium. A bibliography of unclassified literature. United States Atomic energy commission: Technical information service: Oak Ridge, Tenn. 1952. pp.ii.452. [4237.]

L. O. ANDERSON, Ferroelecltrics of barium titanate type, a bibliography. Artices appearing in open literature only, with abstracts. U.S. naval ordnance test station [Inyokern]: China Lake,

Cal. 1952. pp.iv.79. [250.]*

BARIUM bibliography. Food machinery & chemical corporation: New York [1961]. pp.vi.440. [2750.]*

Bergakademie Freiberg.

BIBLIOGRAPHIE der veröffentlichungen aus den instituten der Bergakademie Freiberg Februar 1946 bis Juni 1959. Montanwissenschaftliche literaturberichte (no.3): Berlin 1959. pp.207. [2500.]

Batholiths.

FRANK F[ITCH] GROUT *and others*, A bibliographical list of papers on batholiths and the mechanics of igneous intrusion. [National research council: Washington] 1931. ff.54. [550.]*

Biochemistry.

ANNUAL review of biochemistry. Stanford university.

i. 1932. pp.vii.724. [8000.]
ii. 1933. pp.vii.564. [5500.]
iii. 1934. pp.viii.558. [5500.]
iv. 1935. pp.vii.639. [7000.]
v. 1936. pp.ix.640. [7000.]

vi. 1937. pp.ix.708. [7500.]
vii. James Murray Luck, editor. 1938. pp.ix. 571. [5500.]
viii. 1939. pp.ix.676. [7000.]
ix. 1940. pp.ix.744. [8000.]
x. 1941. pp.ix.692. [7000.]
xi. 1942. pp.ix.736. [8000.]
xii. 1943. pp.ix.704. [7500.]
xiii. 1944. pp.ix.795. [8500.]
xiv. 1945. pp.x.856. [9000.]
xv. 1946. pp.viii.687. [7500.]
xvi. 1947. pp.xi.740. [8000.]
xvii. 1948. pp.ix.801. [8500.]
xviii. 1949. pp.ix.739. [8000.]
xix. 1950. pp.xi.596. [6500.]
xx. 1951. pp.ix.648. [7000.]
— Cumulated index to vol. 11 to 20. Edited by A. C. Griffin. 1952. pp.iv.377.
xxi. 1952. pp.ix.781. [8500.]
xxii. 1953. pp.x.729. [.]
xxiii. 1954. pp.x.636. [.]
xxiv. 1955. pp.xvi.805. [.]

no more published.

FELIX HAUROWITZ, Biochemie des menschen und der tiere seit 1914. Wissenschaftliche forschungsberichte: Naturwissenschaftliche reihe (vol.xii): Dresden &c. 1925. pp.xii.148. [1100.]

— — II. Teil (1924–1931). . . . (vol.xxvi): 1932. pp.x.152. [1250.]

ABSTRACTS of chemical papers. . . . A 3. Biochemistry. Bureau of chemical abstracts: 1937 &c.
details of this work are entered under Chemistry: General, below.

RICHARD KUHN [*and others*], Biochemistry. Office of military government for Germany: Field information agencies technical: Fiat review of german science 1939–1946: Wiesbaden 1947–1948. pp.[xv].218+[iv].241+[iv].201+[iv].318. [3000.]
the text is in german.

РЕФЕРАТИВНЫЙ журнал. Химия-биологическая химия. Академия наук СССР: Институт научной информации: Москва.
1955. pp.80.vii + 84.vii + 77.vii + 71.viii + 90.viii + 82.viii + 96.viii + 95.ix + 105.ix + 109.ix + 109.x + 106.x + 89.ix + 89.ix + 97.x + 103.x + 103.x + 92.x. + 98.x + 104.xii + 110.xi + 111.xii + 114.xiv + 640. [18,004.]
1956. pp.222.vii + 109.vi + 130.vii + 118.vii + 116.viii + 129 + 133 + 117 + 104 + 109 + 125 + 117 + 136 + 105 + 112 + 112 + 100 + 116 + 120 + 120 + 128 + 128 + 141. [23,973.]

1957. pp.136 + 136 + 132 + 133 + 133 + 144 + 133 + 145 + 140 + 161 + 141 + 128 + 120 + 137 + 137 + 137 + 136 + 132 + 125 + 125 + 100 + 105 + 101 + 104. [27,023.]

1958. pp.160 + 152 + 148 + 152 + 152 + 156 + 152 + 148 + 120 + 160 + 160 + 168 + 160 + 168 + 160 + 148 + 141 + 164 + 167 + 160 + 157 + 183 + 201 + 197. [33,207.]

1959. pp.137 + 125 + 149 + 183 + 181 + 196 + 181 + 189 + 196 + 188 + 187 + 193 + 188 + 201 + 204 + 197 + 193 + 205 + 197 + 169 + 168 + 168. [32,818.]

1960. pp.169 + 200 + 201 + 200 + 196 + 209 + 209 + 224 + 201 + 201 + 232 + 208 + 228 + 208 + 200 + 208 + 197 + 180 + 188 + 185 + 188 + 181 + 188 + 208 + 180. [35,182.]

1961. pp.189 + 208 + 196 + 192 + 196 + 184 + 188 + 156 + 194 + 192 + 176 + 168 + 172 + 170 + 168 + 144 + 160 + 153 + 164 + 148 + 144 + 200 + 148 + 200. [37,500.]

1962. pp.168 + 168 + 136 + 176 + 208 + 196 + 192 + 188 + 200 + 212 + 200 +

200 + 200 + 188 + 200 + 200 + 196 + 188 + 168 + 168 + 188 + 212 + 216 + 280. [40,000.]

1963. pp.196 + 156 + 183 + 179 + 184 + 188 + 232 + 216 + 212 + 236 + 214 + 264 + 236 + 196 + 120 + 136 + 164 + 204 + 211 + 220 + 272 + 288 + 156 + 156. [40,000.]

in progress; previously formed part of the sections devoted to chemistry and biology.

E. E. GONCHAROVA, N. M. POLYAKOVA and TS. M. SHTUTMAN, Биохимия нервной системы. Библиографический указатель отечественной литературы 1868–1954. Академия наук Украинской ССР: Институт биохимии: Киев 1957. pp.88. [1000.]

J. A. MCCORMICK, Isotopes in biochemistry and biosynthesis of labeled compounds. A selected list of references. United States atomic energy commission: Technical information service extension: Oak Ridge, Tenn. 1958. pp.89. [2430.]

BIBLIOGRAPHY of papers published by László Zechmeister and co-authors in the fields of chemistry and biochemistry, 1913–1958. Wien 1958. pp.[vii].22. [257.]

JEAN ÉMILE COURTOIS, Biochemistry. French

bibliographical digest (no.27): New York 1959. pp.171. [1200.]

JEAN ÉMILE COURTOIS, Biochimie. Association pour la diffusion de la pensée française: Bibliographie française établie à l'intention des lecteurs étrangers: 1960. pp.152. [1000.]
limited to works published in 1951–1956.

W[ILLIAM] H. FITZPATRICK and W[ILLIAM] HENRY SEBRELL, Biochemistry of the USSR. A 25-year literature search of the journal Biokhimiya. Israel program for scientific translations: [Jerusalem 1961]. pp.[iii].199. [2000.]*

Brass.

EFFECT of iron on brass. Science library: Bibliographical series (no.588): 1943. single leaf. [12.]*

Calculus.

MAURICE LECAT, Bibliographie du calcul des variations depuis les origines jusqu'à 1850, comprenant la liste des travaux qui ont préparé ce calcul. Gand &c. 1916. pp.iv.92. [750.]
—— 1850–1913. 1913. pp.iv.113. [1000.]

Capillarity.

SIGMUND WALDBOTT, References to capillarity to the end of the year 1900, being chapter VII of "A study in pharmacy" by John Uri Lloyd. Lloyd library: Bulletin (no.4): Cincinnati 1902. pp.[ii]. 2.101–212. [665.]

Carbon monoxide.

R. R. SAYERS and SARA J. DAVENPORT, Carbon monoxide literature. Public health service: Public health bulletin (no.150): Washington 1925. pp. iii.53. [200.]

Carboniferous strata.

HENRY PAUL, Schrifttum des unterkarbons. Reichsamt für Bodenforschung: Abhandlungen (n.s., no.198): Berlin 1944. pp.515. [5939.]

Catalysis.

ERWIN SAUTER, Heterogene katalyse. Ergebnisse der neueren entwicklung der chemisch-physikalischen erforschung der durch grenzflächen beschleunigten chemischen reaktionen. Wissenschaftliche forschungsberichte: Naturwissenschaftliche reihe (vol.xxiii): Dresden &c. 1930. pp.x.80. [400.]

I. V. MATVEEVA, Гетерогенных катализ в органической химии. Библиографический указатель . . . литературы (1958–1960). Академия наук СССР: Институт органической химии: Москва 1962. pp.279. [3450.]

Cataphoresis.

CHARLES L. SCHUETTLER, Electrophoresis bibliography. Silver Spring, Md. [1953]. pp.228. [2500.]*

— — [another edition]. Edited by Alfred Henley. [1955]. pp.290. [3000.]*

Cathodes.

PAPERS on cathode sputtering, especially since 1932. Science library: Bibliographical series (no. 537): 1940. ff.3. [50.]*

CATHODIC protection. Public library of South Australia: Research service [Adelaide] 1958. ff.10. [175.]*

Cavitation.

CAVITATION, 1937 to date. Science library: Bibliographical series (no.568): 1942. ff.3. [55.]*

Cerite.

R. J. MEYER, Bibliographie der seltenen erden, ceriterden, yttererden und thorium. Hamburg &c. 1905. pp.79. [780.]

Cerium.

W. H. MAGEE, Indexes to the literatures of cerium and lanthanum. Smithsonian institution: Smithsonian miscellaneous collections (no.971): Washington 1895. pp.[ii].43. [cerium: 750.]

Corticosterone.

THE CHEMISTRY of corticosterone. Science library: Bibliographical series (no.440): 1938. ff.3. [50.]*

Cesium.

B[ORIS] I[OSIFOVICH] KOGAN [*and others*], Редкие щелочные металлы (литий, рубидий и цезний). Библиография. Академия наук СССР: Институт минералогии, геохимии и кристаллохимии редких элементов: Москва 1962. pp.331. [5000.]

Chloramine.

CHLORAMINE and chloramine-T, 1933–7. Science

library: Bibliographical series (no.451): 1939. ff.4. [67.]*

Chromatographic analysis.

FLORENCE EVELYN COZZI, Chromatography, its use in the quantitative analysis of detergent and soap ingredients. An abstract bibliography. QM research & development laboratories. Technical library: Bibliographic series (no.29): Philadelphia [1953]. pp.24. [83.]*

KAREL MACEK and IVO M. HAIS, Bibliografie papírové chromatografie a přehled použití. Česklosvenská akademie věd: Praha 1960. pp.767. [10,290.]

Clouds.

CLOUD formation and dispersal by aeroplanes. 1935–1942. Science library: Bibliographical series (no.603): 1944. f.1. [15.]*

CATALOGUE of meteorological satellite data, Tiros I television cloud photography. Weather bureau: Key to meteorological records documentation (no.5.31): Washington 1961. pp.ii.147.*

CATALOGUE of meteorological satellite data. Tiros IV television cloud photography. Weather bureau: Key to meteorological records documentation (no.5.34): Washington 1963. pp.ii.167.*

CATALOGUE of meteorological satellite data. Tiros v television cloud photography. Weather bureau: Key to meteorological records documentation (no.5.35): Washington 1964. pp.ii.362.*

Cobalt.

FLOTATION of nickel and cobalt minerals. Science library: Bibliographical series (no.101): [1933]. ff.[i].2. [15.]*

— Supplement. . . . (no.438): 1938. single leaf. [12.]*

[FACUNDO ROLF MORRAL], Cobalt and its alloys. A summary of allotropy and phase diagrams. Cobalt information center: [Columbus] 1956.

— — Second edition. 1958. pp.125. [1250.]

Collagen.

RUDOLF HINTERWALDNER, Struktur und eigenschaften des kollagens und seiner umwandlungsprodukte. Garmisch-Partenkirchen 1959. ff.[i]. 103. [438.]*

Colloids.

RAPHAEL E[DUARD] LIESEGANG, Kolloidchemie 1914–1922. Wissenschaftliche forschungsberichte:

Naturwissenschaftliche reihe (vol.vi): Dresden &c. 1922. pp.viii.100. [600.]

— — Zweite . . . auflage. 1926. pp.xii.176. [1250.]

HARRY N[ICHOLLS] HOLMES, Bibliography of colloid chemistry. . . . Preliminary edition. National research council: Washington 1923. ff.[i].iii.135. [2150.]*

RAPHAEL ED. LIESEGANG, Kolloide in der technik. Wissenschaftliche forschungsberichte: Naturwissenschaftliche reihe (vol.ix): Dresden &c. 1923. pp.[viii].158. [1000.]

RAPHAEL ED. LIESEGANG, Biologische kolloidchemie. Wissenschaftliche forschungsberichte: Naturwissenschaftliche reihe (vol.xix): Dresden &c. 1928. pp.xii.128. [600.]

SELECT bibliography of dichromated colloids. Science library: Bibliographical series (no.20): [1931]. ff.[i].3.3. [100.]*

A[NTON] V[LADIMIROVICH] DUMANSKY and I. A. DUMANSKY [iii: and ZAKHAR MARKOVICH VASHCHENKO], Библиографический очерк развития отечественной коллоидной химии. Академия наук Украинской ССР: Институт общей и неорганической химии: Киев 1952–1958. pp.148+240+220. [4000.]

Colour.

VERZEICHNISS sämmtlicher schriften über färberei, druckerei, bleicherei, appretur und farbenfabrikation welche von 1865–1881 im deutschen buchhandel erschienen sind. Gracklauer's fachkatalog (no.19): Leipzig 1881. pp.12. [125.]

THEORY of colour mixture. Science library: Bibliographical series (no.105): 1933. ff.[i].4. [57.]*

COLOUR measurement and matching. Science museum library: Bibliographical series (no.116): 1933. ff.[i].8. [108.]*

ELLERY H[ALE] HARVEY, Bibliography on the lightfastness of coloring matter. Chicago 1942. ff.[i].21. [500.]

COLOUR vision. References from 1930–1940. Science library: Bibliographical series (no.556): 1941. ff.19. [375.]*

INTERNATIONALE bibliographie der farbenlehre und ihrer grenzgebiete. Göttingen.

i. 1940–1949. Bearbeitet . . . von Manfred Richter. 1952. pp.xv.244. [1668.]

Columbium.

REFERENCES to the industrial applications of

columbium (niobium). Science library: Bibliographical series (no.238): 1935. single sheet. [19.]*

FRANK W[EISS] TRAPHAGEN, Index to the literature of columbium: 1801–1887. Smithsonian institution: Miscellaneous collections (no.663): Washington 1888. pp.iii.27. [200.]

Corticosterone.

THE CHEMISTRY of corticosterone. Science library: Bibliographical series (no.440): 1938. ff.3. [50.]*

Cosmic rays. [*see also* **Radiation.**]

RECENT references to cosmic rays. Science library: Bibliographical series (no.68): [1933]. ff.[i]. 4. [48.]*

WALTHER BOTHE, SIEGFRIED FLÜGGE [*and others*], Nuclear physics and cosmic rays. Office of military government for Germany: Field information agencies technical: Fiat review of german science, 1939–1946: Wiesbaden 1948. pp.[viii].230+[iv]. 198. [1250.]

the text is in german.

SELECT list of publications in the Science library relevant to cosmic rays. Science library: Bibliographical series (no.704): 1951. ff.3. [41.]*

ESTHER PEREIRA, Cosmic ray instruments and measuring devices. California institute of technology: Jet propulsion laboratory: Astronautics informations literature search (no.230): Pasadena 1960. pp.vii.83.

M. S. PETUKHOVA, Космические лучи. Библиографический указатель 1956–1960. Академия наук СССР: Физический институт им. П. Н. Лебедева: Библиотека: Москва 1961. pp.195. [3000.]

Cosmology.

N[IKIFOR] R. SAMBORSKY, Живая библіографія послѣднихъ ста лѣтъ. Книга о книгахъ, вышедшихъ въ свѣтъ на русскомъ языкѣ и найболѣе расходящихся въ Россіи, съ 1794–1894 г.г. ч. 1. Космологія. Одесса 1895. pp.[iv].iii.68. [750.]

MILDRED CATHERINE BENTON, Cosmological theories: ancient mythology and modern science. A selected reading list. Naval research laboratory [Bellevue, D.C.]: Bibliography (no.6): Washington 1955. pp.[iii].19. [101.]*

Crystal, Crystallography.

JARED KIRTLAND MORSE, Bibliography of crystal structure. Bulletin of the Crystal structure laboratory, University of Chicago: Chicago 1928. pp. xix.164. [2000.]

GROWTH of crystals. Science library: Bibliographical series (no.161): 1934. ff.[i].16. [303.]*

J[EAN] WYART, Cristallographie [-minéralogie]. Association pour la diffusion de la pensée française: Bibliographie française établie à l'intention des lecteurs étrangers: 1956. pp.68. [700.]
limited to writings published since 1946.

L. V. MATVEEVA, Дислокации в кристаллах. Библиографический указатель. Академия наук СССР: Институт кристаллографии: Москва 1960. pp.116. [1525.]

CRYSTAL imperfections in semiconductors. Bibliography. Library of Congress: Aerospace information division: [Washington] 1962. ff.iii.11. [98.]*

LATTICE dynamics. Annotated bibliography of soviet literature. Library of Congress: Aerospace information division: AID report (no. B-63-101): [Washington] 1963. ff.iii.34.*

Curves.

HEINRICH WIELEITNER, Bibliographie der höheren algebraischen kurven für den zeitabschnitt von 1890–1904. Königl. humanistisches gymnasium zu Speyer: Jahresbericht ... 1904/05: Beilage: Leipzig 1905. pp.58. [1250.]

Cyclones.

ANDRÈS [ANDRÉ] POEY, Table chronologique de quatre cents cyclones qui ont sévi dans les Indes Occidentales et dans l'océan Atlantique nord ... (depuis 1493 jusqu'en 1855). Dépôt des cartes et plans de la Marine: 1862. pp.50. [400.]

— — A chronological table of cyclonic hurricanes [&c.]. 1866. pp.96. [400.]

— — Deuxième édition. Bibliographie cyclonique. 1866. pp.96. [1008.]

Deflectometers.

SOME REFERENCES on deflectometers and extensometers, 1935 to date. Science library: Bibliographical series (no.612): 1944. ff.2. [30.]*

Determinants.

[SIR] THOMAS MUIR, The theory of determinants in the historical order of its development.

i. Determinants in general. Leibnitz (1693) to Cayley (1841). 1890. pp.xi.278. [50.]

— Second edition. 1906. pp.xi.491. [100.]

ii. The period 1841 to 1860. 1911. pp.xvi.475. [300.]

iii. The period 1861 to 1880. 1920. pp.xxvi. 503. [500.]

iv. The period 1880 to 1900. 1923. pp.xxxi. 508. [800.]

— — [continuation].. Contributions to the history of determinants, 1900–1920. 1930. pp.xxiv. 408. [650.]

consists of abstracts.

MAURICE LECAT, Bibliographie des déterminants à plus de deux dimensions. Louvain 1924. pp.16. [200.]

Deuterium.

LAWRENCE M. BROWN, ABRAHAM S. FRIEDMAN and CHARLES W. BECKETT, Bibliography of research on deuterium and tritium compounds, 1945 to 1952. National bureau of standards: Circular (no.562): Washington 1956. pp.ii.85. [2500.]

Deuterum oxide.

J[ACQUELINE] DIRIAN, Constantes physiques des variétés isotopiques de l'eau (^{17}O–^{18}O–D–T).

Bibliographie. Commissariat à l'énergie atomique: Bibliographie (no.15): [Gif-sur-Yvette] 1962. pp. 107. [119.]*

Didymium.

A[RTHUR] C[OMINGS] LANGMUIR, Index to the literature of didymium, 1842–1893. Smithsonian institution: Smithsonian miscellaneous collections ([vol.xxxviii,] no.972): Washington 1894. pp.[ii]. 20. [400.]

Dimensions.

THE METHOD of dimensions. Theory of units mostly excluded. Science library: Bibliographical series (no.563): 1941. ff.2. [30.]*

Dipole moment.

THE DIPOLE moment and polarization of bonds. A selected list of references, 1948–1952. Science library: Bibliographical series (no.710): 1952. ff.3. [53.]*

Dolomite.

LIST of references on dolomite. Library of Congress: Washington 1917. ff.4. [35.]*

DOLOMITIC lines. Science library: Bibliographical series (no.474): 1939. ff.4. [75.]*

Earth pressure.

BRIEF LIST of references on earth pressure. Library of Congress: Washington 1916. ff.3. [34.]*
— [supplement]. 1923. ff.3. [33.]*

Earthquakes, earth currents.

A. POEY, Catalogue chronologique des tremblements de terre ressentis dans les Indes-Occidentales, de 1530 à 1857; accompagné d'une revue bibliographique contenant tous les travaux relatifs aux tremblements de terre des Antilles. [1857]. pp.76. [1500.]

ALEXIS PERREY, Bibliographie séismique. Catalogue des livres, mémoires et notes sur les tremblements de terre et les phénomènes volcaniques. Dijon 1855–1865. pp.183+161+70. [4015.]

MARIO BARATTA, Bibliografia geodinamica italiana. Part I. 1892. pp.[iii].35. [61.]
no more published.

EARTHQUAKES and volcanoes. A list of books with references to periodicals in the Brooklyn public library. Brooklyn 1906. pp.13. [150.]
— Second edition. 1909. pp.28. [400.]

VERZEICHNIS der bibliothek des Zentralbureau

der Internationalen seismologischen assoziation. Würzburg 1909. pp.22. [300.]

[COUNT FERNAND DE MONTESSUS DE BALLORE], Notice sur les travaux sismologiques du c^te^ de Montessus de Ballore. Chartres [printed] [1910]. pp.11. [50.]

CATALOGUE of books and papers relating to internal dynamics of the earth mainly for the year 1911. International association for seismology: 1914. pp.vii.54. [400.]

C[OUNT] F[ERNAND] DE MONTESSUS DE BALLORE, Bibliografía general de temblores y terremotes. Sociedad chileña de historia y geografía: Santiago de Chile 1915–1919. pp.115+117–253+255–502 + 503–613 + 615–765 + 767–998 + 999–1174 + 1175–1516. [9140.]

LIST of references on new Madrid earthquake 1811–1812. Library of Congress: Washington 1916. ff.2. [16.]*

BULLETIN bibliographique trimestriel. Union géodésique et géophysique internationale: Section de séismologie [*afterwards:* International association of seismology and the physics of the earth's interior]: 1922 &c.
in progress.

ERNEST A. HODGSON, Bibliography of seismology. Dominion observatory: Publications: Ottawa 1929–1946. pp.368+436+329. [6142.]

ALFREDO ROSENZWEIG, Bibliografía sismológica. Instituto geológico del Peru: Boletín (no.2). Lima 1945. pp.26. [223.]

MIROSLAVA D. UZELAC, Bibliographie séismologique de Yougoslavie, 1672–1947. Institut séismologique [Seismološki zavod]: Beograd 1948. pp. 93. [546.]

a slip has been pasted on the titlepage, correcting the date to 1667–1947.

WILLARD ROUSE JILLSON, A bibliography of early printed books, pamphlets, articles and maps pertaining to the geology, paleontology and seismology of Kentucky, 1744–1854. Frankfort, Ky. 1950. pp.53. [150.]

CONTENTS of the publications of the Earthquake research institute, Tokyo university, 1926–1951. Tokyo 1952. pp.[ii].72. [1000.]

EDWARD P. HOLLIS, Bibliography of engineering seismology 1954. Earthquake engineering research institute: [Berkeley] 1954. pp.[v].iv.146. [1500.]*

— — Second edition. [San Francisco] 1958. pp. ix.144. [2000.]

C[HARLES] MARTIN DUKE, Bibliography of effects of soil conditions on earthquake damage. Earthquake engineering research institute: [San Francisco] 1958. pp.iii.47. [350.]*

W. D. MCLEAN and T. W. CALESS, A bibliography of seismology for the VELA uniform program. University of Michigan: Institute of science and technology: Acoustics and seismics laboratory: Ann Arbor 1961. pp.v.205. [1000.]*

SOVIET seismology and seismometry. A preliminary bibliography. [Library of Congress:] Air information division: [Washington] 1961. ff. [iii]. 77. [900.]*

UNDERGROUND communications. Annotated bibliography of soviet-bloc literature for the per-

Eclipse.

SKY illumination during eclipse. Annotated bibliography of soviet-bloc literature for the period 1940–1963. Library of Congress: Aerospace information division: AID report (no.B-64-20): [Washington] 1964. ff.iii.68. [116.]*

iod 1960–1962 (preliminary). Library of Congress: Aerospace information division: AID report (no. B-63-99): [Washington] 1963. ff.[ii].19. [54.]*

Emulsions.

EMULSIFICATION by supersonic cavitation, 1940–1952. Science library: Bibliographical series (no. 714): 1952. ff.2. [41.]*

EMULSIONS — non aqueous. Public library of South Australia: Research service: [Adelaide] 1957. ff.5. [72.]*

Epidiascopes.

EPIDIASCOPES, 1926–1936. Science library: Bibliographical series (no.255): 1936. ff.2. [28.]*

Euclid.

GEORG MATHIAS BOSE, Schediasma literarivm qvo contenta Elementorvm Evclidis envnciat, simvl de variis horvm editionibvs post Fabricium nonnulla disserit m. George Mathias Bose. Lipsiae 1737. pp.[viii].24. [60.]

GUSTAVE [GUSTAF] ENESTRÖM, Notice bibliographique sur les traductions en suédois des Éléments d'Euclide. Rome 1885. pp.13. [100.]

[PIETRO RICCARDI], Le prime edizioni degli Elementi di Euclide. Bologna 1886. pp.16. [4.]

PIETRO RICCARDI, Saggio di una bibliografia euclidea. Bologna 1887–1893. pp.125+25+60+58. [2500.]

CHARLES THOMAS-STANFORD, Early editions of Euclid's Elements. Bibliographical society: Illustrated monographs (no.xx): 1926. pp.viii.67. [86.]

SIR STEPHEN GASELEE, The Euclid collection in the National library of Wales. Aberystwyth 1939. pp.8. [100.]

Evaporation. [*see also* **Liquids.**]

GRACE J. LIVINGSTON, An annotated bibliography of evaporation. [Weather bureau: Washington 1910]. pp.121. [1100.]

EVAPORATION from land and water surfaces, 1930–1948. Science library: Bibliographical series (no.658): 1948. ff.7. [110.]*

Flocculation.

FLOCCULATING agents. Public library of South Australia: Research service: [Adelaide] 1957. ff.8. [111.]*

Friction.

DRY friction, 1930–1936. Science library: Bibliographical series (no.308): 1937. ff.5. [100.]*

Functions.

TABLES of Bessel functions in the Science library. Science library: Bibliographical series (no.2): [1931]. ff.[i].4. [39.]*

J[AMES] A[LEXANDER] SHOHAT, EINAR HILLE and JOSEPH L[EONARD] WALSH, A bibliography on orthogonal polynomials. National research council: Bulletin (no.103): Washington 1940. pp.ix. 204. [2000.]

Galilei, Galileo.

PIETRO RICCARDI, Bibliografia galileiana. Modena 1872. pp.[iv].coll.72. [1500.]

ANTONIO FAVARO, Per la edizione nazionale delle opere di Galileo Galilei. Firenze 1888. pp.57. [87.]
comprises a list of works omitted from the collected edition of Galilei's works.

[ANTONIO FAVARO], Per la edizione nazionale delle opere di Galileo Galilei. . . . Indice alfabetico e topografico del commercio epistolare. Firenze [printed] 1889. pp.23. [400.]

ANTONIO FAVARO, Per la edizione nazionale delle opere di Galileo Galilei. . . . Materiali per un indice dei manoscritti e documenti galileiani non posseduti dalla Biblioteca nazionale di Firenze. Venezia 1894. pp.[ii].127. [1255.]

ANTONIO FAVARO, Per la edizione nazionale delle opere di Galileo Galilei. . . . Indice cronologico del carteggio galileiano. Firenze 1896. pp.101. [4202.]

A[LARICO] CARLI and A[NTONIO] FAVARO, Bibliografia galileiana (1568–1895). Ministero della pubblica istruzione: Indice e cataloghi (vol.xvi): Roma 1896. pp.viii.403. [2130.]

— — Supplemento. Bibliografia galileiana, 1896–1940 . . . da Giuseppe Boffito . . . (vol.xviii): 1943. pp.lii.310. [1850.]

ANTONIO FAVARO, Vent' anni di studi galileiani. Firenze [printed] 1896. pp.26. [100.]

— — [another edition]. Per la edizione nazionale delle opere di Galileo Galilei. . . . Trent'anni di studi galileiani. 1907. pp.29. [160.]

lists of Favaro's own publications.

ANTONIO FAVARO, Bibliografia galileiana . . . nel terzo centenario della morte di Galileo Galilei. Reale istituto veneto di scienze, lettere ed arti: Venezia 1942. pp.vii.39. [260.]

DINO CINTI, Biblioteca galileiana raccolta dal principe Giampaolo Rocco di Torrepadula. Biblioteca bibliografica italica (vol.15): Firenze 1957. pp.3–396. [194.]

ANGIOLO PROCISSI, La collezione galileiana della Biblioteca nazionale di Firenze. Ministero della pubblica istruzione: Indici e cataloghi (new ser.v): [Rome] [1959]– . pp.xxiv.331+ .
in progress.

Gallic acid.

PRODUCTION of gallic acid. Science library: Bibliographic series (no.678): 1949. single sheet. [9.]*

Gallium.

PHILIP E[MBURY] BROWNING, Index to the literature of gallium, 1874–1903. Smithsonian institution: Miscellaneous collections (vol.xlvi): Washington 1904. pp.12. [100.]

PIERRE DE LA BRETÈQUE, Gallium bibliographie. Aluminium-industrie-aktiengesellschaft: Neuhausen am Rheinfall 1958. pp.[250]. [2000.]

— — [supplement]. Gallium, bulletin bibliographique. Société française pour l'industrie de l'aluminium: Marseille 1960. pp.[98]. [600.]*

PIERRE DE LA BRETÈQUE [*and others*], Gallium. Propriétés principales. Bibliographie. Société française pour l'industrie de l'aluminium: Institut de recherches: Marseille 1962. pp.xxxii.[342]. [2500.]

Gamma function.

MAURICE GODEFROY, La fonction gamma: théorie, histoire, bibliographie. Faculté des sciences de Paris: Thèses (ser. A, no.10): 1901. pp.vii.94. [50.]

Gauge.

PIRANI gauges. Science library: Bibliographical series (no.253): 1936. ff.2. [26.]*

Geometry.

I[GNAZ] ROGG, Handbuch der mathematischen literatur vom anfange der buchdruckerkunst. . . . Erste abtheilung, welche die arithmetischen und geometrischen wissenschaften enthält. Tübingen 1830. pp.[ii].vi.579. [7500.]

no more published.

— CHASLES, Rapport sur les progrès de la géométrie. Ministère de l'instruction publique: Recueil de rapports sur les progrès des lettres et des sciences en France: 1870. pp.[iii].389. [1000.]

BARON FRIEDRICH VON LÖFFELHOLZ-COLBERG, Forstliche chrestomathie. . . . III. 2. Abtheilung. Die literatur der geometrie, stereometrie, trigonometrie und höheren mathematik überhaupt. Berlin 1873. pp.viii.859–1092. [2500.]

[PIERRE JEAN BAPTISTE] H[ENRI] BROCARD, Notes de bibliographie des courbes géométriques. Bar-le-Duc [printed] 1887–1899. pp.[xxii].296.xxvi+viii.243. [1022.]
reproduced from handwriting.

A LIST of books and pamphlets in the National art library . . . on drawing, geometry, and perspective. South Kensington [*now* Victoria and Albert] Museum: 1888. pp.92. [1000.]

C. ALASIA, Saggio terminologico-bibliografico sulla recente geometria del triangolo. Bergamo 1902. pp.iv.43. [200.]

DUNCAN M['LAREN] Y[OUNG] SOMMERVILLE, Bibliography of non-euclidean geometry. University of St. Andrews: London 1911. pp.xii.404. [4500.]

PAULINE SPERRY, Bibliography of projective differential geometry. University of California: Publications in mathematics (vol.ii, no.6): Berkeley 1931. pp.[ii].119–127. [200.]

V. M. GERASIMOVA, Указатель литературы по геометрии Лобачевского и развитию ее идей. Геометрия Лобачевского и развитие ее идей (vol.vii): Москва 1952. pp.192. [1750.]

ILONA PÁLFFY and MIKLÓS PÁLFFY, Bibliographia bolyaiana 1831–1960. Uj bibliográfiai füzetek (no.v): Budapest 1962. pp.67. [500.]

Geophysics.

1. *Periodicals*

OTTO [FRANZ] MEISSER and ULRICH GERHARDT, Periodica geophysica . . . in bibliotheken der Deutschen Demokratischen Republik. Bücherei des dokumentalisten (vol.4): Berlin 1956. pp. [viii].184. [1510.]*

2. *General*

JAHRBUCH der astronomie und geophysik. Enthaltend die wichtigsten fortschritte auf den gebieten der astrophysik, meteorologie und physikalischen erdkunde. Leipzig 1890–1913.

details of this work are set out under Astronomy, above.

ABSTRACTS of current articles on geophysical prospecting. Bureau of mines: Mining division (G.A. no.1-86): Washington 1928–1936. pp.1848. [2250.]*

previously issued as part of the Information circular *of the Geological survey; continued as:*

Geophysical abstracts. Geological survey bulletin (no.87 &c.).

lxxxvii. July–December 1936. Compiled by W[ladimir] Ayvazoglou . . . (no.887) pp.ii.98. [290.]

lxxxviii–xci. 1937. . . . (no.895). pp.ii.203. [579.]

xcii–xcv. 1938. . . . (no.909). pp.[ii].222. [645.]

xcvi–xcix. 1939. . . . (no.915). pp.[ii].195. [550.]

c–ciii. 1940. . . . (no.925). pp.[ii].200. [578.]

civ–cvii. 1941. . . . (no.932). pp.[ii].183. [500.]

cviii–cxi. 1942. . . . (no.939). pp.[ii].138. [407.]

cxii–cxv. 1943. . . . (nos.7256, 7257, 7267, 7273). pp.45+53+51+78. [469.]

cxvi–cxix. 1944. . . . (nos.7285, 7292, 7303, 7310). pp.54+49+52+77. [533].

cxx–cxxiii. 1945. . . . (nos.7324, 7338, 7349, 7355). pp.53+48+49+78. [541.]

cxxiv–cxxvii. 1946. . . . (nos.7384, 7386, 7400,

7414). pp.60+62+63+86. [636.]

cxxviii–cxxxi. 1947. . . . (no.957). pp.iii.265. [745.]

cxxxii–cxxxv. 1948. . . . (no.959). pp.[ii].388. [1052.]

cxxxvi–cxxxix. 1949. . . . (no.966). pp.iii.333. [942.]

cxl–cxliii. 1950. . . . (no.976). pp.iii.301. [835.]

cxliv–cxlvii. 1951. . . . (no.981). pp.iii.281. [770.]

clviii–cli. 1952. . . . (no.991). pp.iii.287. [900.]

clii–clv. 1953. . . . (no.1002). pp.iii.284. [843.]

in progress; the issues for 1942–1946 were published as Information circulars of the Bureau of mines.

CARLOS DE MELLO, Les lois de la géographie. . . . Bibliographie systématique de la géophysique. Berlin 1902. pp.viii.360. [5000.]

RICHARD AMBRONN, Methoden der angewandten geophysik. Wissenschaftliche forschungsberichte: Naturwissenschaftliche reihe (vol.xv): Dresden &c. 1926. pp.xii.258. [1750.]

[E. LANCASTER-JONES, Current bibliography of applied geophysics]. [1929–1932]. ff.[160]. [2000.]

[*continued as:*]

Current bibliography of applied geophysics. Science museum library: Current bibliography series (no.1): 1933–1947.*

in progress; issued about eight times a year, each issue containing about fifty titles.

C[ARL] A[UGUST] HEILAND and DART WANTLAND, A selected list of books and references on geophysical prospecting. Colorado school of mines: Quarterly (vol.xxvi, no.3): Golden 1931. pp.24. [200.]

J[ESSE] HARLAN JOHNSON, Bibliography of geophysical principles, apparatus and methods applied to prospecting. [Golden *c.*1933]. pp.[7]. [150.]

ZENTRALBLATT für geophysik, meteorologie und geodesie. Berlin.

i. Redaktion: Karl Jung. 1938. pp.iv.490. [2000.]
ii. 1938. pp.iv.499. [2000.]
iii. 1939. pp.iv.486. [2000.]
iv. 1939. pp.iv.493. [2000.]
v. 1940. pp.iv.470. [1750.]

W[LADIMIR] AYVAZOGLOU and V[SEVOLOD LAWROVITCH] SKITSKY, Bibliography of russian literature on geophysical exploration, 1929–1941. Bureau of mines: Information circular (no.7323): [Washington] 1945. pp.44. [521.]*

JULIUS BARTELS [*and others*], Geophysics. Office of military government for Germany: Field information agencies technical: Fiat review of german science, 1939–1946: Wiesbaden 1948. pp. [viii].237+[iv].322. [1750.]

the text is in german.

V. LAURSEN, Bibliography for the second international polar year 1932–33. International meteorological organization: Copenhagen 1951. pp.3–253. [2000.]

SELECCIÓN de reseñas bibliográficas meteorológicas y geofísicas. Servicio meteorológico nacional: Serie G: Buenos Aires 1951 &c.

details of this work are entered unter Meteorology, below.

BOOKS and pamphlets on soil mechanics. South african council for scientific and industrial research: Library and information division: Library class list (no.3): [Pretoria 1952]. pp.20. [150.]*

— Supplement. [1954]. pp.9. [60.]*

[EUGENIUSZ RYBKA *and others*], Polska bibliografia analityczna. Astronomia-geodezja-geofizyka. Polska akademia nauk: Ośrodek bibliografii i dokumentacji naukowej: Warszawa.

i. 1955. pp. . [246.]
ii. 1956. pp.48.88. [328.]
iii. 1957. pp.159. [356.]
no more published.

BIBLIOGRAFÍA de la biblioteca del Instituto geofísico de los Andes colombianos sobre geología y geofísica de Colombia. Secunda edición . . . por Jesús Emilio Ramírez. Banco de la república: Archivo de la economía nacional: Bogotá 1956. pp.3–521. [3900.]

[LEE ANNA EMBREY, MARGUERITE CHRISTESEN and ELIZABETH HUNT], A bibliography for the International geophysical year. National science foundation: [Washington 1957]. pp.vi.51. [400.]

THE EARTH sciences. Books to observe and interpret the International geophysical year. United States information agency: Information center service: Subject bibliography (no.28): [New York] 1957. pp.iii.32. [150.]*

V. K. STEPANOVA, Международный геофизический год. Рекомендательный список литературы. Ленинград 1957. ff.25.*

РЕФЕРАТИВНЫЙ журнал. Физика: геофизика. Академия наук СССР: Институт научной информации: Москва.

1957. pp.113 + 108 + 144 + 176 + 164 + 144 + 128 + 121 + 113 + 125 + 84 + 88. [10,750.]
1958. pp.124 + 112 + 112 + 104 + 96 + 128 +97+132+119+120+120+136. [9650.]
1959. pp.140 + 160 + 168 + 151 + 157 + 164 + 156 + 140 + 172 + 144 + 140 + 116. [12,650.]
1960. pp.154 + 160 + 160 + 180 + 199 + 220 + 220 + 228 + 214 + 224 + 216 + 264+222*+128*. [16,510.]
1961. pp.47.72.80.40 + 48.64.64.56 + 62.48.88.48 + 40.56.80.56 + 43.52.87.38 + 44.56.68.39 + 40.64.92.52 + 40.72.64.44 + 36.56.64.32 + 32.56.48.40 + 44.48.56.40 +48.64.80.32+248*+122*. [20,000.]
1962. pp.44.84.67.40 + 40.72.52.36 + 36.74.56.36 + 36.68.60.44 + 44.60.60.36 + 40.64.67.40 + 40.72.76.40 + 48.84.76.36 + 56.76.68.36 + 64.64.64.40 + [282] + [248] +117*+120*. [20,000.]
1963. pp.[232]+[264]+[252]+[242]+[228] + [226] + [256] + [236] + [208] + [232] +[236]+[256]+ + .

in progress; this subseries previously formed part of the series devoted to Physics.

МЕЖДУНАРОДНЫЙ геофизический год... Библиографический указатель литературы на русском языке. Академия наук СССР: Москва.

1954–1957. [By] Р. Ф. З. pp.44. [452.]

1958. [By] Р. Ф. Затрутина и З. М. Михайлова. pp.64. [616.]

1959. [By] Р. Ф. Затрутина и Л. С. Рубина. pp.88. [793.]

1960. pp.128. [1172.]

AN INTERIM bibliography on the International geophysical year. National academy of sciences: Washington 1958. pp.vii.56. [650.]*

GEORGE E. TARBOX, Bibliography of graduate theses on geophysics in U.S. and canadian institutions. Colorado school of mines: Quarterly (v.53, no.1): Golden 1958. pp.vi.55. [450.]*

BIBLIOGRAFIA MRG. Polska akademia nauk: Komisija miedzynarodowego roku geofizycznego: Warszawa.*

i. 1958. ff.[ix].98.[vi]. [271.]

no more published.

B. B. G. O. A. Boletín bibliográfico de geofísica y oceanografía americanas. Parte geofísica. Instituto panamericano de geografía e historia: México.

i. 1958. pp.viii.160. [500.]

ii. 1959. pp.[v].365. [600.]

in progress?

BIBLIOGRAPHY of publications. University of Alaska: Geophysical institute: College 1960. ff.[i].i.11. [225.]*

JANE CLAPP, Museum publications. Part II. Publications in biological and earth sciences. 1962. pp.610. [9231.]*

DEVI MITRA and JOHN E. WILLSON, A bibliography on geomechanics. Utah engineering experiment station: Bulletin (no.119): Salt Lake City 1962. pp.[v].210. [1630.]*

A. I. PAVLOVA, Строение коры и верхней части мантии земли по геофизическим данным. Библиографический указатель 1937–1961. Академия наук СССР: Сектор сети специальных библиотек: Москва 1962. pp.92. [1138.]

Germanium.

PHILIP E[MBURY] BROWNING, Index to the literature of germanium, 1886–1903. Smithsonian institution: Miscellaneous collections (vol.xlvi): Washington 1904. pp.8. [34.]

A[LLEN] W. SCHOLL and ROY W. ARRICK, Germanium bibliography [Huntington, W. Va.] 1955. pp.vii.131. [1027.]*

Glaciology. [*see also* **Ice.**]

[DANIEL] DOLLFUSS-AUSSET, Matériaux pour l'étude des glaciers.... Tome premier.... Auteurs. 1864–1865. pp.[iv].676+[iv].304. [5000.]

pages 285–304 of the second volume are printed in red.

W[ILLIAM] JEROME HARRISON, A bibliography of midland glaciology. Birmingham 1895. pp.86. [250.]

W[ILLIAM] JEROME HARRISON, A bibliography of Norfolk glaciology. 1897. pp.91. [350.]

LIST of publications in the Science library on glaciology. Science library: Bibliographical series (no.698): 1950. ff.4. [60.]*

FRANK A. COOK, Selected bibliography on periglacial phenomena in Canada. Annotations and abstracts. Department of mines and technical surveys: Geographical branch: Bibliographical series (no.24): Ottawa 1960. pp.[ii].23. [93.]*

Globes, geographic.

WILHELM BONACKER, Das schrifttum zur globenkunde. Leiden 1960. pp.[iii].58. [660.]

Glumiflorae.

JOHN W. SCHERMERHORN and MAYNARD W. QUIMBY, The [Eldin Verne] Lynn index. A bibliography of phytochemistry. Monograph IV. Order, *glumiflorae*. Massachusetts college of pharmacy: Boston 1960. pp.70. [769.]

Glutamic acid.

PREPARATION of glutamic acid and its derivatives from cereals. Science library: Bibliographical series (no.646): 1947. single leaf. [15.]*

Graphs, symbolic.

LIST of references on graphic methods of presenting facts. Library of Congress: Washington 1916. ff.3. [28.]*

— [supplement]. 1924. ff.5. [52.]*

Gravity.

GRAVIMETERS. Instruments, excluding pendulums, for the determination of total gravity. Science library: Bibliographical series (no.216): 1936. ff.2. [33.]*

Halogens.

ANTISEPTIC and bactericidal properties of the halogen derivatives of the phenols, naphthols and xylenols. Science library: Bibliographical series (no.189): 1935. ff.[i].3. [54.]*

Helium.

BIBLIOGRAPHY of scientific literature relating to helium. National bureau of standards: Circular (no.81): Washington 1919. pp.21. [400.]

— Second edition. 1923. pp.32. [600.]

A SELECTED list of Bureau of mines publications . . . dealing with petroleum, natural gas, helium, oil shale, and allied subjects. Bureau of mines: Petroleum experimental station: Bartlesville, Okla. 1930. pp.66. [500.]*

— [another edition]. 1952. pp.x.125.xxviii. [1000.]*

BIBLIOGRAPHY of aeronautics. Part 40 — Helium. Works progress administration: [New York] 1937. pp.[v].45. [500.]*

HENRY P. WHEELER, The properties and uses of helium (including a comprehensive bibliography, 1933–1945). Bureau of mines: [Washington] 1946. pp.66. [912.]*

HENRY P. WHEELER and LOUISE B. SWENARTON, Helium: bibliography of technical and scientific literature from its discovery (1868) to January 1, 1947. Bureau of mines: Bulletin (no.484): Washington 1952. pp.xii.76. [2616.]

Humidity.

THE INFLUENCE of atmospheric humidity upon health (excluding literature purely on industrial hygiene and ventilation engineering). Science library: Bibliographical series (no.283): 1935 [1936]. ff.2. [37.]*

FRANCIS D. HORIGAN, Humidity studies. A literature survey. QM research & development laboratories: Technical library: Bibliographic series (no.19): Philadelphia [1951]. pp.[iii].34. [119.]*

Hygrometry.

HYGROMETERS, 1949–1953. Science library: Bibliographical series (no.724): 1954. ff.6. [100.]*

Ice.

A JÉGHIÁNY. Fővárosi könyvtár: Aktuális kérdések irodalma (no.4): [Budapest] 1910. pp.[2]. [25.]

BIBLIOGRAPHY on ice of the northern hemisphere. Hydrographic office: Publication (no.240): Washington 1945. pp.xii.179. [1700.]

D. C. PEARCE, A bibliography on snow and ice. National research council: Division of building research: Bibliography (no.1): Ottawa 1951. ff. [ii].69. [800.]*

FRANCIS D. HORIGAN, Electrical characteristics of snow and ice. A literature search. QM research & development laboratories: Technical library: Bibliographic series (no.28): Philadelphia [1953]. pp.[ii].15. [39.]*

SELECTED bibliography on sea ice distribution in the coastal waters of Canada. Department of mines and technical surveys: Geographical branch: Bibliographical series (no.18): Ottawa 1957. pp. [iii].50. [500.]*

FRANK A. COOK, Selected bibliography on canadian permafrost. Department of mines and technical surveys. Geographical branch: Bibliographical series (no.20): Ottawa 1958. pp.[iii].23. [102.]*

Indium.

PHILIP E[MBURY] BROWNING, Index to the literature of indium, 1863–1903. Smithsonian institution: Miscellaneous collections (vol.xlvi): Washington 1905. pp.15. [100.]

HERBERT A[UGUST] POTRATZ and JOHN B[ERNARD] EKELEY, A bibliography of indium, 1863–1933. University of Colorado: Boulder 1934. pp. 151–187. [600.]

MARIA THOMPSON LUDWICK, A bibliography of indium, 1934–1940. Utica, N.Y. [1940]. pp.22. [300.]

— — Supplement. [1941]. pp.8. [100.]

Influence lines.

INFLUENCE lines. Science library: Bibliographical series (no.269): 1936. ff.4. [72.]*

Infra-red. [*see also* **Ultra-violet.**]

DRYING by means of infra-red radiation. Science library: Bibliographical series (no.527): 1940. ff.2. [20.]*

URSULA FLECK, Infrared in relation to skin and underlying tissue. A bibliography. Library of Congress: Reference department: Technical information division: Washington 1952. ff.vii.20. [195.]*

CLEMENT R[APHAEL] BROWN [*and others*], Infrared. A bibliography. Library of Congress: Technical information division [ii: Department of commerce: Office of technical services]: Washington 1954–1957. pp.xi.374+[ii].vi.150. [7141.]*

HERBERT M[ALCOLM] HERSHENSON, Infrared absorption spectra. Index for 1945–1957. New York &c. 1959. pp.xvi.111. [12,500.]

R. J. P. LYON, Minerals in the infrared. A critical bibliography. Stanford research institute: Menlo Park 1962. pp.xii.76. [440.]*

E. D. PALIK, A far infrared bibliography. Naval research laboratory: NRL bibliography (no.21): Washington 1963. pp.iv.50. [650.]*

Integraphs.

A READING list upon integraphs, exclusive of planimeters and the like. Science library: Bibliographical series (no.542): 1940. single leaf. [15.]*

Ionosphere.

LAURENCE A. MANNING, A survey of the literature of the ionosphere. 1947.

— — [third edition]. Bibliography of the ionosphere. An annotated survey through 1960. Stanford 1962. pp.xv.613. [3000.]*

Isostasy.

ADOLPH KNOPF, Bibliography of isostasy. National research council: Division of geology and geography: Washington 1924. ff.[ii].39. [462.]*

Isotopes. [*see also* **Radiation.**]

ISOTOPES. A three-year summary of distribution, with extensive bibliography. Atomic energy commission: Washington 1949. pp.iv.201. [1866.]

ELIZABETH JENNINGS, Subject index to radioisotope research and techniques, including use of radioactive calcium, fluorine, phosphorus, sodium and strontium. [San Francisco 1949]. ff.[ii].113. [1104.]

J. A. MCCORMICK, Bibliography on uses of radioactive and stable isotopes in industry. A list of selected references. Atomic energy commission: Technical information service extension: Oak Ridge, Tenn. 1957. pp.27. [455.]*

J. A. MCCORMICK, Isotopes. A bibliography of United States research and application, 1955–1957. Atomic energy commission: Technical information service extension: [Oak Ridge] 1958. pp.vi. 261. [3500.]*

J. A. MCCORMICK, Radioisotopes in medicine and human physiology. A selected list of references. Atomic energy commission: Technical information service: Oak Ridge, Tenn. 1958. pp. 124. [2750.]*

J. A. MCCORMICK, Isotopes in biochemistry and biosynthesis of labeled compounds. A selected list of references. United States atomic energy commission: Technical information service extension: Oak Ridge, Tenn. 1958. pp.89. [2430.]*

J. A. MCCORMICK, Radioisotopes in animal physiology. A literature search. Atomic energy commission: Technical information service: Oak Ridge, Tenn. 1958. pp.118. [983.]*

J. A. MCCORMICK, Isotope techniques in biological sciences. A selected list of references. United States atomic energy commission: Technical information service extension: Oak Ridge, Tenn. 1958. pp.44. [807.]*

G. M. BEGUN, Isotope separation and isotope exchange. A bibliography with abstracts. Oak Ridge national laboratory: Oak Ridge, Tenn. 1959. pp. vi.235. [2497.]*

J. A. MCCORMICK, Radioisotopes in agriculture; animal husbandry, bacteriology, fertilizer uptake, plant physiology, photosynthesis, and entomology. A selected list of references. Atomic energy commission: Technical information service: Oak Ridge, Tenn. 1959. pp.67. [1500.]*

J. A. MCCORMICK, Utilization of radioisotopes in physical and biological sciences—general topics. A literature search. Atomic energy commission: Technical information service: Oak Ridge, Tenn. 1959. pp.30. [492.]*

VĚRA KLIMEŠOVÁ, Radioizotopy z zemědělství. Universita: Knihovna: Výběrový seznam (no.43): Brně 1959. pp.[iii].38. [115.]*

RADIOISOTOPE applications in industry . . . with selected reference to the international literature. International atomic energy agency: Vienna 1963. pp.131.

Keratin.

KERATIN, 1933–37. Science library: Bibliographical series (no.434): 1938. ff.3. [55.]*

Lanthanum.

W. H. MAGEE, Indexes to the literature of cerium and lanthanum. Smithsonian institution: Smithsonian miscellaneous collections (no.971): Washington 1895. pp.[ii].43. [Lanthanum: 250.]

Laser.

PATRICIA WARREN BECKER and STANLEY EDWARD BECKER, Bibliography on lasers. Oak Ridge, Tenn. 1962. pp.51. [514.]*

MASERS and lasers. An ASTIA report bibliography. Armed services technical information agency: Arlington, Va. 1962. pp.v.69. [300.]*

PERSONALITIES and organizations active in soviet manser and laser research. (Preliminary). Library of Congress: Aerospace information division: [Washington] 1962. ff.[iii].ii.28. [275.]*

LASER abstracts. New York.*

i. By A. K. Kamal. 1964. pp.[vii].177. [743.]

in progress.

Lea, Isaac.

GEORGE W[ASHINGTON] TRYON, Publications of Isaac Lea on recent conchology. Philadelphia [printed] 1861. pp.13. [125.]

[I. LEA], A catalogue of the published works of Isaac Lea. Philadelphia 1876. pp.22. [223.]

NEWTON PRATT SCUDDER, The published writings of Isaac Lea. Smithsonian institution: United States national museum: Bulletin (no.23 = Bibliographies of american naturalists, vol.ii): Washington 1885. pp.lix.278. [279.]

Le Verrier, Urbain Jean Joseph.

[U. J. J. LE VERRIER], Exposé sommaire des recherches astronomiques de m. U.-J. Le Verrier. [1843]. pp.8. [11.]

— — [another edition]. [1845]. pp.22. [17.]

LE VERRIER et son temps. Exposition. . . . Catalogue. Observatoire de Paris: [1946]. pp.[vii].79. [425.]

Levulinic acid.

LITERATURE references on levulinic acid (B). A. E. Staley manufacturing company: Decatur, Ill. [1941]. ff.[i].29. [317.]*

— [another edition]. Levulinic acid. [*c.*1943]. pp.38. [423.]

Liesegang rings.

KURT H. STERN, Bibliography of Liesegang rings. University of Arkansas: Fayetteville 1955. ff.70.*

Light. [*see also* **Optics.**]

ALFRED TUCKERMAN, Bibliography of the chemical influence of light. Smithsonian institution: Miscellaneous collections (no.785): Washington 1891. pp.iii.22. [400.]

LIST of references on the effect of light on human beings, physiological, psychological and therapeutic. Library of Congress: Washington 1913. ff.5. [56.]*

SUBJECT list of works on sound and light (including music, musical instruments, and general optical instruments). Patent office library: Subject lists (n.s. GG–GP): 1914. pp.iv.134. [1250.]

SELECTED list of books on fuels and lighting. Public library of New South Wales: Sydney 1919. pp.12. [250.]

W. H. MARTIN, Light scattering; bibliography. University of Toronto: Chemical laboratories: Papers (no.135): Toronto 1923. pp.276–281. [100.]

LIGHT reflected from the eyes of animals. Science library: Bibliographical series (no.78): [1933]. single sheet. [7.]*

TIME integration of light radiation. Science library: Bibliographical series (no.242): 1936. single sheet. [18.]*

SCATTERING of light by small particles, with reference to particle-size and wavelength, 1920–1937. Science library: Bibliographical series (no. 431): 1938. ff.3. [50.]*

THE EFFECT of light on germination. Science library: Bibliographical series (no.700): 1950. ff.4. [66.]*

РЕФЕРАТИВНЫЙ сборник по люминесцентному анализу. Академия наук СССР: Комиссия по люминесценции [ii– : Отделение физико-математических наук]: Москва.

i. [By] M[ariya] A[leksandrovna] Konstantinova-Shleringer. 1951.

ii. 1954. pp.104. [180.]

W[ILLIAM] H[ENRY] PLANT, Night vision and dark adaptation, 1946–1951, a selected reading list. Research and development board: [Washington] 1952. ff.[i].10. [100.]*

ABSTRACTS of the literature on semiconducting and luminescent materials and their application. Battelle memorial institute [Columbus]: New York &c. 1953 &c.*

details of this work are entered under Semiconduction, below.

MYER M. FISHMAN, Light scattering by colloidal systems. An annotated bibliography. Technical service laboratories: River Edge, N.J. [1957]. pp. [iii].i.84. [858.]*

ELSIE F. DUPRÉ and LEO H. DAWSON, Transmission of light in water. An annotated bibliography. U. S. Naval research laboratory: NRL bibliography (no.20): Washington 1961. pp.viii.80. [657.]*

YA. M. BOGUSLANSKY and S. I. VASILOV, Библиографический справочник литературы по люминесцентному анализу в медицине и биологии (с 1947 г. по июнь 1961 г.). Читинский государственный медицинский институт: Чита 1961. pp.77. [509.]

SELECTED bibliography on light scattering (after 1945). Philadelphia [1963]. pp.16. [500.]

LUMINESCENCE of complex molecules. Annotated bibliography based on soviet open literature. Library of Congress. Aerospace information division: AID report (no. B-63-125): [Washington] 1963. ff.iii.74. [145.]*

Limnology.

DIE LIMNOLOGISCHE literatur des jahres. Internationale vereinigung für theoretische und angewandte hymnologie: Archiv für hydrobiologie: Literatur-supplement (no.i &c.): Stuttgart.

1923. Herausgegeben von August Thienemann. . . . (no.i): 1925. pp.[iv].165. [2000.]

1924–1925. . . . (no.ii): 1927. pp.[vi].317. [4000.]

1926–1927. Herausgegeben von G. Vereščagin [Gleb Yurevich Vereshchagin] . . . (no.iii): 1931. pp.xiii.638. [3097.]

[*continued as:*]

Bericht über die literatur der binnengewässerkunde. Zentralverwaltung des Hydrometeorologischen dienstes der USSR: Moskau.

1928. 1943. pp.xii.500.4. [3471.]

no more published.

Lipides.

INTRANUCLEAR lipides. Science library: Bibliographical series (no.785): [1963]. pp.5. [100.]*

Lipids.

ÉMILE F. TERROINE, Lipides (décembre 1936–décembre 1937). Physiologie, revue annuelle (no.6 = Actualités scientifiques et industrielles, no.636): 1938. pp.59. [275.]

LIPIDS in lungs. Science library: Bibliographical series (no.782): [1963]. pp.5. [100.]*

Linseed.

LIST of references on linseed oil industry. Library of Congress: [Washington] 1919. ff.3. [20.]*

Linselles.

TH[ÉODORE DÉSIRÉ JOSEPH] LEURIDAN, Inventaire sommaire des archives communales antérieures à 1790. . . . Commune de Linselles. Lille 1881. pp. [iii].xiii.11.5.31.4.4.9.19.3.4.3.9.6. [15,000.]

Lintner, Joseph Albert.

EPHRAIM PORTER FELT, Memorial of the life and entomologic work of Joseph Albert Lintner. New York state museum: Bulletin (vol.v, no.24): Albany 1899. pp.[ii].303–400. [1250.]

Linz.

LAMBERT GUPPENBERGER, Bibliographie des clerus der diöcese Linz von deren gründung. Linz 1893. pp.ix.270. [3000.]

Liouville, Roger.

[R. LIOUVILLE], Notice sur les travaux scientifiques de m. Roger Liouville. 1886. pp.8. [11.]

— — [another edition]. 1897. pp.24. [46.]

Lippe.

O. WEERTH and E[RNST] ANEMÜLLER, Bibliotheca lippiaca. Uebersicht über die landeskundliche und geschichtliche litteratur des fürstenthums Lippe. Detmold 1886. pp.[vi].88. [926.]

WILHELM HANSEN, Lippische bibliographie. . . . Mit hinweisen auf die buchbestände der Lippischen landesbibliothek. Landesverband Lippe: Detmold 1957. pp.xxxii.coll.1640. [20,000.]

Lippmann, Gabriel.

[G. LIPPMANN], Notice sur les travaux scientifiques de m. G. Lippmann. 1882. pp.18. [39.]

ERNEST LEBON, Gabriel Lippmann. Biographie, bibliographie analytique des écrits. Savants du jour: 1911. pp.viii.70. [212.]

Lipsius, Justus.

JUSTE LIPSE ou Lipsius. Liste sommaire et provisoire de ses œuvres. Université de l'état: Gand [1885]. pp.[ii].57. [750.]

[FERDINAND VANDERHAEGHEN, THOMAS JAMES I. ARNOLD and R. VANDENBERGHE], Bibliographie lipsienne. Œuvres de Juste Lipse. [Université de

Gand:] Gand 1886–1888. pp.[iv].xxviii.598+709 +[iv].418. [1750.]

also issued as vols.xv–xvii of the first series of Bibliotheca belgica.

Liquids. [*see also* **Drinks.**]

ABRIDGMENTS of specifications. Class 32. Distilling, concentrating, evaporating, and condensing liquids. Patent office.

1855–1866. pp.xii.126. [500.]
1867–1876. pp.xii.166. [600.]
1877–1883. pp.viii.98. [400.]
1884–1888. pp.ix.92. [400.]
1889–1892. pp.ix.94. [400.]
1893–1896. pp.x.96. [400.]
1897–1900. pp.x.92. [400.]
1901–1904. pp.xiv.140. [500.]
1905–1908. pp.xii.140. [500.]
1909–1915. pp.xii.209. [750.]
1916–1920. pp.xi.52. [500.]
1921–1925. pp.xiii.240. [800.]
1926–1930. pp.xvi.349. [1250.]

no more published.

ABRIDGMENTS of specifications. Class 46. Filtering and otherwise purifying liquids. Patent office.

1855–1866. pp.xii.66. [200.]

1867–1876. pp.xiii.104. [300.]
1877–1883. pp.xi.74. [250.]
1884–1888. pp.xii.122. [400.]
1889–1892. pp.xi.88. [300.]
1893–1896. pp.xii.108. [350.]
1897–1900. pp.xvi.154. [500.]
1901–1904. pp.xvi.194. [600.]
1905–1908. pp.xii.238. [750.]
1909–1915. pp.xiv.299. [1000.]
1916–1920. pp.ix.104. [300.]
1921–1925. pp.xii.190. [600.]
1926–1930. pp.xvi.208. [600.]
no more published.

ABRIDGMENTS of specifications. Class 102. Pumps and other means for raising and forcing liquids, excepting rotary pumps. Patent office.

Lithium.

ANNOTATED bibliography on the use of organolithium compounds in organic synthesis. Lithium corporation of America: Minneapolis [1949]. pp. [80]. [223.]*

— Supplement.*
i. [1949]. pp.[60]. [136.]
ii. [1949–1952]. pp.[96]. [218.]
iii. [1952–1953]. pp.[80]. [173.]

iv. [1954–1955]. pp.[124]. [271.]
v. [1956–1957]. pp.[170]. [451.]
vi. [1958]. Compiled by Florence Emert and Robert W. Gibson. pp.[152]. [328.]
vii. [1959]. Compiled by F. Emert, R. W. Gibson and Ralph L. Darby. pp.[198]. [487.]

B[ORIS] I[OSIFOVICH] KOGAN [*and others*], Редкие щелочные металлы (литий, рубидий и цезний). Библиография. Академия наук СССР: Институт минералогии, геохимии и кристаллохимии редких элементов: Москва 1962. pp.331. [5000.]

Loess.

D[ONALD] T[HOMAS] DAVIDSON, T. Y. CHU and J. B. SHEELER, A bibliography of the Loess. Iowa state college: Bulletin (vol.l, no.21 = Engineering report no.8 of the Iowa engineering experiment station): Ames 1951. pp.15. [275.]

Magnesite.

LIST of references on magnesite. Library of Congress: Washington 1918. ff.6. [73.]*

Magnetism, terrestrial.

R. BILLWILL, Klimatologie. Mit einem anhang: Erdmagnetismus, von J. Maurer. Centralkommission für schweizerische landeskunde: Bibliographie der schweizerischen landeskunde (section IV.4): Bern 1927. pp.xv.140.9. [terrestrial magnetism: 60.]

LIST of publications of the Department of terrestrial magnetism of the Carnegie institution. Washington.

1904–1927. pp.39. [658.]
1928. pp.11. [90.]
1929. pp.11. [118.]
1930. pp.11. [124.]*
1931. pp.13. [154.]*
1932. pp.9. [103.]*
1933. pp.8. [100.]*
1934. pp.8. [100.]*
1935. pp.6. [64.]*
1936. pp.10. [106.]
1937. pp.11. [140.]*
1938. pp.11. [140.]*
1939. pp.12. [130.]
1940. pp.13. [140.]*
1941. pp.11. [100.]
1942. pp.8. [60.]

1943. pp.8. [60.]
1944. pp.6. [40.]
1945.
1946. pp.7. [49.]
1947.
1948. pp.7. [47.]
1949.
1950.
1951. pp.5. [25.]
1952. pp.6. [36.]
1953. pp.7. [42.]
1954.
1955.
1956. pp.[9]. [85.]
1957. pp.[8]. [49.]
1958.
1959. pp.7. [40.]

in progress.

EDDA BARBER, Methods of measuring magnetic fields. California institute of technology: Jet propulsion laboratory: Astronautics information literature search (no.195): Pasadena 1960. pp.v.18. [134.]

Magnetohydrodynamics.

C. W. J. WILSON, Bibliography on magnetohydrodynamics. Royal aircraft establishment: Library bibliography (no.190): Farnborough 1957. pp.15. [201.]

LIST of scientific reports on plasma physics, magnetohydrodynamics and thermonuclear reactions in the CERN library. European organization for nuclear research: Library: CERN bibl[iography] (no.1): Genève [19]58. ff.11. [200.]*

L. G. NAPOLITANO and G. CONTURSI, Magneto-fluid-dynamics. Current papers and abstracts. North Atlantic treaty organization: Advisory group for aeronautical research and development: AGARD bibliography (vol.i): Oxford &c. 1962. pp.xxi.251. [2038.]

Magnetostriction.

MAGNETOSTRICTION, 1930 to date. Science library: Bibliographical series (no.566): 1942. ff.4. [65.]*

Manganese.

H[ENRY] CARRINGTON BOLTON, Index to the literature of manganese, 1596–1874. Salem 1876. pp.44. [2000.]

HENRY P[AUL] TALBOT and JOHN W[ESLEY]

BROWN, A bibliography of the analytical chemistry of manganese, 1785–1900. Smithsonian institution: Miscellaneous collections (no.1313): Washington 1902. pp.viii.124. [750.]

F. E. DEARBORN, The arsenates of manganese sa insecticides. (A review of the literature). Bureau of entomology and plant quarantine: Division of insecticide investigations: Washington 1937. pp. 2–27. [89.]*

Manila Observatory.

MIGUEL SELGA, The publications of the Observatory of Manila. Government of the Philippine Islands: Manila 1929. pp.21. [400.]

Mars.

ATMOSPHERE of Mars. Review of soviet literature. [Library of Congress]: Aerospace information division: AID report (no.61–138): [Washington] 1961. pp.[i].ii.31. [25.]*

Maser.

MASERS and lasers. An ASTIA report bibliography. Armed services technical information agency: Arlington, Va. 1962. pp.v.69. [300.]*

PERSONALITIES and organizations active in soviet maser and laser research. Preliminary. Library of Congress: Aerospace information division: [Washington] 1962. ff.ii.28. [98.]*

PATRICIA WARREN BECKER and STANLEY EDWARD BECKER, Bibliography on masers 1954–1961. Oak Ridge, Tenn. 1962. pp.37.

— — Supplement.

i. 1963. pp.63. [549.]

in progress.

Mason, William.

[JOHN] PHILIP [WELLESLEY] GASKELL, The first editions of William Mason. Cambridge bibliographical society: Monograph (no.1): Cambridge 1951. pp.xiv.41. [52.]

112.viii + 91.vi + 102.vi + 136 + 44. [6481.]

1956. pp.116.v + 103.v + 89.iv + 101 + 112 + 109 + 109 + 93 + 89 + 99 + 114 +101+167+52. [8755.]

1957. pp.199 + 155 + 172 + 165 + 180 + 167 + 164 + 173 + 180 + 160 + 189 + 305. [13,649.]

1958. pp.188 + 136 + 144 + 177 + 172 +

137 + 140 + 152 + 180 + 172 + 252 + 321. [13,733.]
1959. pp.156 + 136 + 176 + 175 + 200 + 180 + 174 + 172 + 194 + 238 + 271 + 347. [14,750.]
1960. pp.188 + 160 + 160 + 168 + 176 + 216 + 192 + 184 + 184 + 193 + 212 + 257+404. [17,065.]
1961. pp.120.72 + 96.64 + 96.64 + 808.64 + 124.72 + 112.64 + 104.72 + 104.72 + 103.64 + 148.68 + 144.72 + 320 + 112. [20,000.]
1962. pp.132.72 + 140.76 + 152.80 + 148.76 + 144.76 + 140.68 + 132.76 + 140.68 + 116.72 + 144.72 + 24.120.76 + 20.120.84+360+135. [20,000.]
1963. pp.20.116.92 + 28.115.80 + 32.131.84 + 28.132.80 + 32.144.83 + 28.142.92 + 24.140.80 + 28.148.87 + 24.136.80 + 32. 152.96 + 32.144.116 + 32.140.96 + + . [20,000.]

in progress.

[JERZY NOWIŃSKI *and others*], Polska bibliografia analityczna. Mechanika. Polska akademia nauk: Ośrodek bibliografii i dokumentacji naukowej: Warszawa.

i. 1955. pp. . [513.]
ii. 1956. pp.88.102. [587.]
iii. 1957. pp.99. [767.]
no more published.

WOJCIECH URBANOWSKI, Polska bibliografia mechaniki 1945–1954. Polska akademia nauk: Instytut podstawowych problemów techniki: Warszawa 1960. pp.315. [2084.]

G[URY] N[IKOLAEVICH] SAVIN and V[ALENTINA] V[LADIMIROVNA] GEORGIEVSKAYA, Развитие механики на Украине за годы советской власти. Академия наук Украинской ССР: Институт механики: Киев 1961. pp.284. [5000.]

Mechanics. [*see also* **Mathematics: Applied.**]

MECHANICS. International catalogue of scientific literature (Section B): Royal society.
i. 1902. pp.xiv.128. [908.]
ii. 1903. pp.viii.132. [841.]
iii. 1905. pp.viii.148. [1022.]
iv. 1906. pp.viii.115. [663.]
v. 1907. pp.viii.200. [1459.]
vi. 1908. pp.viii.105. [653.]
vii. 1909. pp.viii.137. [1098.]
viii. 1910. pp.viii.125. [1055.]
ix. 1911. pp.viii.156. [1501.]

x. 1912. pp.viii.124. [1236.]
xi. 1913. pp.viii.140. [1438.]
xii. 1914. pp.viii.112. [1202.]
xiii. 1915. pp.viii.84. [623.]
xiv. 1918. pp.viii.134. [1324.]
no more published.

CATALOGUE of scientific papers, 1800–1900. Subject index. Volume II. Mechanics. [Edited by] Herbert McLeod. Royal society: Cambridge 1909. pp.lxxiii.356. [20,000.]

ZENTRALBLATT für mechanik. Berlin.
i. Schriftleitung: W. Flügge und O. Neugebauer. 1934. pp.[iii].462. [2000.]
ii. 1935. pp.[iii].462. [2000.]
iii. 1935. pp.[iii].466. [2000.]
iv. 1936. pp.[iii].465. [2000.]
v. 1937. pp.[iii].494. [2000.]
vi. 1938. pp.[iii].468. [2000.]
vii. Redaktion: W. Flügge und I. Flügge-Lotz. 1938. pp.[iii].471. [2000.]

METHODS of analogy in physics and mechanics (exclusive of photoelasticity) 1930 to date. Science library: Bibliographical series (no.629): [1946]. ff.2. [25.]*

РЕФЕРАТИВНЫЙ журнал. Механика. Академия наук СССР: Институт научной информации: Москва.

1953–1954. coll.172.pp.vi+coll.168.pp.vi+coll.164.pp.vii + pp.84.vi + 95.vii + 84.vi + 80.vi + 95.v + 91.v + 80.v + 76.v + 72.vi + 72.iv + 84.v + 148 + 44. [5890.]

1955. pp.117.vi + 92.vi + 90.vi + 104.viii + 110.vii + 106.x + 106. vii +120.viii +

Mercury.

MERCURY: occurrence, concentration and extraction, 1930–1941. Science library: Bibliographical series (no.576): 1942. ff.6. [120.]*

HUGH E. VORESS and NAOMI K. SMELCER, Mercury toxicity. A bibliography. U. S. atomic energy commission: Technical information service extension: Oak Ridge 1957. pp.iii.52. [1500.]*

Mercury, planet.

EDWARD S[INGLETON] HOLDEN, Index-catalogue of books and memoirs on the transits of Mercury. Library of Harvard university: Bibliographical contributions (no.1): Cambridge, Mass. 1878. pp.6. [175.]

Meson.

BRUNO VITALE, A bibliography on heavy mesons and hyperons. Amsterdam 1960. pp.vii.142. [2250.]

Meteorites, meteors.

OTTO BUCHNER, Versuch eines quellenverzeichniss zur literatur der meteoriten. Frankfurt a. M. 1861. pp.18. [1500.]

— — Zweites quellenverzeichniss [&c.]. 1863. pp.19. [1500.]

E. A. WÜLFING, Die meteoriten in sammlungen und ihre literatur. Tübingen 1897. pp.xlvii.461. [12,500.]

PAUL COGELS, Céraunies et pierres de foudre. Histoire et bibliographie. [Académie royale d'archéologie de Belgique: Bulletin (vol.i, supplement):] Anvers 1907. pp.406. [250.]

HARRISON [SCOTT] BROWN, *ed.* A bibliography on meteorites. Chicago [1953]. pp.viii.686. [9000.]*

METEOR trail observations. Annotated bibliography. [Library of Congress:] Aerospace information division: AID report (no.62–151): [Washington] 1962. ff.[ii].7. [33.]*

Metric system.

LIST of references on the use of the metric system in the United States. Library of Congress: Washington 1907. ff.3. [15.]*

CLIFTON HILDEBRAND, Metric literature clues. A list of references . . . on metric standardization of weights and measures. World trade club: San Francisco [1919]. pp.64. [750.]

Microscopy.

[FREEMAN CLARKE SAMUEL ROPER], Catalogue of works on the microscope, and of those referring to microscopical subjects, in the library of Freeman C. S. Roper. 1865. pp.vii.5–102. [1000.]
privately printed.

CATALOGUE of the books in the library of the Quekett microscopical club. [1876]. pp.12. [150.]
— [another edition]. 1929. pp.vii.178. [3550.]
— — Brochure containing lists of recent additions to the library [&c.]. 1932. pp.10. [75.]

THE ELECTRONE microscope, mainly 1939–41. Science library: Bibliographical series (no.567): 1942. ff.2. [35.]*

V[ERNON] E[LLIS] COSSLETT, Bibliography of electron microscopy. Institute of physics: 1950.

pp.350. [3000.]

CLAIRE MARTON [*and others*], Bibliography of electron microscopy. National bureau of standards: Circular (no.502): Washington 1950. pp. iv.87. [1923.]

THE APPLICATION of the electron microscope to the study of paints and paint materials. References from 1947 to 1951. Science library: Bibliographical series (708): 1952. ff.4. [63.]

THE INTERNATIONAL bibliography of electron microscopy. New York society of electron microscopists: New York.

i. 1950–1955. [1959]. pp.vii.166. [4054.]*

in progress; first published quarterly as Bibliography of electron microscopy, *on cards.*

Molecules.

THEORY of molecular attraction. Science library: Bibliographical series (no.155): 1934. ff.[i].2. [26.]*

BIBLIOGRAPHY of molecular association in gases and vapours, excluding papers on quantum mechanics. Science library: Bibliographical series (no.166): 1934. ff.[2]. [19.]*

MAX ULMANN, Molekülgrössenbestimmungen

hochpolymerer naturstoffe. Wissenschaftliche forschungsberichte: Naturwissenschaftliche reihe (vol.xxxix): Dresden &c. 1936. pp.xi.194. [1000.]

R. M. VILENSKAYA, Библиографический указатель работ научных сотрудников института высокомолекулярных соединений А. Н. СССР, 1949–1959 гг. Академия наук СССР: Институт высокомолекулярных соединений А. Н. СССР: Ленинград 1961. pp.104. [651.]

LUMINESCENCE of complex molecules. Annotated bibliography based on soviet open literature. Library of Congress: Aerospace information division: AID report (no.B-63-125): [Washington] 1963. ff.iii.74. [145.]*

Moon, the.

FICTIONAL accounts of trips to the moon, 160–1901 (A.D.). Syracuse university: Library: [Syracuse 1959]. pp.14. [11.]

BIBLIOGRAPHY of the moon. Department of the army: Office of the chief of engineers: Intelligence and mapping division: Washington 1960. pp.[iii]. 170. [2000.]*

ALEXANDER POGO [*and others*], Annotated bibliography of physical observations of the moon 1920–1960. California institute of technology: Division of the geological sciences: Pasadena 1961. ff.83.*

LUNAR dimensions. Annotated bibliography of soviet-bloc literature. Library of Congress: Aerospace information division: [Washington] 1963. ff.vi.49.*

National bureau of standards.

PUBLICATIONS of the National bureau of standards, 1901 to June 30, 1947. National bureau of standards: Circular (no.460): Washington 1948. pp.[ii].375. [4000.]

National physical laboratory.

ABSTRACTS of papers published in the year. National physical laboratory.

1936. pp.iii.65. [140.]
1937. pp.iii.71. [141.]
1938. pp.iii.63. [119.]

National research council of Canada.

LIST of publications of the National research council of Canada. National research council of Canada (no.808): [Ottawa] 1938. pp.38. [773.]

— [another edition]. Publications [&c.] . . . (no.1638): 1947. [1412.]

National science foundation.

PUBLICATIONS resulting from the National science foundation research grants. Washingon [1957]. pp.v.38. [1200.]

Nebulae.

EDWARD S. HOLDEN, Index catalogue of books and memoirs relating to nebulae and clusters, &c. Smithsonian institution: Smithsonian miscellaneous collections (no.311): Washington 1877. pp.[iii].ix.109. [1500.]

Neutron.

J[ACQUELINE] DESDEVISES and D[ANIEL] DIDIER, Spectrométrie des flux de neutrons intermédiaires et rapides. Commissariat à l'énergie atomique: Centre d'études nucléaires de Saclay: Service de documentation: Série bibliographies (no.7): Gif-sur-Yvette 1960. pp.39. [185.]

J[ACQUELINE] DESDEVISES, Sources étalons de neutrons. Commissariat à l'énergie atomique: Centre d'études nucléaires de Saclay: Service de documentation: Série bibliographies (no.11): Gif-sur-Yvette 1960. pp.22. [78.]

[SERGEI G. RODOVNICHENKO], Effects of neutron irradiation in non-fissionable metals and alloys. International atomic energy agency: Bibliographical series (no.6): Vienna 1962. pp.171. [735.]

Nicotine.

SOLUBILITY and steam-distillation of nicotine. Science library: Bibliographical series (no.206): 1935. ff.2. [19.]*

R. L. BUSBEY and N. E. MCINDOO [part ii &c.: N. E. MCINDOO, R. C. ROARK and R. L. BUSBEY], A bibliography of nicotine. Bureau of entomology and plant quarantine: Washington.*

i. Chemistry of nicotine. 1936. pp.257. [851.]
ii. The insecticidal uses of nicotine and tobacco. 1936. pp.[v].198+[v].199–358+[v].359–628. [2497.]

NICOTINE acid. Annotated bibliography. Merck & co.: Rahway, N.J. [1940]. ff.xi.88. [305.]*
— [another edition]. Nicotine acid (and nicotinamide). 1941. ff.[ii].130. [350.]*
— — Supplement. 1941. pp.[iii].19. [85.]
— — Supplement. 1942. pp.[iii].56. [300.]

Nomography.

NOMOGRAPHY, 1929–1939. Books and papers, exclusive of particular applications. Science library: Bibliographical series (no.525): 1940. ff.4. [70.]*

— 1939 to date. . . . (no.616): 1944. ff.3. [50.]*

FRANCIS D. HORIGAN and CARY R. SAGE, A bibliography of selected references on nomographic or alignment charts. QM research and development laboratories: Technical library: Bibliographic series (no.7): Philadelphia [1949]. ff.[ii].21. [242.]*

DOUGLAS P[AYNE] ADAMS, An index of nomograms. Massachusetts institute of technology [&c.]: New York &c. [1950]. pp.ix.174. [1250.]

Nuclear physics.

WALTHER BOTHE, SIEGFRIED FLÜGGE [*and others*], Nuclear physics and cosmic rays. Office of military government for Germany: Field information agencies technical: Fiat review of german science, 1939–1946: Wiesbaden 1948. pp.[viii].230+[iv]. 198. [1250.]

the text is in german.

PARTICLE counting and analysis. Science library: Bibliographical series (no.719): 1953. ff.5. [93.]*

Nucleic acids.

NUCLEIC acids. Schwarz laboratories: New York 1946. pp.48. [300.]

Numbers.

LEONARD EUGENE DICKSON, History of the theory of numbers. Carnegie institution: Publication (no.256): Washington 1919–1923. pp.xiii.486 +xxv.803+v.313. [6000.]

consists of abstracts.

DERRICK HENRY LEHMER, Guide to tables in the theory of numbers. National academy of sciences: National research council: Bulletin (no.105 = Committee on mathematical tables and aids to computation, Report 1): Washington 1941. pp. xiv.177. [500.]

ANITA RIESS, Number readiness in research. A survey of the literature. Chicago &c. [1947]. pp.70. [153.]

Oceanography.

1. *Periodicals*

PUBLICATIONS périodiques consultées pour la

préparation de la bibliographie. Bibliographia oceanographica: Venezia 1930. pp.24. [500.]

CH. VIDONNE, Inventaires des publications périodiques reçues à la bibliothèque. Fondation Albert Ier: Institut océanographique: Bulletin (no.1122): Monaco 1958. pp.40. [1000.]

2. *General*

XAVIER DA CUNHA, A Bibliotheca nacional de Lisboa na exposição de oceanographia. Catálogo summario. Coimbra 1904. pp.16. [110.]

BOOKS of the sea. Compiled by the Seafarers' education service. Third edition. National book council: Book list (no.69): 1930. pp.[4]. [125.]

G[IOVANNI] MAGRINI, M. PICOTTI and P[AOLO] REVELLI, Partecipazione italiana alla Mostra oceanografica internazionale di Siviglia (1929). Consiglio nazionale delle ricerche: Genova 1934. pp.[iv].215. [500.]

ESSAI d'une bibliographie générale des sciences de la mer. Venezia.

[i]. 1928. [By] Giovanni P. Magrini. 1929. pp.iv.48+iv.80+iv.127.39. [3076.]

[*continued as:*]

Bibliographia oceanographica. Venetiis [vol.vii

&c.: Consiglio nazionale delle ricerche, Romae].
ii. 1929. Edidit 'Johannes Magrini'. pp.viii. 216.43. [2290.]
iii. 1930. 1932. pp.vii.514. [3647.]
iv. 1931. 1933–1934. pp.iv.329. [3866.]
v. 1932. 1935. pp.vii.292. [3484.]
vi. 1933. 1934. pp.vi.384. [4544.]
vii. 1934. [Directed by Gustavus Brunelli]. 1935. pp.iii–xi.328. [4015.]
viii. 1935. 1936. pp.vii.350. [4153.]
ix. 1936. Delegatio Italica commissionis internationalis mari Mediterraneo scientiae causa explorando curavit. 1937. pp.viii.333. [4164.]
x. 1937. 1938. pp.iii–xi.317. [4155.]
xi. 1938. 1939. pp.vii.325. [4288.]

[*continued as:*]

Bibliographia maris mediterranei . . . extracta de Bibliographia oceanographica.
xii. 1939. 1940. pp.vii.319. [4078.]
xiii. 1940. 1941. pp.vii.286. [3588.]
xiv. 1941. 1942. pp.249. [3155.]

[*continued as:*]

Bibliographia oceanographica.
xv. 1942. 1944. pp.vii.211. [2505.]
xvi. 1943. 1949. pp.[v].85. [714.]
xvii. 1944. 1955. pp.ix.202. [1472.]

xviii. 1945. 1956. pp.viii.101. [784.]
xix. 1946. 1950. pp.123. [1147.]
xx. 1947. 1952. pp.170. [1431.]
xxi. 1948. 1953. pp.246. [1723.]
xxii. 1949. 1954. pp.xii.347. [2268.]
xxiii. 1950. 1956. pp.xxiv.523. [2453.]
xxiv. 1951.
xxv. 1952.
xxvi. 1953.
xxvii. 1954.
xxviii. 1955. 1958. pp.xxvii.939. [5140.]

in progress?

[MARY CATHERINE GRIER], References on the physical oceanography of the western Pacific ocean. Hydrographic office: Publication (no.238): Washington 1946. pp.ix.174. [1227.]

[—] — [another edition]. 1953. pp.ix.168. [1200.]

AUGUSTO CONTE Y LACAVE, Exposición del libro antiguo del mar. Catálogo. Casino gaditano: [Cadiz 1953]. pp.[ix].63. [886.]

450 copies printed.

WILLIAM MARCUS INGRAM and CLARENCE M. TARZWELL, Selected bibliography of publications relating to undesirable effects upon aquatic life by algicides, insecticides, weedicides. Public health service: Public health bibliography series (no.13): Cincinnati 1954. pp.[iii].28. [163.]*

BIBLIOGRAPHY of New Zealand oceanography, 1949–1953. New Zealand oceanographic institute: Geographical memoir (no.4): Wellington 1955. pp.19. [190.]

CONTRIBUTIONS from the Woods Hole oceanographic institution, 1933–1956 [Woods Hole] 1957. pp.126. [1100.]

B. B. G. O. A. Boletín bibliográfico de geofísica y oceanografía americanas. Parte oceanográfica. Instituto panamericano de geografía e historia: México.

i. 1958. pp.vi.161–471. [200.]
ii. 1959–1961. pp.[iii].365–539. [250.]

in progress.

CATALOGUE of IGY oceanographic data in IGY world date center A: oceanography. College Station, Tex. 1959. ff.[i].46. [250.]*

CURRENT bibliography for aquatic sciences and fisheries. Food and agriculture organization: Fisheries division: Biology branch: Rome 1959 &c.*

in progress; details of this work are entered under Fisheries.

RESEARCH catalogue of the American geographical society [New York]. Polar regions, oceans, tropics. Boston 1962. pp.xx.10,049–10,436. [8134.]

this is a photographic reproduction of catalogue cards; the figure in square brackets refers to these cards.

ESTHER E. THOMPSON, Oceanography. A report bibliography. Defense documentation center: Arlington, Va. 1963. pp.ix.355. [2000.]*

Optics.

[FRANÇOIS NAPOLÉON MARIE] MOIGNO, Répertoire d'optique moderne ou analyse complète des travaux modernes relatifs aux phénomènes de la lumière. 1847–1850. pp.[iii].iv.404+[iii].ii.401–902+[iii].ii.897–1301+[iii].ii.1295–1856. [750.]

ABRIDGMENTS of specifications. Class 97 (i). Optical systems and apparatus. Patent office.

1909–1915. pp.xxii.520. [2000.]
1916–1920. pp.xix.293.
1921–1925. pp.xxiv.474.
1926–1930. pp.xxv.466.

no more published.

PATENTS for inventions. Fifty years subject index, 1861–1910. Class 97 (i). Optical systems and apparatus. Patent office: 1917. pp.20. [2750.]

JOHN H. SUTCLIFFE, Library and museum catalogue. British optical association: 1932–1935. pp. xv.335+[vii].138. [3000.]

CATALOGUE of the Adolph Lomb optical library. University of Virginia: Bibliographical series (no.7): Charlottesville 1947. pp.203. [2250.]

AN ANNOTATED bibliography of form perception. Department of commerce: Office of technical services: [Washington 1954]. pp.[ii].iv.28. [30.]*

Oscillator.

ASYNCHRONOUS actions. Annotated bibliography: Library of Congress: Aerospace information division: Report (no. B-63-84): [Washington] 1963. ff.ii.10. [28.]*

Oximes.

CHARLES V. BANKS and RICHARD FULLERTON, Annotated bibliography of diaminoethanedionedioxime. Department of commerce: Office of technical services: [Washington] 1956. pp.14. [16.]*

π

S[YLVESTER] C[LARK] GOULD, Bibliography of the polemic problem: what is the value of π? Manchester, N.H. 1888. pp.32. [100.]

Pacific ocean and region.

MARY C[ATHERINE] GRIER, Oceanography of the north Pacific ocean, Bering sea and Bering strait: a contribution toward a bibliography. University of Washington: Publications: Library series (no. 2): Seattle 1941. pp.xxiii.290. [2929.]

Particle size.

A BIBLIOGRAPHY on particle size and surface area determination, with brief abstracts. Research council of the British whiting federation: Bibliography (no.1): 1950–1951. pp.[ii].45+[ii].55. [550.]*

Periodicity.

BIBLIOGRAPHIE sur la recherche des périodicités dans les séries temporelles. [1961]. pp.19. [400.]

Permeability.

CLARENCE J[AY] WEST, WALTER B[RAMBLE] KUNZ and GEORGE R[OBERSON] SEARS, Permeability of organic materials to gases. Institute of paper chemistry: Bibliographic series (no.169): Appleton, Wis. 1948. ff.[iii].153+[ii].154–292. [565.]*

— — [supplement]. Permeability . . . to gases and vapours. [By] Sally Wilkinson, Curtis L. Brown and Jack Weiner. 1956. ff.[ii].172. [435.]*

Phenanthrene.

HÅKAN SANDQVIST, Tabellarische uebersicht der in den jahren 1902–1912 in der literatur erwähnten phenanthrenderivate. Arbeiten utgifna med understöd af W. Ekmans universitetsfond (no.14): Uppsal [1913]. pp.vii.106. [1000.]

Phenols.

ANTISEPTIC and bactericidal properties of phenol and the naphtols, their derivatives and homologues (excluding inorganic derivatives), 1935–1937. Science library: Bibliographical series (no.387): 1938. ff.4. [85.]*

DETECTION and determination of phenols. Science library: Bibliographical series (no.671): [1949]. ff.3. [50.]*

Phlegraean fields.

R[OBERT WILLIAM] T[HEODORE] GÜNTHER, A bibliography of topographical and geological works on the phlegræan fields. Royal geographical society: 1908. pp.vii.100. [2200.]

Photosynthesis.

ФОТОСИНТЕЗ. Указатель отечественной иностранной литературы. Московский... университет: Научная библиотека им. А. М. Горького: Научно-библиографический отдел: [Москва].

i. 1951–1958. [By] Н. В. Арциховская. 1961. pp.387.403.507. [8995.]

in progress.

Planetaria.

PLANETARIA. Science library: Bibliographical series (no.610): 1944. ff.2. [35.]*

Plasma physics.

LIST of scientific reports on plasma physics, magnetohydrodynamics and thermonuclear reactions in the CERN library. European organization for nuclear research: Library: CERN bibl[iography] (no.1): Genève [19]58. ff.11. [200.]*

JAMES D. RAMER, Bibliography on plasma physics and magnetohydrodynamics and their application to controlled thermonuclear reactions. University of Maryland: Engineering and physical sciences library: College Park [1959]. pp.[iii].105. [1707.]

Plasticity.

DIMITRI KECECIOGLU, Bibliography on plasticity: theory and application. American society of mechanical engineers: [New York] 1950. pp. [iv].191. [1845.]*

Platinum.

JA[ME]S LEWIS HOWE, Bibliography of the metals of the platinum group, platinum, palladium, iridium, rhodium, osmium, ruthenium, 1784–1896. Smithsonian institution: Smithsonian miscellaneous collections (no.1084): Washington 1897. pp.318. [2500.]

— — 1748–1917. By J. L. Howe and H[endrick] C[oenraad] Holtz. Geological survey: Bulletin (no.694): 1919. pp.558. [4000.]

LIST of references on platinum. Library of Congress: Washington 1904. ff.3. [16.]*

J[ACQUES] S[IMHA] NEGRU, Literature and patent references to platinum group metals. International nickel company: New York.

i. 1930. ff.153. [400.]

ii. 1931. ff.38*.pp.39–98. [700.]

iii. March–June 1932. pp.32. [300.]

no more published.

JAMES LEWIS HOWE [*and others*], Bibliography of the platinum metals. Newark, N.J.

1918–1930. [1947]. pp.138. [3250.]

1931–1940. [1949]. pp.248. [8000.]

no more published; the first volume includes a supplement to the Bibliography *of 1919.*

Polarization.

THE DIPOLE moment and polarization of bonds. A selected list of references, 1948–1952. Science library: Bibliographical series (no.710): 1952. ff.3. [53.]*

Polarography.

BIBLIOGRAPHY of polarographic literature, 1922–1941. E. H. Sargeant & co: Chicago 1941. pp.[ii]. 98. [699.]

— 1922–1955. [1956]. pp.[iv].192. [6815.]*

SELECT bibliography on polarography. Central library: Research bibliographies (new ser., no.8): Sheffield [1948]. ff.3. [20.]*

GIOVANNI SEMERANO, Bibliografia polarografica (1922–1948). Consiglio nazionale delle ricerche: 1948. pp.124. [1858.]

— — Seconda edizione. [Edited by Elena Gagliardo]. 1949. pp.145. [2175.]

— — Supplemento.

ii. 1948. pp.10. [97.]
iii. 1949. pp.10. [94.]
iv. 1951. pp.63. [900.]
v. A cura . . . Maria Menegus-Scarpa, Elena Gagliardo. 1953. pp.94. [1290.]
vi. 1954. pp.56. [741.]
vii. A cura . . . M. Menegus-Scarpa, Bianca Tosini. 1955. pp.49. [579.]
viii. A cura . . . B. Tosini. 1956. pp.59. [775.]
ix. 1957. pp.64. [848.]
x. 1958. pp.80+96. [1053.]
xi. 1959. A cura . . . Anita Delmarco, Luciana Griggio. pp.86. [1152.]

in progress; intermediate supplements to the original works are regarded as the first supplement.

WILL AUGUSTO H[ARMANN] BRASE, Bibliografia antartica internaçionale. [Braunschweig 1956]. ff.[36]. [200.]

BIBLIOGRAPHY of publications dealing with the polarographic method in 1959. Československá akademie věd: Sekce chemická: Praha 1961. pp. 96. [1000.]

Polymers.

ROBERT SIMHA [*and others*], Bibliography of recent research in the field of high polymers. National bureau of standards: Circular (no.498): Washington 1950. pp.iii.56. [1300.]

ПОЛИМЕРЫ. Информация о статьях, опубликованых в зарубежных научных журналах. Всесоюзнауя государственная библиотека иностранной литературы: Москва 1958. pp.78. [500.]*

I. I. RESHETINSKY, Полимеры. Материалы неограниченных возможностей. Рекомендательный обзор литературы. Государственная ордена Ленина библиотека СССР имени В. И. Ленина: Центральная политехническая библиотека: Новости техники (no.16): Москва 1958. pp.36. [66.]

R. M. VILENSKAYA, Библиографический указатель работ научных сотрудников института . . . 1949–1959 гг. Академия наук

СССР: Институт высокомолекулярных соединений: Ленинград 1961. pp.104. [651.]

E. R. YESCOMBE, Book exhibition on polymer science, rubber and plastics technology. List. National college of rubber technology: 1963. ff.9. [300.]*

Polynomials.

J[AMES] A[LEXANDER] SHOHAT [*and others*], A bibliography on orthogonal polynomials. National academy of sciences: National research council: Bulletin (no.103): Washington 1940. pp.ix. 204. [2080.]

Porphin.

CHARLES V. BANKS and SAMUEL ANDERSON, Annotated bibliography alpha, beta, gamma, delta-tetraphenylporphine. Iowa state college: Ames laboratory: [Ames] 1957. pp.17. [28.]*

Powders.

HYDRAULIC classification of powders, 1933–7. Science library: Bibliographical series (no.473): 1939. ff.3. [46.]*

Propolis.

PROPOLIS. Science library: Bibliographical series (no.697): 1950. ff.2. [25.]*

Quantum theory.

A. LANDÉ, Fortschritte der quantumtheorie. Wissenschaftliche forschungsberichte: Naturwissenschaftliche reihe (vol.v): Dresden &c. 1922. pp.xi.92. [125.]

the second edition, 1926, is of a general nature.

NONLINEAR field theories. Annotated bibliography. Library of Congress: Aerospace information division: AID report (no. B-63-115): [Washington] 1963. ff.iii.36. [83.]*

Quaternions.

ALEXANDER MACFARLANE, Bibliography of quaternions and allied systems of mathematics. International association for promoting the study of quaternions: Dublin 1904. pp.86. [1250.]

Queuing.

VERA RILEY, Bibliography on queuing theory. Johns Hopkins university: Operations research office: BRS [Bibliography reference series] (no.3): Chevy Chace, Md. 1955. pp.27. [175.]*

Radiation.

RECENT references to cosmic rays. Science library: Bibliographical series (no.68): [1933]. ff.[i].4. [48.]*

RADIATION from building materials at normal temperatures. Science library: Bibliographical series (no.213): 1936. ff.2. [21.]*

Radioactivity. [*see also* **Radium.**]

RICHARD LUCAS, Bibliographie der radioaktiven stoffe. Hamburg &c. 1908. pp.[iv].92. [1300.]

BIBLIOGRAPHY on radioactive springs. Science library: Bibliographical series (no.53): [1932]. ff. [i].2. [25.]*

WISSENSCHAFTLICHE veröffentlichungen aus dem Institut für strahlenforschung der Humboldt-universität zu Berlin, 1926–1954. Berlin [1955]. pp.27. [296.]

HANS KÖNIG, Les récepteurs physiques de radiation. Informations bibliographiques . . . 1939–1949. Union internationale de physique pure et appliquée: Commission internationale d'optique: Document (no.4): 1950. pp.39.

J. A. MCCORMICK, Radiation detection. A literature search. Atomic energy commission: Technical information service: Washington 1950. pp.22. [335.]*

RADIATION sterilization. Review of the literature in selected fields. Quartermaster food and container institute for the armed forces: [Chicago] 1955. pp.[v].74. [250.]*

A[LEKSEI] N[IKOLAEVICH] BUCHENKOV, Меченые атомы и их применение в народном хозяйстве. Рекомендательный обзор литературы. Государственная... библиотека СССР имени В. И. Ленина: Новости техники (no.4): Москва 1956. pp.14. [25.]

ROGER WALLACE, Bibliography of technical reports on the effects of fallout. University of California: Radiation laboratory: [Berkeley] 1958. pp.70. [447.]*

RICHARD S. CUTTER, Biological effects of nonionizing radiation on humans and higher animals: Selected references in english, 1916–1957. National library of medicine: Washington 1958. pp. [iii].10. [118.]*

BIBLIOGRAPHY and index of literature on uranium and thorium and radioactive occurrences in the United States. Geological society of America: Special paper (no.67 &c.): York 1958 &c.

in progress.

WILLIAM E. BOST [*and others*], Radioactive fallout. A literature search. Atomic energy commission: Technical information service: Washington 1959. pp.[ii].41. [400.]*

L[UCIE] ROULE and [ÉLISÉE] MOULET, Principales publications relatives à la radiotoxologie du ^{239}Pu. Commissariat à l'énergie atomique: Centre d'études nucléaires de Saclay: Service de documentation: Série bibliographies (no.6): Gif-sur-Yvette 1959. pp.76. [356.]

WILLIAM E. BOST, Radioactive decontamination. A literature search. Atomic energy commission: Technical information service: Washington 1959. pp.[ii].34. [300.]*

V. G. NEISHILD and B. N. PANOVKIN, Радиоастрономия. Аннотированный библиографический указатель отечественной и иностранной литературы 1932–1958 гг. Академия наук СССР: Физический институт им. П. Н. Лебедева: Библиотека: Москва 1960. pp.216. [2750.]

I. P. SABANEEVA and R. S. ELIASBERG, Применение геофизических методов для поисков и разведки месторождений радиоактивных элементов. Библиографический указатель, 1945–1957 гг. Министерство геологии и охраны недр: Москва 1960. pp.96. [569.]

RADIOAKTIVITA ve vědě a technice. Výběr publikací ze Státní vědecke knihovny. Státní vědecká knihovna: Publikace (6th ser., no.7): [Ostrava] 1960. ff.17. [70.]*

INDUCTION de mutations chez les plantes supérieures d'intérêt agricole par les rayonnements bêta, gamma, les neutrons et les protons, 1955–1960. Commissariat à l'énergie atomique: Isotopes, rayonnements, agriculture: Études bibliographiques (no.1): Gif-sur-Yvette 1962. pp.xv. 282. [439.]

JUDITH HAYES, Television, photogrammetry, photometry, and radiometry adaptable to space reconnaissance. California institute of technology: Jet propulsion laboratory: Astronautics information literature search (no.490): Pasadena 1963. pp.v.303.

RALPH L. WENTWORTH and MARIE P. CANFIELD,

Patent literature in radiation chemistry. Cambridge, Mass. [1964]. pp.[iii].81. [500.]*

Radioisotopes.

EDUARD PETRŮ, Radioisotopy a jejich praktické využito. (Soupis knižní literatury z fondů universitvú knihovny). Universita: Knihovna: Publikace (1960, no.21): Olomouc 1960. pp.28. [110.]*

Radium.

LIST of references on radium, thorium, yttrium and aluminium. Library of Congress: Washington 1904. ff.4. [36.]*

LIST of references on radium and radioactivity. Library of Congress: Washington 1914. ff.3. [23.]*
— [another edition]. 1921. ff.13. [143.]*

LE JOURNAL de physique et le radium. . . . Revue bibliographique. 1920 &c.
in progress; details of this work are entered under Physics, above.

ABSTRACTS of selected articles on radium and radium therapy. Compiled by American institute of medicine for United States radium corporation. New York 1922. pp.[ix].164. [150.]

BIBLIOGRAPHY on radium, its uses and results.

Compiled by American institute of medicine for United States radium corporation. New York 1922. pp.[vii].132. [2500.]

— Supplement 1. 1922. pp.[v].47. [400.]

— Supplement no.2 [1924]. pp.[vii].55. [550.]

DETECTION of poisoning by radium salts and radon. Science library: Bibliographical series (no.373): 1938. ff.2. [32.]*

Relativity.

MAURICE LECAT and M. LECAT-PIERLOT, Bibliographie de la relativité. Bruxelles 1924. pp.xii. 291.47. [4500.]

MILDRED CATHERINE BENTON, The clock problem (clock paradox) in relativity. . . . An annotated bibliography. Naval research laboratory [Bellevue, D.C.]: Bibliography (no.15): Washington 1959. pp.[ii].v.48. [241.]*

Research.

LIST of references on scientific research. Library of Congress: Washington 1917. ff.8. [107.]*

ADMINISTRATION of research. A selective bibliography. Forest service: Library: Washington 1938. ff.[iii].26. [200.]*

READING list on the organisation of research, 1935–1942. Science library: Bibliographical series (no.575): 1942. ff.3. [35.]*

THE ADMINISTRATION of research: a selective bibliography. Library of Congress: Washington 1947. pp.105. [417.]*

RESEARCH training in american universities: a selected list of references. Library of Congress: Washington 1947. ff.7. [22.]*

HELEN DUDENBOSTEL JONES, Fiscal and budgetary phases of research. A selected list of references. Library of Congress: General reference and bibliography division: Washington 1949. pp.[ii]. iv.33. [134.]

GEORGE POLLOCK BUSH, Bibliography on research administration. Washington [1954]. pp. iv.146. [1100.]

RESEARCH and development (and related aspects) in foreign countries. A selected list of titles. Department of the army: Army library: Special bibliography (no.8): [Washington] 1956. pp. v.171+v.145. [1500.]*

A SELECTED bibliography of research and development and its impact on the economy. National science foundation: Office of special studies: Washington 1958. pp.v.21. [250.]*

[THEODORE SURANYI-UNGER], Bibliography on the economic and social implications of scientific research and development. National science foundation: Office of special studies: [Washington 1959]. pp.v.53. [200.]

HERMAN W. MILES, Management of scientific research. . . . An ASTIA report bibliography. Armed service technical information agency: Arlington, Va. 1962. pp.[v].77. [375.]*

CONSEIL national de la politique scientifique. Bibliographies. Bruxelles.*

i. Le financement de la recherche scientifique. 1962. ff.[i].24. [167.]
ii. Le personnel scientifique. 1962. ff.[i].26. [166.]
iii. Les chercheurs. 1962. ff.[i].26. [209.]

Rheology.

RHEOLOGY abstracts. A survey of world literature. British society of rheology.

i. 1958. [Edited by J. H. C. Vernon]. pp.32.28. 32.48.xviii. [587.]
ii. 1959. pp.48.36.24.52.xix. [546.]

iii. 1960. pp.36.24.36.28.xix. [533.]
iv. 1961. pp.25.32.32.32.xxiii. [628.]
v. 1962. [Edited by R. W. Whorlow]. pp.36. 35.36.40.xxi. [634.]
vi. 1963. pp.123.xvi. [523.]

in progress.

Rhodium.

ANN E. REA and MARGUERITE BEBBINGTON, Annotated bibliography on ruthenium, rhodium and iridium as catalysts. Selected references with abstracts, 1881 to June, 1959. International nickel company: New York 1959. pp.[iii].vii.192. [350.]*

Royal society.

[SIR] A[RTHUR] H[ERBERT] CHURCH, The Royal society. Some account of the 'classified papers' in the archives. Oxford 1907. pp.38. [1000.]

—— Some account of the 'letters and papers' of the period 1741–1806 in the archives. Oxford 1908. pp.78. [2000.]

Royal society of Canada.

JOHN GEORGE BOURINOT, Bibliography of the members of the Royal society of Canada. [Ottawa] 1894. pp.79. [3000.]

Rubrene.

RUBRENE. Science library: Bibliographical series (no.128): 1934. ff.[i].3. [36.]*

Ruthenium.

ANN E. REA and MARGUERITE BEBBINGTON, Annotated bibliography on ruthenium, rhodium and iridium as catalysts. Selected references with abstracts, 1881 to June, 1959. International nickel company: New York 1959. pp.[iii].vii.192. [350.]*

Sand.

SAND, its occurrence, properties and uses. A bibliography. Carnegie library: Pittsburgh 1918. pp.72. [600.]

H[ENRY] E[RNEST] HAFERKORN, Sand movement & beaches. Including references on bars, bays, coast changes, currents, erosion, estuaries, shorelines, tides, waves and wave action. Engineer school: Fort Humphreys, Va. 1929. pp.[iii].114. [vii]. [1101.]

Scientific instruments.

ABRIDGMENTS of specifications relating to

optical, mathematical, and other philosophical instruments; including nautical, astronomical, and meteorological instruments, A.D. 1636–1866. Commissioners of patents for inventions: 1875. pp.xxvii.552. [750.]

ABRIDGMENTS of specifications. Class 97. Philosophical instruments, (including optical, nautical, surveying, mathematical, and meteorological instruments). Patent office.

1855–1866. pp.xv.118. [500.]
1867–1876. pp.xvi.158. [600.]
1877–1883. pp.xii.105. [400.]
1884–1888. pp.xvi.172. [750.]
1889–1892. pp.xv.178. [750.]
1893–1896. pp.xviii.222. [900.]
1897–1900. pp.xx.244. [1000.]
1901–1904. pp.xxii.302. [1250.]
1905–1908. pp.xxviii.472. [2000.]

[*continued as:*]

i. [*entered under Optics*].
ii. Surveying, nautical, and astronomical instruments.

1909–1915. pp.ix.165. [500.]
1916–1920. pp.viii.134. [400.]
1921–1925. pp.vii.113. [350.]
1926–1930. pp.vii.87. [250.]

iii. Thermometers, photometers, meteorological and mathematical instruments, and miscellaneous philosophical instruments.

1909–1915. pp.xii.254. [750.]

1916–1920. pp.x.161. [500.]

1921–1925. pp.x.190. [600.]

1926–1930. pp.x.172. [500.]

no more published.

PATENTS for inventions. Fifty years subject index, 1861–1910. Class 97. Patent office.

i. [*entered under Optics*].

ii. Surveying, nautical, and astronomical instruments. 1916. pp.8. [1000.]

iii. Thermometers, photometers, meteorological and mathematical instruments, and miscellaneous philosophical instruments. 1916. pp.11. [1500.]

AUSZÜGE deutscher patent-anmeldungen. Instrumente. Klasse 42, 51. [Berlin 1947–1948]. pp.1011.3. [7098.]

covers the period 1939–1945; issued by the allied occupation forces.

INSTRUMENT abstracts. British scientific instrument research association: Chislehurst.

i. 1946. pp.200. [748.]

ii. 1947. pp.357. [1372.]
iii. 1948. pp.377. [1887.]
iv. 1949. pp.376. [2494.]
v. 1950. pp.426. [2369.]
vi. 1951. pp.460. [2574.]
vii. 1952. pp.468. [2787.]
viii. 1953. pp.440. [2776.]
ix. 1954. pp.508. [3374.]
x. 1955. pp.452. [2914.]
xi. 1956. pp.410. [2396.]
xii. 1957. pp.387. [2073.]
xiii. 1958. pp.396. [2293.]
xiv. 1959. pp.625. [4829.]
xv. 1960. pp.824. [5990.]
xvi. 1961. pp.1125. [8365.]
xvii. 1962. pp.1203. [8641.]
xviii. 1963. pp. . [7981.]

in progress.

JULIAN F[RANCIS] SMITH, Instrumentation literature and its use. A guide and source list. National bureau of standards: Washington 1952. pp.[iii]. 129. [1000.]*

W[ILLIAM] G[EORGE] BROMBACHER, JULIAN F. SMITH and LYMAN VAN DER PYL, Guide to instrumentation literature. National bureau of standards: Circular (no.567): [Washington] 1955. pp. iv.156. [1500.]*

INSTRUMENTATION. Government-owned inventions available for license. Department of commerce: Small business administration: Patents abstracts (no.1): Washington [1954]. pp.iii.65. [775.]*

Sedimentation.

CARL B[ARRIER] BROWN and FARRELL F[RANCIS] BARNES, Selected annotated bibliography on sedimentation as related to soil conservation and flood control. Department of agriculture: Soil conservation service: Washington [1939]. ff.[i].40. [101.]*

LIST of publications and reports on sedimentation. Department of agriculture: Soil conservation service: Technical letter (Sed.-15e): Washington 1952. pp.22. [315.]

[ANDRÉ VATAN *and others*], Bibliographie des travaux récents de sédimentologie. Association internationale de sédimentologie: Symposium: Sedimentology and the oil industry: [Paris 1960]. pp.164. [2500.]

BIBLIOGRAPHY of published reports on sedimentation, Arkansas-White-Red river basins. AWRBIAC [Arkansas-White-Red basins interagency committee]: Muskogee, Okla. 1962. ff.[i].iv.pp.21. [350.]*

Selenium.

LIST of references on selenium. Library of Congress: Washington 1913. ff.4. [44.]*

— Additional references. 1915. ff.2. [17.]*

MARION FOSTER DOTY, Selenium. A list of references, 1817–1925. Public library: New York 1927. pp.[iii].114. [1665.]

Shock.

J. N[ORTON] BRENNAN, *ed.* Bibliography on shock and shock excited vibrations. Pennsylvania state university: College of engineering and architecture: Engineering research bulletin (no.68–69): University Park 1957–1958. pp.[iii].vi.348+xi. 181.32. [4000.]*

BENJAMIN W[ILLIAM] ZWEIFACH, Annotated bibliography on shock, 1950–1962. National research council: Publication (no.1182): Washington 1963. pp.[ix].206. [944.]*

WENDELL L. HERCULES, Shock and vibration environment. Armed services technical information agency: Arlington, Va. 1962. pp.v.135. [800.]*

WENDELL HERCULES, Shock and vibration. Data analysis and application. A report bibliography. Defense documentation center: Alexandria, Va. 1963. pp.v.44. [250.]*

Sky.

SKY illumination during eclipse. Annotated bibliography of soviet-bloc literature for the period 1940–1963. Library of Congress: Aerospace information division: AID report (no. B-64-20): [Washington] 1964. ff.iii.68. [116.]*

Slide rule.

SELECT list of references on the slide rule. Library of Congress: Washington 1912. ff.2. [17.]*

Slovenská akadémia vied.

JURAJ KUBÍNSKY, Les publications de l'Académie slovaque des sciences, 1953–1955. . . . Traduction par J. Vránek et A. Mišíková. Bratislava 1956. pp.190. [500.]

translations into other languages were also issued.

Smithsonian institution.

WILLIAM J[ONES] RHEES, Catalogue of publications of the Smithsonian institution 1846–1882. Smithsonian institution: Smithsonian miscellaneous collections (no.437): Washington 1882. pp.[ii].328. [3000.]

— — [another edition]. List of publications . . .

1846–1903.... (no.1376): 1903. pp.[vii].99. [1500.]

MAMIE [RUTH] TANQUIST MILLER, An author, title and subject check list of Smithsonian institution publications relating to anthropology. University of New Mexico: Bulletin (no.405 = Bibliographical series, vol.i, no.2): Albuquerque 1946. pp.218. [1250.]

Snow.

D. C. PEARCE, A bibliography on snow and ice. National research council: Division of building research: Bibliography (no.1): Ottawa 1951. ff. [ii].69. [800.]*

FRANCIS D. HORIGAN, Electrical characteristics of snow and ice. A literature search. QM research development laboratories: Technical library: Bibliographic series (no.28): Philadelphia [1953]. pp. [ii].15. [39.]*

Sodium.

POST-WAR papers on sodium arsenite and its applications. Science library: Bibliographical series (no.157): 1934. ff.[i].2. [32.]*

BIBLIOGRAPHY of the literature on sodium and iodine in relation to plant and animal nutrition, Chilean nitrate educational bureau: New York.

i. 1948. pp.[ii].123. [1000.]
no more published.

Sorbitol.

SORBITOL, 1933–7. Science library: Bibliographical series (no.463): 1939. ff.3. [50.]*

Spectroscopy.

ALFRED TUCKERMAN, Index to the literature of the spectroscope. Smithsonian institution: Miscellaneous collections (no.658): Washington 1888. pp.x.423. [4000.]

— — 1887–1900 . . . (no.1312): 1902. pp.iii.373. [3000.]

LIST of references on spectroscopy. Library of Congress: Washington 1916. ff.8. [93.]*

BIBLIOGRAPHY of magnetic spectra. Science library: Bibliographical series (no.4): [1930]. ff.[i].3. [49.]*

ADAM HILGER, LTD. Recent applications of spectrophotometry. 1932. pp.44. [750.]

D[ONALD] M[URGATROYD] SMITH, Bibliography of literature on spectrum analysis. British nonferrous metals research association: Research monographs (no.2, supplement): 1935. pp.20. [195.]

— — Second edition. Bibliography of spectrochemical analysis. 1940. pp.3–57. [510.]
this edition was published out of series.

RECENT spectroscopic research on combustion, with special reference to internal-combustion engines, 1934 and onwards. Science library: Bibliographical series (no.285): 1935 [1937]. ff.2. [24.]*

SPECTROCHEMICAL abstracts.
- [i]. 1933–1937. By F[rank] Twyman. pp.52. [239.]
- ii. 1938–1939. By Ernest H. S. Van Someren. pp.39. [98.]
- iii. 1940–1945. pp.112. [468.]
- iv. 1946–1951. By E. H. S. Van Someren and F. Lachman. pp.179. [827.]
- v. 1952–1953. pp.66. [258.]
- vi. 1954–1955. pp.100. [361.]
- vii. 1956–1959. pp.[vii].104. [484.]

O[SWALD] J[AMES] WALKER, Absorption spectrophotometry and its applications; bibliography and abstracts 1932 to 1938. 1939. pp.68. [866.]

F[RANK] TWYMAN, Analyse spectrochimique en 1938. 1939. pp.[iii].74. [100.]

WILLIAM F[REDERICK] MEGGERS and BOURDON F[RANCIS] SCRIBNER, Index to the literature on spectrochemical analysis, 1920–1937. American society for testing materials: Philadelphia [1939]. pp.[ii].59. [932.]

— — Second edition. Index . . . 1920–1939. [1941]. pp.[ii].94. [1446.]

— — — [supplement].

1940–1945. [1947]. pp.iii.180. [1018.]

1946–1950. [1954].

J[ESSE] W[ILBUR] STILLMAN and E[DWARD] C. DUNLOP, Bibliography of photoelectric spectrophotometric methods of analysis for inorganic ions. American society for testing materials: Special technical publication (no.125): Philadelphia [1952]. pp.[iii].25. [750.]

— — Supplement.

1959. By Melvin Guy Mellon and D. D. Bly.

SPECTROGRAPHIC abstracts. Abstracts of literature on infra-red spectroscopy. Ministry of supply.*

1–2. 1954. 1954–1955. pp.[iv].82+[iv].138. [655.]

3. 1955. 1956. pp.[iv].151. [410.]

4–5. 1956–1957. 1958–1961. pp.[ii].168+[ii].258. [1337.]

in progress.

[G. I. KIBISOV *and others*], Спектральный анализ. Аннотированный указатель советских работ . . . 1931–1950. Академия наук СССР: Комиссия по спектроскопии: Москва 1955. pp.183. [920.]

DMS current literature list. 1958 &c.*
in progress.

L. S. MAYANTS [*and others*], Рассеяние света и инфракрасная спектроскопия. Библиографический указатель 1928–1940. Академия наук СССР: Комиссия по спектроскопии: Москва 1961. pp.452. [4618.]

MOLECULAR formula list of compounds, names, and references to published infrared spectra. American society for testing and materials: ASTM special technical publication (no.331): Philadelphia [1962]. pp.iv.312. [43,500.]

SOVIET infrared stellar spectroscopy. Annotated bibliography of the period 1947–1962. [Library of Congress:] Aerospace information division: AID report (no.62–153): [Washington] 1962. ff.[iii].46. [126.]*

X-RAY and optical spectra. Annotated bibliography of soviet literature. (Preliminary). Library of Congress: Aerospace information division: AID report (no.B-63-10): [Washington] 1963. ff.viii.127. [70.]*

Speleology.

É[DOUARD] A[LFRED] MARTEL, Bibliographie spéléologique, 1895–1897. 1897. pp.71. [250.]

É. A. MARTEL, La spéléologie au XXe siècle, Revue et bibliographie des recherches souterraines de 1901 à 1905. 1905. pp.810. [700.]

SELECT list of references on caves. Library of Congress: Washington 1911. ff.4. [25.]*

BOOKS of [*sic*] the geology of caves from *c.*1850 to date. Works of purely local reference excluded. Science library: Bibliographical series (no.298): 1938. single sheet. [8.]*

BENNO WOLF, Animalium cavernarum catalogus. Gesellschaft für höhlenforschung und höhlenkunde zu Berlin: 's-Gravenhage 1934–1938. pp.[iii].xxiii.616+[iii].918. [15,000.]

INTERNATIONALE bibliographie für speläologie. Landesverein für höhlenkunde in Wien und Niederösterreich: Die höhle: Wissenschaftliche beihefte (no.2 &c.): Wien.*

1950. Von Hubert Trimmel. . . . (no.2): 1954. pp.62. [924.]

1951. . . . (no.2): 1956. pp.72. [1072.]

1952. . . . (no.4): 1958. pp.72. [1183.]

1953. . . . (no.5): 1959. pp.80. [1245.]

1954. . . . (no.6): 1959. pp.95. [1429.]

ESSAI d'index bibliographique sur les grottes du Maroc. Société spéléologique du Maroc: [Rabat?] 1956. ff.8. [80.]*

Stellite.

LIST of references on stellite. Library of Congress: Washington 1917. ff.2. [12.]*

Steroids.

FRANCIS D. HORIGAN, Psychiatric studies on adrenocortical steroids. A survey of the literature. National institutes of health: Psychiatric abstracts series (no.4): Bethesda, Md. [1960]. ff.66.

Sterols.

ÉMILE F. TERROINE, Stérides (décembre 1936–décembre 1937). Physiologie, revue annuelle (no.7 = Actualités scientifiques et industrielles): 1938. pp.27. [108.]

Stratigraphy. [*see also* **Geology.**]

W[ILLIAM] E[MERSON] PUGH and B[OBBY] G[LYNN] PRESTON, Bibliography of stratigraphic traps. Seismograph service corporation: Tulsa [1951]. pp.iii–xxxi.198. [1500.]

B[ORIS] K[ONSTANTINOVICH] LINKHAREV, *ed.* Стратиграфический словарь СССР. Всесоюзный научно-исследовательский геологический институт: Москва 1956. pp.1284. [3500.]

Stratosphere.

BIBLIOGRAPHY of aeronautics. Part 50. Stratospheric flight. . . . Compiled from the index of aeronautics of the Institute of the aeronautical sciences. Works progress administration: [New York] 1937. ff.iii].35. [400.]*

Strontium.

A BIBLIOGRAPHY related to radioactive strontium. Public health service: [Washington *c.*1955]. pp.38. [428.]*

Sulphur. [*see also* **Chemistry: Technical.**]

LIST of references on sulphur and the sulphur industry. Library of Congress: [Washington]

1921. ff.11. [117.]*

P. BORGSTROM, R. W. BOST and D. F. BROWN, Bibliography of sulfur compounds (1871–1929).

American petroleum institute: Baltimore [printed] 1930. pp.[v].187. [3500.]

ALLEN W. SCHOLL and RICHARD L. YOUNG, Sulfur. A bibliography of analytical methods. [Huntington, W. Va. 1954]. pp.324. [3400.]*

Sun.

SUNLIGHT literature. A selected list prepared by the Sunlight league. [National book council:] Bibliography (no.101): 1928. pp.[2]. [40.]

A READING list on solar power plants, 1930–39. Science library: Bibliographical series (no.553): 1941. f.1. [11.]*

— 1939 to date. . . . (no.606): 1944. ff.2. [27.]*

— 1943 to date. . . . (no.659): 1948. single leaf. [14.]*

BIBLIOGRAPHY on domestic and industrial applications of solar heating. Engineering societies library: Bibliography (no.7): New York [1950]. ff.20.*

EFFECTS of sunspots on weather. Science

library: Bibliographical series (no.725): 1954. ff.4. [67.]*

T. BRIMELOW, Solar energy technology: 1954–1959. Library association: Special subject list (no.34): 1961. pp.32. [327.]

DIRECT energy conversion in the USSR: solar cell research. Comprehensive report. Library of Congress: Aerospace information division: [Washington] 1963. ff.[iii].47.12. [176.]*

Syenites.

C. J. KOENIG, Literature abstracts pertaining to nepheline syenites. Ohio state university: Engineering experiment station: Bulletin (vol.xvi, no.4 = no.130): Columbus 1947. pp.iv.33. [150.]

SOME references to the syenites of Leicestershire. Science library: Bibliographical series (no.657): 1948. ff.2. [28.]*

Tannoids.

J[ENS] CHRISTIAN BAY, Bibliography of the tannoids. [St. Louis] 1893. pp.27. [400.]

Thallium.

MARTHA DOAN, Index to the literature of thallium, 1861–1896. Smithsonian institution: Miscellaneous collections (no.1171): Washington 1899. pp.26. [300.]

FRANCIS F[ARNHAM] HEYROTH, Thallium. A review and summary of medical literature. Public health service: Public health reports: Supplement (no.197): Washington 1947. pp.[ii].23. [298.]

Thermocouples.

THERMOCOUPLES, 1932–1936. Science library: Bibliographical series (no.305): 1937. ff.7. [125.]*

Thermodynamics.

[ ] BERTIN, Rapport sur les progrès de la thermodynamique en France. Ministère de l'instruction publique: Recueil de rapports sur les progrès des lettres et des sciences en France: 1867. pp.[iii].84. [1000.]

ALFRED TUCKERMAN, Index to the literature of thermodynamics. Smithsonian institution: Miscellaneous collection (no.741): Washington 1890. pp.v.239. [2500.]

Thorium.

CAVALIER H[ARGRAVE] JOÜET, Index to the literature of thorium, 1817–1902. Smithsonian institution: Miscellaneous collections (no.1374): Washington 1903. pp.154. [1123.]

R. J. MEYER, Bibliographie der seltenen erden, ceriterden, yttererden und thorium. Hamburg &c. 1905. pp.79. [780.]

THORIUM. A bibliography of unclassified report literature. Atomic energy commission: Technical information service extension: Oak Ridge, Tenn. 1956. pp.vi.9. [74.]*

BIBLIOGRAPHY and index of literature on uranium and thorium and radioactive occurences in the United States. Geological society of America: Special paper (no.67 &c.): New York 1958 &c.
in progress.

[E. M. EL SHAZLY], Geology of uranium and thorium. International atomic energy agency: Bibliographical series (no.4): Vienna 1962. pp.134. [892.]

Tides.

LIST of references on tidal water power. Library of Congress: Washington 1915. ff.2. [25.]*

J[OSEPH] PROUDMAN, Tidal bibliography. Association d'océanographie physique: Publication scientifique (no.2): Helsingfors 1932. pp.17. [200.]

TIDAL power, 1920 to date. Science library: Bibliographical series (no.579): 1942. ff.5. [100.]*

[JOSEPH PROUDMAN *and others*], Bibliography on tides, 1665–1939. Association d'océanographie physique: Publication scientifique (no.15): Bergen 1955. pp.220. [2500.]

Titanium.

TITANIUM: occurrence, mining, extraction, metallurgy (excluding alloys), 1925–1938. Science library: Bibliographical series (no.406): 1938. ff.11. [220.]*

NON-FERROUS titanium alloys, 1925–1937. Science library: Bibliographical series (no.468): 1939. ff.3. [40.]*

PRODUCTION of titanium carbide (excluding alloys). Science library: Bibliographical series (no.536): 1940. ff.2. [30.]*

CHARLES A[RMIE] BROPHY [*and others*], Titanium bibliography, 1900–1951. Battelle memorial institute: Columbus, O. 1952. pp.[iv].iv.197. [4000.]

— — Supplement. 1952 &c.

JEAN RICHARDS CARPENTER and GWENDOLYN WERTH LUTTRELL, Bibliography on titanium. Geological survey: Circular (no.87): Washington 1953. pp.i.19. [331.]

TITANIUM abstract bulletin. A summary of published information on titanium and its alloys. Imperial chemical industries: Metals division: Birmingham.

i. 1955–1956. pp.[ii].761. [1403.]
ii. 1956–1957.
iii. 1957–1958.
iv. 1958–1959.
v. 1959–1960.
vi. 1960–1961.
vii. 1961–1962. pp.[vi].316. [840.]

in progress.

E[RNA] L[AZAREVNA] ZILBERBERG, Титан. Библиографический указатель. Центральная научно-техническая библиотека: Москва 1957. pp.140. [1100.]

JEAN LOMBARD, Sur la géochimie et les gisements du titane. A la recherche du rutile. Étude bibliographique. La chronique des mines d'outre-mer et de la recherche minière (no.253/254): 1957.

BIBLIOGRAPHY of principal articles on titanium that have appeared in the Welding journal starting with the year 1938. American welding society: A. W. S. bibliographies: New York [1958]. pp.3. [75.]*

Triangle.

C. ALASIA, Saggio terminologico-bibliografico sulla recente geometria del triangolo. Bergamo 1902. pp.vi.43. [200.]

J[EAN BAPTISTE JOSEPH] NEUBERG, Bibliographie de triangles spéciaux. Bruxelles [printed] 1924. pp.54. [250.]

Trigonometry.

MAURICE LECAT, Bibliographie des séries trigonométriques. Avec un appendice sur le calcul des variations. Louvain &c. 1921. pp.viii.ff.168. [2562.]

— — Compléments. 1924. pp.16. [266.]

Tritium.

LAWRENCE M. BROWN, ABRAHAM S. FRIEDMAN and CHARLES W. BECKETT, Bibliography of research on deuterium and tritium compounds, 1945 to 1952. National bureau of standards: Circular (no.562): Washington 1956. pp.ii.85. [2500.]

Ultra-violet rays.

WILLIAM B. BEHRENS, Ultra-violet rays. References to material in the New York public library. New York 1915. pp.20. [317.]

MEASUREMENT of ultra-violet fluorescence, papers 1930–1936. Science library: Bibliographical series (no.309): 1937. ff.3. [55.]*

REFLECTION and emission of infra-red and ultra-violet radiation from the surfaces of animals. Science library: Bibliographical series (no.505): 1939. ff.3. [55.]*

INFRA-RED and ultra violet ray apparatus for medical use, 1930–1940. Science library: Bibliographical series (no.526): 1940. ff.14. [265.]*

Uranium.

H[ENRY] CARRINGTON BOLTON, Index to the literature of uranium, 1789–1885. Washington 1885. pp.36. [400.]

FRED[ERICK] E[MORY] CROXTON, Uranium and its compounds. A bibliography of unclassified literature. Union carbide and carbon corporation: Oak Ridge, Tenn. 1951. pp.708. [3970.]*

S[TEWART] M. LANG, An annotated bibliography of selected references on the solid-state reactions of the uranium oxides. National bureau of standards: Circular (no.535): Washington 1953. pp.iv.95. [325.]

ROBERT EDWARD ALLEN, Uranium and its compounds. A bibliography of unclassified literature. Atomic energy commission: Technical information service: Oak Ridge, Tenn. 1953. pp.126. [2000.]

URANIUM production. Public library of South Australia: Research service: [Adelaide] 1955. ff.3. [42.]*

URANIUM recovery from raw materials. A bibliography of unclassified report literature.

Atomic energy commission: Technical information service extension: Oak Ridge, Tenn. 1956. pp.39.

GIFFORD A. YOUNG, Enriched uranium fuel materials. A bibliography of unclassified literature. Atomic energy commission: Technical information service extension: Oak Ridge, Tenn. 1956. pp.vii.33. [257.]*

BIBLIOGRAPHY and index of literature on uranium and thorium and radioactive occurences in the United States. Geological society of America: Special paper (no.67 &c.): New York 1958 &c. *in progress.*

CHARLES E. MAKEPEACE, The non-atomic uses of uranium. A bibliography of metallurgical abstracts. Ottawa 1960. pp.xxxvii.165. [299.]*

HENRY H[ERMAN] HAUSNER and HELEN C. FRIEDEMANN, Bibliography on uranium carbides. Rentte/Tyrol 1961. pp.[ii].ii.36. [209.]*

[E. M. EL SHAZLY], Geology of uranium and thorium. International atomic energy agency: Bibliographical series (no.4): Vienna 1962. pp.134. [892.]

Vanadium.

WILHELM PRANDTL, Die literatur des vanadins, 1804–1905. Hamburg &c. 1906. pp.117. [589.]

VANADIUM: occurrence, mining, extraction, metallurgy (excluding patent references). Science library: Bibliographical series (nos.389, 394): 1938. ff.8+8. [261.]*

Vapour pressure.

MAURICE [MARIE ALBERT] LECAT, L'azéotropie: la tension de vapeur des mélanges de liquides. Bibliographie. Bruxelles 1932–1942. pp.viii.ff.264 +pp.x.ff.123.pp.124–128.v. [4000.]

Viscometers.

VISCOMETERS based on the principle of time of fall of a solid body in a cylinder of liquid. Science library: Bibliographical series (no.344): 1937. ff.4. [67.]*

COMPARISON of viscosity results obtained with various viscometers, 1927–1947. Science library: Bibliographical series (no.669): 1949. single leaf. [17.]*

Vision.

J[OSEPH] PLATEAU, Bibliographie analytique des principaux phénomènes subjectifs de la vision, depuis les temps anciens jusqu'à la fin du XVIIIe siècle, suivie d'une bibliographie simple pour la partie écoulée du siècle actuel. Académie royale des sciences, des lettres et des beaux-arts de Belgique: Mémoires (vol.xlii, [nos.1–6] &c.): Bruxelles 1878.

[i. Persistance des impressions sur la rétine]. pp.[ii].iv.59. [400.]

ii. Couleurs accidentelles ordinaires de succession. pp.59. [150.]

iii. Images qui succèdent à la contemplation d'objets d'un grand éclat ou même d'objets bien éclairés. pp.26. [75.]

iv. Irradiation. pp.44. [100.]

v. Phénomènes ordinaires de contraste. pp.55. [150.]

vi. Ombres colorées. [Supplément à l'ouvrage entier, comprenant l'année 1877]. pp.43. [150.]

—— Deuxième supplément. . . . (vol.xliii, [no.11]): 1882. pp.17. [150.]

—— Troisième supplément. . . . (vol.xlv, [no.4]): 1884. pp.20. [175.]

REPORTS of the Committee upon physiology of vision. Privy council: Medical research council: Special report series (no.104 &c.).

i. Illumination and visual capacities. (A review of recent literature). By R. J. Lythgoe . . . (no.104): 1926. pp.80. [500.]

ii. Dark adaptation. (A review of the literature). By Dorothy Adams . . . (no.127): 1929. pp.138. [750.]

COLOUR vision. References from 1930–1940. Science library: Bibliographical series (no.556): 1941. ff.19. [375.]*

NIGHT VISION: a list of recent references. Library of Congress: Washington 1943. ff.7. [71.]*

JOHN F[ARQUHAR] FULTON, PHEBE M. HOFF and HENRIETTA T. PERKINS, A bibliography of visual literature, 1939–1944. Yale medical library: Historical library: Publication (no.11): Washington 1945. pp.x.117. [3400.]

JACK WEINER and MORRIS C[ECIL] LEIKIND, Visibility: a bibliography. Library of Congress: Reference department: Technical information division: Washington 1952. pp.vi.90. [2008.]*

AN ANNOTATED bibliography of form perception. Department of commerce: Office of technical services: [Washington 1954]. pp.iv.28. [40.]

Volcanoes.

[A.] SILVESTRI, Bibliografia generale riguardante la vulcanologia, mineralogia, geologia, paleontologia e paleoetnologia della provincia di Catania e delle isole vulcaniche adiacenti alla Sicilia. Bologna 1881. pp.vii.64. [950.]

EARTHQUAKES and volcanoes. A list of books with references to periodicals in the Brooklyn public library. Brooklyn 1906. pp.13. [150.]

— Second edition. 1909. pp.28. [400.]

HENRY JAMES JOHNSTON-LAVIS, Bibliography of the geology and eruptive phenomena of the more important volcanoes of southern Italy. . . . Second edition, compiled by miss B. M. Stanton. 1918. pp.xxiv.375. [7349.]

SELECT list of references on volcanoes. Library of Congress: Washington 1920. ff.6. [51.]*

LIST of references on volcanic ash. Library of Congress: [Washington] 1921. ff.5. [40.]*

Waves.

S. FLÜGGE and A. KREBS, Experimentelle grundlagen der wellenmechanik. Wissenschaftliche forschungsberichte: Naturwissenschaftliche reihe (vol.xxxviii): Dresden &c. 1936. pp.x.236. [600.]

CLAUDE C. LEE, Waves and wave action. A bibliography . . . from 1687 through December 1943. Vicksburg 1944. ff.[xiv].xviii.91. [700.]*

Wind. [*see also* **Meteorology.**]

LIST of references on wind power. Library of Congress: Washington 1919. ff.3. [33.]*

UTILISATION of wind power. Papers published in England, 1920–1936. Science library:

Bibliographical series (no.266): 1936. single sheet. [18.]*

— Supplement . . . 1935 to date, not confined to England . . . (no.606): 1944. ff.2. [27.]*

Yttrium.

R. J. MEYER, Bibliographie der seltenen erden, ceriterden, yttererden und thorium. Hamburg &c. 1905. pp.79. [780.]

The Besterman World Bibliographies

The Besterman World Bibliographies

Physical Sciences

A BIBLIOGRAPHY OF BIBLIOGRAPHIES

By Theodore Besterman

VOLUME ONE

TOTOWA, N. J.
Rowman and Littlefield
1971

Published by
Rowman and Littlefield
A Division of
Littlefield, Adams & Co.
81 Adams Drive
Totowa, N. J. 07512

*

Printed in the United States of America

*

Typography by George Hornby
and Theodore Besterman

ISBN 0-87471-054-5

Contents

Preface

I have explained in the Introduction to the successive editions of *A World bibliography of bibliographies* why I decided to arrange it alphabetically by specific subjects. Since that decision was taken, and after prolonged experience of the book in use, I have had no reason to regret it, nor among the many letters I have received from librarians has there been a single one complaining of the alphabetical form of the *World bibliography*.

The *World bibliography of bibliographies* covers all subjects and all languages, and is intended to serve reference and research purposes of the most specific and specialised kind. Yet contained in it are broad and detailed surveys which, if relevant entries throughout the volumes are added to them, can serve also the widest reference inquiries, and be useful to those who seek primary signposts to information in varied fields of inquiry.

Therefore I can only thank Rowman and Littlefield for having gathered together all the titles in some of the major fields found throughout the 6664 columns of the fourth edition (1965-1966) of *A World bibliography of bibliographies*.

These fields are:

1. Bibliography
2. Printing
3. Periodical Publications
4. Academic Writings
5. Art and Architecture
6. Music and Drama
7. Education
8. Agriculture
9. Medicine
10. Law
11. English and American Literature
12. Technology
13. Physical Sciences
14. Biological Sciences
15. Family History
16. Commerce, Manufactures, Labour
17. History
18. Geography

Of course these categories by no means exhaust the 117,000 separately collated volumes set out in the *World bibliography*, and the above titles will be added to if librarians wish for it.

Th. B.

Notes on the Arrangement

An Alternative to critical annotation

Consider what it is we look for in a normal bibliography of a special subject. Reflection will show, I think, that we look, above all, for completeness, just as we do in a bibliography of bibliographies. We desire completeness even more than accuracy (painfully uncongenial though it is for me to make such a statement); for in most cases a bibliography is intended to give us particulars of publications to which we wish to refer; thus we can always judge for ourselves (waiving gross errors) whether the bibliographer has correctly described these publications. On the other hand, anything that is omitted is lost until rediscovered.

The question is, therefore, whether it is possible to give some indication of the degree of completeness of a bibliography without indulging in the annotation which is impossible in a work of the present scope and scale. It seemed to me that this could be achieved, to a considerable extent, by recording the approximate number of entries set out in it. This method is, of course, a rough-and-ready one, but experience shows that it is remarkably effective: and I hope that its novelty will not tell against it.

The recording of the number of works set out in a bibliography has another advantage in the case of serial publications: it displays in statistical form the development of the subject from year to year—often in a highly significant manner.

This procedure, then, is that which I have adopted, the number of items in each bibliography being shown in square brackets at the end of the entry. This, I may add, is by no means an easy or mechanical task, as can be judged from the fact that this process, on the average, just about doubles the time taken in entering each bibliography.

Supplementary information in footnotes

I have said that this method of indicating the number of entries is intended to replace critical treatment; but it is not possible to exclude annotation altogether, for a certain minimum of added information is indispensable. Consequently many of my entries will be seen to have footnotes, in which the following types of information are recorded: a few words of explanation where the title is inaccurate, misleading, obscure, or insufficiently informative; a statement to that effect where a work is in progress, where intermediate volumes in a series have not been published, or where no more have been published; an attempt to clarify complicated series; a note that a book was privately printed or in a limited number of copies, where this does not exceed 500, or in some abnormal manner, as on one side of the leaf, on coloured paper, or in a reproduction of handwriting, or with erratic pagination; when I have come across copies containing manuscript or other added matter, I have recorded the fact; substantial corrections and additions to bibliographies are sometimes published in periodicals, and I have noted a good many of these—but without aiming at anything even remotely approaching completeness, the attainment of which would be impossible. Various minor types of information are also occasionally to be found in the footnotes.

Owing to the great increase in the number of bibliographies reproduced directly from typewritten copy, such publications are designated by an asterisk at the end of the entry; this device saves a good deal of space.

Place of publication

The place of publication is not shown when it is London in the case of an english book and

Paris in the case of a french one. In the case of a series or sequence of entries, however, the absence of a place of publication means that it is the same as the place last shown in the series. The same applies to the names of editors and compilers. The place of publication is given as it appears on the titlepage, but prepositions are omitted even if violence is done to grammatical construction.

The Order of entries

Under each heading the order of the entries is chronological by date of publication; in the case of works in successive volumes or editions the chronological order applies to the first volume or edition. In suitable long headings an additional chronological order by period covered has been created; see, for instance, France: History, or Drama: Great Britain.

Method of collating

An effort has been made, so far as space allows, to give detailed and accurate information of the kind more usually found in small bibliographies. For instance, I have paid special attention to the collation of bibliographies in several (or even numerous) parts or volumes. It is, in fact, difficult to understand why it is usually considered necessary to give collations of works in a single volume, where difficulties seldom occur (from the point of view of systematic bibliography), but not of a work in several volumes, where confusion much more frequently arises. An occasional gap in the collations of such publications will be noticed. This is because, naturally enough, I have not been able in every case to see perfect sets; and I have thought it better to leave a very small number of such blanks rather than to hold up the bibliography indefinitely.

Serial publications

Where successive issues of a serial publication are set out, the year or period shown is usually that covered by the relevant issue; in such cases no date of publication is given unless publication was abnormal or erratic in relation to the period covered.

Bibliographies in more than one edition

Where a bibliography has gone into more than one edition I have tried (though I have not always been able) to record at least the first and latest editions. Intermediate editions have also been recorded wherever it seemed useful to do so, that is, for bibliographies first published before 1800, and for those of special interest or importance; but in general intermediate editions, though examined, have not been recorded.

Notes

Transcription of titles

Titles have been set out in the shortest possible form consistent with intelligibility and an adequate indication of the scope of the bibliography; omissions have of course been indicated. The author's name, generally speaking, is given as it appears on the titlepage, amplified and interpreted within square brackets where necessary.

Anonymous bibliographies

Far too large a proportion of bibliographical work is published anonymously. This is due, in part, to the all too common practice of library committees and similar bodies of suppressing altogether or of hiding in prefaces the names of those who have compiled bibliographies and catalogues for them. I have spent a good deal of time in excavating such and other evidences of authorship, and the result may be seen in the large number of titles preceded by names enclosed within square brackets.

Th. B.

Science.

[*works entered in section 3.iii are not repeated in section 3.iv.*]

Science

1. *Bibliographies*

[E. WYNDHAM HULME and CHARLES KINZBRUNNER], Class catalogue of current serial digests and indexes of the literature of pure and applied science exhibited at the Liverpool meeting of the Library association. 1912. pp.40. [350.]

CLASS LIST of current serial digests and indexes of the literature of science, technology and commerce. Library association: Technical and commercial libraries committee: Final report (appendix A): 1917. pp.13. [75.]

RUTH COBB, Periodical bibliographies and abstracts for the scientific and technological journals of the world. National research council: Bulletin (vol.i, part 3, no.3): Washington 1920. pp.131–154. [250.]

BIBLIOGRAPHY of current periodical abstracts and indexes published in the British Commonwealth and contained in the Science museum library. [1948]. pp.12. [125.]

A LIST of periodicals and bulletins containing abstracts published in Great Britain. With appendix giving partial list of journals containing abstracts published in the british commonwealth. Royal society: 1949. pp.64. [135.]

— Second edition. 1950. pp.80. [134.]

limited to science and technology.

W. SCHOLTEN and M. ELSEN, Bronnen voor het literatuuronderzoek op technisch en natuurwetenschappelijk gebied. Nederlands instituut voor documentatie en registratuur: Publ[icatie] (no. 313): 's-Gravenhage [1952]. pp.[iii].48. [1000.]

JOHN L[EONARD] THORNTON and R[OBERT] I[AN] J[AMES] TULLY, Scientific books, libraries and collection. A study of bibliography and the book trade in relation to science. Library association: 1954. pp.x.288. [750.]

CURRENT research and development in scientific documentation. National science foundation: Science information service: Washington.

i. 1957.

ii. 1958.

iii. 1958. pp.v.76. [200.]

iv. 1959. pp.viii.85. [250.]
v. 1959. pp.viii.102. [300.]
vi. 1960.
vii. 1960. pp.vii.153. [300.]
viii. 1961. pp.x.193. [300.]
ix. 1961. pp.xii.270. [500.]
x. 1962. pp.xv.383. [600.]

in progress.

GERTRUDE SCHUTZE, Bibliography of guides to the s-t-m literature, scientific, technical, medical. Contributions to s-t-m library literature (no.1): New York 1958. ff.[i].iv.64. [654.]*

ВОПРОСЫ отраслевой библиографии по естественным наукам. Академия наук СССР: Библиотека: Ленинград 1960. pp.145.

NADEZHDA YAKOVLEVNA YUVACHEVA, Научно-техническая информация и пропаганда. Библиографический указатель литературы. Ленинград 1961. pp.68.

2. *Dictionaries*

GLOSSARIES of technical definitions (exclusive of botany and zoology). Science Library: Bibliographical series (no.114): 1933. ff.[i].3. [52.]*

— [fifth edition]. Technical glossaries and dictionaries. . . . (no.707): 1952. pp.82. [1300.]*

— — Supplement no.1. 1953. pp.16. [250.]*

J[OHN] E[DWIN] HOLMSTROM, Bibliography of interlingual scientific and technical dictionaries. Unesco [United nations educational, scientific and cultural organization]: Paris 1951. pp.231. [1044.]

[—] — Fourth edition. 1961. pp.xxxvi.236. [2500.]

[KURT GASSEN *and others*], Fachwörterbücher. Naturwissenschaften, land- und forstwirtschaft, technik, medizin. Öffentliche wissenschaftliche bibliothek: Bibliographische mitteilungen (no.4): [Berlin 1954]. pp.32. [250.]

EUGEN WÜSTER, Bibliographie de vocabulaires scientifiques et technologiques monolingues. Unesco: Documentation et terminologie scientifiques: [1955–1959]. pp.219 + 146. [2800.]

SCIENTIFIC & technical dictionaries, encyclopedies & handbooks. Linda hall library: Bibliographic bulletin (no.13): Kansas City 1956. pp. [i].51. [500.]*

GLOSSARIES. A preliminary survey of selected titles of technical and scientific, domestic and foreign terms and definitions employed by the Department of defense. Department of the army:

Army library: Special bibliography (no.5): [Washington] 1956. pp.iv.27. [250.]

3. *Periodicals*

i. *History*

A LIST of current periodicals on the history of science. Science library: Bibliographical series (no.721): 1953. ff.3. [70.]*

ii. *General*

[G. R. CROTCH], List of scientific journals, with abbreviated titles, compiled for the use of the recorders. Third issue. Zoological record association: 1873. pp.16. [450.]

HENRY CARRINGTON BOLTON, A catalogue of scientific and technical periodicals (1665 to 1882), together with chronological tables and a library check-list. Smithsonian institution: Miscellaneous collections (no.514): Washington 1885. pp.x.773. [5105.]

— — Second edition . . . (no.1076): 1897. pp. vii.1247. [8603.]

LIST of journals. International catalogue of scientific literature: Royal society: 1903. pp.xv. 312. [4666.]

— Supplementary list. 1904. pp.vi.68. [800.]

[ADOLF LAW VOGE], Bibliography of periodics [*sic*] & serials & society publications of natural science & technology that are regularly excerpted for the card-index of electrochemistry & allied subjects by the Concilium bibliographicum. Zurich 1909. ff.[656]. [600.]
loose-leaf slips.

ENRIQUE SPARN, Catálogo universal de revistas de ciencias exactas, físicas y naturales. Academia nacional de ciencias: Miscelanea (no.1): Córdoba [Argentina] 1920. pp.255. [4500.]
— — Primer suplemento . . . (no.6): 1922. pp.71. [1250.]
— — Segundo suplemento . . . (no.9): 1925. pp.67. [1000.]
— — Suplemento III . . . (no.20): 1932. pp.76. [1000.]
— — Cuarto y último suplemento. 1939. pp.39. [500.]

MITTEILUNGEN aus den gebieten der naturwissenschaften und der technik herausgegeben von Friedr. Vieweg & sohn. [Heft 1–28]. Braunschweig 1922–1932. [10,000.]

A WORLD list of scientific periodicals published

in the years 1900–1921. 1925–1927. pp.xii.ff.499 +pp.xii.ff.344. [25,000.]

— 1900–1933. Second edition. 1934. pp.xiv. 780. [36,000.]

— 1900–1950. Third edition. Edited by William Allan Smith and Francis Lawrence Kent. 1952. pp.xvii.1058. [50,000.]

— — Fourth edition. Edited by Peter Brown and George Burder Stratton. 1963- . pp. xxv.531+ .

in progress.

AMERICAN and foreign periodicals on science and technology. With an appendix . . . containing publications of the important learned societies. New York [1952]. pp.114.

CHARLES HARVEY BROWN, Scientific serials. Characteristics and lists of most cited publications in mathematics, physics, chemistry, geology, physiology, botany, zoology, and entomology. Association of college and reference libraries: ACRL monograph (no.16): Chicago 1956. pp. xv.189. [750.]

LISTE des périodiques analysés par le Bulletin signalétique. Centre national de la recherche scientifique: Centre de documentation: 1959. pp.[iii]. 54. [4000.]*

CHOIX de périodiques en cours pour les bibliothèques universitaires. Sciences. Direction des bibliothèques de France: 1963. pp.[ii].ii.83.ix. [722.]*

iii. *Countries &c.*

America, latin

KATHERINE LEONORE MORGAN, Journals dealing with the natural, physical and mathematical sciences published in latin America. A tentative directory. Pan american union: Division of intellectual cooperation: Washington 1944. ff.[i].ii.62. [283.]*

GUIDE to latin american scientific and technical periodicals. An annotated list. Pan american union: Division of science development: Washington 1962. pp.xii.188. [1141.]

Asia

LIST of scientific and technical journals published in southeast Asia (Philippines, Indonesia, Malaya, Thailand, Indo-China). [United nations educational, scientific and cultural organization:] East Asia scientific cooperation office: Manila 1950. ff.[i].28. [125.]*

— [revised edition]. 1951. ff.xiii.32. [142.]*

— — Supplement . . . (China). 1951. ff.[i]. xiii.28. [120.]*

— Second [*sic*] edition. Scientific and technical journals of east and south east Asia (Hong Kong, Japan, Indonesia, Macao, Malaya, north Borneo, the Philippines, Sarawak, Thailand, Vietnam). 1953. pp.[v].ii.233. [800.]*

LIST of scientific & other periodicals published in the indo-pacific area. 2nd edition. Indo-pacific fisheries council: Bangkok 1953. pp.50. [400.]

Canada

DIRECTORY of canadian scientific and technical periodicals. National research council: Ottawa 1961. pp. . [361.]*

— Second edition 1962. ff.iv.34. [423.]*

China

CHINESE scientific and technical serial publications in the collections of the Library of Congress. Library of Congress: Science and technology division: Washington 1955. pp.vii.55. [700.]*

— Revised edition. 1961. pp.v.107. [1400.]

JOURNALS in science and technology published in Japan and mainland China. A selected list. Library of Congress: Science and technology division: Washington 1961. pp.iii.47. [400.]*

LIST of communist chinese scientific and technical periodicals. [Library of Congress:] Aerospace information division: AID report (no.61–21): [Washington] 1961. ff.[ii].63. [247.]*

Denmark

LIST of scientific periodicals published in Denmark. Compiled by the Royal danish academy of science. United nations educational, scientific and cultural organization: [Paris] 1952. pp.12. [107.]*

Europe

M[ANUEL] SANCHEZ, *ed.* Selected list of european scientific and technical journals. . . . Compiled from the Library of Congress union list of european journals in the District of Columbia. Washington 1942. loose-leaf.*

LIST of irregular serials received from eastern Europe and the USSR. Department of scientific and industrial research: Lending library unit.*

1959. pp.[ii].ii.90. [500.]
1961. pp.[ii].ii.111. [600.]

Germany

KURZTITELVERZEICHNIS technisch-wissenschaftlicher zeitschriften. Reichsgemeinschaft der technisch-wissenschaftlichen arbeiten: Berlin [1937]. pp.39. [1200.]

Great Britain

A LIST of british scientific publications reporting original work or critical reviews. Royal society: Information services committee: [1950]. pp.95. [212.]

limited to periodical publications.

India

LIST of scientific periodicals published in India. Unesco: Paris 1952. pp.18.

Ireland

T. PERCY C. KIRKPATRICK, The periodical publications of science in Ireland. Bibliographical society of Ireland (vol.ii, no.3): Dublin 1921. pp.[ii].33–58. [50.]

Italy

I PERIODICI scientifico-tecnici italiani. Consiglio nazionale delle ricerche: Bologna 1928. pp.viii. 375. [1600.]

— II. edizione. Periodici italiani che se occupano delle scienze e delle loro applicazioni. [By] Giovanni Magrini. 1929. pp.xv.286. [1600.]

— Quinta edizione. Periodici italiani scientifici, tecnici e di cultura generale. Roma 1939. pp.169+534+595. [1444.]

Japan

A LIST of scientific periodicals from japanese learned organizations. [Mombushō: Daigaku gakujutsukyoku]: Tokyo 1949. pp.[iii].58. [1158.]*

reproduced from handwriting.

LIST of scientific periodicals published in Japan. No.1. Association for science documents information: Meguroku 1953. pp.[ii].14. [800.]*

BIBLIOGRAPHICAL list of japanese learned journals, natural sciences. Department of education: Higher education and science bureau: Tokyo 1957 &c.

GEORGE S[CHLEGEL] BONN, Japanese journals in science and technology. An annotated checklist. Public library: New York 1960. pp.xv.119. [660.]

JOURNALS in science and technology published in Japan and mainland China. A selected list. Library of Congress: Science and technology division: Washington 1961. pp.iii.47. [400.]*

JAPANESE scientific and technical serial publications in the collections of the Library of Congress: Washington 1962. pp.v.247. [1500.]*

DIRECTORY of japanese scientific periodicals, 1962. Natural sciences, medical sciences and industry. National diet library: Tokyo 1962. pp.229.

Mexico

ANNITA MELVILLE KER, A survey of mexican scientific periodicals. Harvey Bassler foundation: Baltimore [printed] 1931. pp.xv.105. [100.]

LISTA de revistas científicas mexicanas. Centro de documentación científica y técnica: México [1953]. pp.8. [120.]

Netherlands

A[UGUSTUS] GORTER, Scientific periodicals in the Netherlands. A classified list of the more important current periodicals and serial publications . . . published in the Netherlands. Central organisatie T.N.O. [voor toegepast-naturwetenschappelijk onderzoek]: The Hague 1947. pp.36. [400.]

Pakistan

SCIENTIFIC & technical periodicals of Pakistan.

Pakistan national scientific and technical documentation centre: Pansdoc bibliography (no.303): Karachi [1961]. ff.ii.12. [70.]*

A. R. GHANI and A. GHANI, Scientific & technical periodicals of Pakistan: a preliminary survey of holdings in foreign libraries. Pakistan national scientific and technical documentation centre: Karachi 1963. pp.[ii].4.xc. [50.]*

Poland

PIOTR GRZEGORCZYK, Czasopisma naukowe w Polsce. Życia nauki: Kraków 1947. pp.16. [150.]

SPIS polskich czasopism i wydawnictw ciagłych matematyczno — przyrodniczych i techicznych ukazujących sie bieżąco. Ósrodek bibliografii i dokumentacji naukowej P. A. N.: Warszawa 1958. pp.[ii].50. [307.]

Russia

SCIENTIFIC and technical periodicals received from the U.S.S.R. [Survey of the issues for 1939–1942]. Association of special libraries and information bureaux: Aslib war-time guides to british sources of specialised information (no.4): 1942. ff.[i].9. [334.]*

LIST of scientific and technical periodicals received from the U.S.S.R., 1933–1942. British museum: [1943]. ff.10. [162.]*

GUIDE to russian scientific periodical literature. Brookhaven national laboratory: Upton, N.Y. 1947 &c.

SCIENTIFIC and technical serial publications, Soviet Union, 1945–1953. Library of Congress: Science and technology: Washington 1954. pp. viii.118. [2000.]*

— 1945–1960. Prepared by Nikolay T[ikhonovich] Zikeev. 1963. pp.vii.347. [5091.]*

CARL [ALGOT PETER] BJÖRKBOM, Samkatalog över ryska tidskrifter och seriepublikationen inom ren och tillämpad naturvetenskap 1939–1954: svenska bibliotek. Kungl. tekniska högskol: Bibliotek: Stockholm 1955. pp.24.

L[ÉON] NAVEZ, Liste des périodiques scientifiques, techniques et agricoles russes existant dans les bibliothèques belges. Bruxelles 1957. ff.[ii].48. [400.]*

TITLES of current periodicals from the USSR and "cover-to-cover" translations. Department of scientific and industrial research: Lending library unit.*

1957. pp.[i].iii.26. [225.]
1958. pp.[ii].iii.28. [300.]
1959. pp.[ii].iii.76. [600.]
1960. pp.[ii].iii.82. [700.]

ROLAND C[LIFFORD] STEWART and AGNES N[ELLIE] TYSSE, Current russian science and technology. Selected list of scientific journals and bibliographies availablc in the university library of the university of Michigan. Ann Arbor 1959. ff.[ii].52. [500.]*

PÉRIODIQUES scientifiques et techniques soviétiques. Union française des organismes de documentation: 1960. pp.210. [1557.]

LIST of russian scientific journals available in english. National science foundation: Office of science information service: Washington 1960. pp.13. [80.]*

— [another edition]. 1961. pp.[ii].39. [126.]*

UNION list of russian scientific and technical periodicals available in european libraries. Editor: L. J. van der Wolk and S. Zandstra. Amsterdam 1963 &c.

in progress.

Sweden

HANS BAUDE and LARS ERIK SANNER, Current

swedish scientific and technical periodicals. Swedish institute: Stockholm [1959]. ff.[i].50. [307.]*

United States

RALPH C. MARTIN and WAYNE JETT, Guide to scientific and technical periodicals. A selected and annotated list of those publishing in english. [Denver 1963]. pp.ix.170. [320.]*

Uruguay

[JULIO GARRIDO, *ed.*], Catálogo de las revistas científicas, técnicas y económicas publicadas en Uruguay desde 1850. Biblioteca nacional: Misión de asistencia técnica de la Unesco: Centro de documentación científica, técnica y económica: Montevideo 1954. pp.29. [358.]

iv. *Library and exhibition catalogues*

Africa

SERIAL publications in the library. Royal society of South Africa: Capetown 1935. pp.41. [387.]

[G. ANNESLEY, *ed.*], Scientific and technical periodicals held in the principal libraries of British east Africa. East Africa commission: Paper (no.2): [Nairobi] 1949. pp.75. [1000.]*

CATALOGUE of periodicals. South african council for scientific and industrial research: Pretoria 1950. ff.60. [750.]*

LIST of current journals received in the library. Council for scientific and industrial research: Library and information division: Pretoria [1953]. pp.19. [900.]*

Argentina

RICHARD GANS, Catálogo de revistas de ciencias exactas, naturales y de ingeniería existentes en bibliotecas argentinas. Buenos Aires 1917. pp.xi.45. [1100.]

REINALD VANOSSI, Catálogo de las publicaciones periodicas existentes en la biblioteca de la Sociedad científica argentina. [Buenos Aires] 1927. pp.201. [2000.]

LISTA de publicaciones que se reciben en la biblioteca de la Sociedad científica argentina. Buenos Aires 1938. pp.73. [1000.]

[ERNESTO GUSTAVO GIETZ], Catálogo de publicaciones científicas y técnicas recibidas en las bibliotecas de las instituciones adheridas al Comité. Sociedad científica argentina: Comité argentino de bibliotecarios de instituciones científicas y técnicas: Buenos Aires 1942. pp.xx.343. [7387.]

— — Secundo edición. Catálogo colectivo de publicaciones periódicas existentes en bibliotecas científicas y técnicas argentinas. 1962. pp.xxix. 1726.

Asia

S[ĪGĀṚI] R[ĀMĀṚITA] RANGANATHAN [RAṄGĀNĀTHAN] [*and others*], Union catalogue of learned periodical publications in south Asia. Volume I. Physical and biological sciences. India library association: English series (no.7): [Delhi] &c. 1953. pp.391. [6000.]

Australia

W. T. DAYTON, Catalogue of the scientific literature in the following libraries in Sydney, N.S.W., viz.: the Australian museum, Free public library, Linnean society of N.S.W., Observatory, Parliamentary library, Royal geographical society of N.S.W., Royal society of N.S.W., Technological museum and the University of Sydney. Compiled under the direction of [sir] T[homas] P[eter] Anderson Stuart. 1889. pp.v.146. [1250.]

T[HOMAS] S[ARGEANT] HALL, Catalogue of the scientific and technical periodical literature in the libraries in Melbourne. Melbourne 1899. pp.vii. 225. [2250.]

[GEORGE HENRY PITT], Alphabetical list of serial

publications: scientific, technical, and historical. Public library, museum, and art gallery of South Australia: Adelaide 1914. pp.120. [1500.]

ERNEST R[OLAND] PITT, Catalogue of the scientific and technical periodicals in the libraries of Australia. Council for scientific and industrial research: Melbourne 1930. pp.xxiv.1208. [12,500.]

[—] — [another edition]. Scientific serials in australian libraries. Edited by Adelaide L. Kent. Commonwealth scientific and industrial research organization: 1958.

Austria

VERZEICHNIS der von mitte April 1909 bis mitte April 1910 an die mathematisch-naturwissenschaftliche Akademie der wissenschaften gelangten periodischen druckschriften. [Vienna 1910]. pp.36. [1000.]

FRANZ MAYR, Verzeichnis der zeitungen und periodika an wissenschaftlichen anstalten Tirols und Vorarlbergs. Naturwissenschaftlich-medizinischer verein: Innsbruck [1927]. pp.316. [4826.]

Belgium

[JEAN TIMMERMANS], Répertoire des périodiques scientifiques existant dans les bibliothèques belges.

Science

Fédération belge des sociétés de sciences mathématiques, physiques, chimiques, naturelles, médicales et appliquées: Gand 1927. pp.xx.377. [6500.]

CATALOGUE des périodiques. Musée royal du Congo belge: Bibliothèque: Tervuren 1960. pp. [ii].273. [3000.]*

Canada

A UNION list of magazines, chiefly of a scientific and technical character, in the libraries of certain government offices in Ottawa. Ottawa Carnegie library: [Ottawa] 1921. ff.iv.38. [800.]*

GERHARD R[ICHARD] LOMER and MARGARET S. MACKAY, A catalogue of scientific periodicals in canadian libraries. McGill university: Montreal 1924. pp.iii–xx.255. [5500.]

ERNEST HESS, A catalogue of scientific periodicals in libraries of the maritime provinces. Nova Scotian institute of science: Halifax 1936. pp.82. [3000.]

UNION list of scientific serials in canadian libraries. National research council: Library: Ottawa 1957. pp.xiii.805. [20,000.]

Ceylon

M[OHAMED] U[VAIS] S[IDDEEK] SULTANBAWA, List

of the scientific periodicals in the libraries of Ceylon. Ceylon association for the advancement of science: [Colombo] 1953. pp.143. [3000.]

China

LIST of scientific periodicals in the library of the Shanghai science institute. [Shanghai] 1934. pp. [ii].2.83. [600.]

Czechoslovakia

SEZNAM odborných časopisů objednaných na rok 1956 v zemích s neplánovaným hospodařstvím. Ustřední technicka knihovna ČSR: Praha 1956. pp.[iii].212.31. [5000.]*

Denmark

TIDSSKRIFTFORTEGNELSE omfaltende medicin, odontologi, almindelig naturvidenskab, kemi og fysik i statsbiblioteket, hospitaler og universitetsinstitutter i Århus. Århus 1955. pp.58. [1500.]

Egypt

UNION catalogue of scientific periodicals in Egypt. Fouad I national research council: Cairo 1951. pp.x.383. [3150.]

Finland

PAAVO SUOMALAINEN, MARJATTA HAVU and

Science

EERO K. NEUVONEN [vol.ii: MARIA WIDNÄS], Yhteisluettelo Suomen tieteellisissä kirjastoissa olevisto eksaktisten tieteiden, luonnontieteiden, tekniikan ja maatalous-metsätieteiden aikakaus- ja sarjajulkaisuista. Suomalaisen tiedeakatemia: Helsinki 1950–1953. pp.xii.699+viii.181. [7500.]

France

AUGUSTE [FRANÇOIS] LE JOLIS, Catalogue de la bibliothèque de la Société nationale des sciences naturelles et mathématiques de Cherbourg. . . . Première partie. Publications des académies, sociétés savantes et établissements scientifiques, recueils périodiques. Cherbourg 1870. pp.46. [292.]

— — 2e édition. 1881. pp.84. [750.]

LISTE des périodiques français et étrangers qui se trouvent à la bibliothèque de la Faculté des sciences de Marseille. Marseille 1898. pp.32. [600.]

[GEORGES] FLEURY and [MAURICE] GODEFROY, Catalogue des périodiques français et étrangers de la bibliothèque de l'université d'Aix-Marseille. Section des sciences à Marseille. Faculté des sciences de Marseille: Annales (vol.xvi, supp.1): 1907. pp.3–57. [1000.]

LÉON BULTINGAIRE, Inventaire des périodiques scientifiques des bibliothèques de Paris. Dressé sous la direction de m. Alfred Lacroix. Académie des sciences: 1924–1925. pp.xv.976+979–1103. [17,500.]

— — Supplément I. 1929. pp.xv.283. [2500.]

— — Supplément II. 1939. pp.xvi.532. [4000.]

R[AYMOND] BOURRIAU, Inventaire des périodiques de la bibliothèque scientifique de La Rochelle. La Rochelle 1927. pp.40. [750.]

BULLETIN bibliographique. Publications périodiques reçues pendant l'année. Institut de France: Académie des sciences.

1928. pp.21. [450.]
1929. pp.20. [400.]
1930. pp.20. [400.]
1931. pp.20. [400.]
1932. pp.20. [400.]
1933. pp.20. [400.]
1934. pp.24. [500.]
1935. pp.27. [550.]
1936. pp.27. [550.]
1937. pp.28. [550.]
1938. pp.28. [550.]
1939. pp.28. [550.]
1940. pp.19. [400.]

1941–1944. pp.11. [250.]
1945–1947. pp.23. [500.]
1948–1949. pp.23. [500.]
1950. pp.20. [500.]
1951. pp.23. [500.]
in progress?

CATALOGUE des périodiques étrangers scientifiques et techniques reçus par les principaux centres de documentation et bibliothèques de Paris. Secrétariat d'état aux forces armées (guerre): Direction des études et fabrications d'armement: Service technique: 1954. pp.[ii].xxxi.266. [4000.]*

— — 3e édition. Catalogue de périodiques étrangers scientifiques, techniques, industriels, reçus par les bibliothèques, instituts techniques et centres de documentation de Paris. Par [Geneviève] Dollfus et [Marcelle Olimpe] Lesur. École nationale supérieure des mines: 1962. pp.414. [8000.]

LISTE des revues étrangères reçues à la section scientifique et technique. État-major de l'armée: 1956. pp.7. [200.]*

LISTE des périodiques figurant à la bibliothèque. Centre d'études nucléaires de Saclay: Service de documentation: Gif-sur-Yvette 1961. pp.61. [800.]*

LISTE des publications étrangères et internationales reçues à la bibliothèque centrale. École polytechnique: Direction des études: Bibliothèque centrale: 1961. pp.41. [400.]*

Germany

[ARTHUR LUTHER, HANS RUPPERT and EWALD ROLLENBLECK], Verzeichnis der in der Deutschen bücherei zu Leipzig vorhandenen laufenden zeitschriften aus den gebieten der naturwissenschaften und medizin. Leipzig 1922. pp.120. [1178.]

[L. KLAIBER], Verzeichnis der naturwissenschaftlichen, mathematischen und medicinischen zeitschriften der universitäts-bibliothek und der institute der universität Freiburg i. Br. Freiburg i. Br. 1930. pp.[iv].113. [1333.]

RENATE RAUCH, Verzeichnis der periodischen schriften der bibliothek der Kaiserlich leopoldinisch-carolinisch deutschen akademie der naturforscher, sitz in Halle. Halle 1937. pp.xii.186. [3653.]

BESTAND an deutschen und ausländischen zeitschriften. Bibliothek der Technischen hochschule: München 1939. pp.124.

VERZEICHNIS der seit 1945 vorhandenen periodica. Technische hochschule: Bibliothek: [Dresden 1953]. pp.24. [900.]

VERZEICHNIS der laufend gehaltenen zeitschriften aus den gebieten: naturwissenschaften, technik, volkswirtschaft. Hessische landes- und hochschulbibliothek: Darmstadt 1953. pp.iv.38. [500.]

ZEITSCHRIFTEN-VERZEICHNIS. Battelle-Institut: Bibliothek: Frankfurt/Main 1960. pp.[i].67. [1000.]

ZEITSCHRIFTENVERZEICHNIS seit 1945 gehaltener zeitschriften. Technische hochschule: Bibliothek: Stuttgart 1962. pp.[xiii].471. [2500.]*

Great Britain

[BAILEY], Catalogue of transactions of societies, periodicals, and memoirs, available for the use of students in the reading room of the Radcliffe library at the Oxford museum. Third edition. Oxford 1876. pp.vi.45. [700.]

[—] — Fourth edition. 1887. pp.vii.95. [1250.]

CATALOGUE of the scientific books in the library of the Royal society. Transactions — journals — observations and reports — surveys — museums. 1881. pp.[iii].262. [4000.]

CHARLES W. E. LEIGH, List of the current scientific serial publications received by the principal libraries of Manchester. Literary and philosophical society: Manchester 1898. pp.vi.52. [741.]

LIST of short titles of current periodicals in the Science library. 1904.

— Fifth edition. Hand-list [&c.]. Part I. Alphabetical. Edited by S[amuel] C[lement] Bradford. 1938. pp.262. [9000.]*

— Eighth edition. 1956. pp.[ii].iv.426. [11,000.]

[LUXMOORE NEWCOMBE and L. ELLSTON], Catalogue of the periodical publications in the library of the Royal society of London. 1912. pp.viii.456. [1811.]

[E. WYNDHAM HULME and CHARLES KINZBRUNNER], Class catalogue of current serial digests and indexes of the literature of pure and applied science exhibited at the Liverpool meeting of the Library association. 1912. pp.40. [350.]

R. T. LEIPER, Periodicals of medicine and the allied sciences in British libraries. British medical association: [1923]. pp.vi.194. [5500.]

the scope is much wider than is suggested by the title.

INTERCHANGE of technical periodicals. Union list of scientific, technical and commercial periodicals in the libraries of members of the group

interchange system. Sheffield 1934 &c.
details of this work are entered under Trade, below.

LIST of technical, scientific & commercial periodicals taken by certain libraries in Manchester. Compiled by the Lancashire & Cheshire branch of A.S.L.I.B. [Association of specia libraries and information bureaux]. [Manchester 1936]. ff.[i].45. [900.]*

LIST of scientific and technical periodicals in D.S.I.R. libraries. Department of scientific and industrial research: 1952. pp.96. [2500.]

UNION list of scientific and technical periodicals in northern libraries. Library association: Reference and special libraries section: Northern group: [Newcastle-upon-Tyne] 1955. pp.39. [2250.]*

PERIODICALS held by the library. 2nd edition. Atomic energy research establishment: Library: Harwell 1955. pp.61. [900.]*

CATALOGUE of periodicals available in CSIR organisations: Council of scientific & industrial research: New Delhi 1955. pp.[iii].108. [1500.]

C. J. SPITTAL [*and others*], A union-catalogue of current scientific and technical periodicals in some group libraries. Library association: Reference and special libraries section: Western group: Bristol 1955. ff.[i].ix.72. [1696.]*

— — First supplement. 1957. ff.[i].ii.31. [300.]*

A UNION-CATALOGUE of current scientific and technical periodicals in some group libraries. Library association: Reference, special and information section: Western group: Bristol 1955.

— 2nd edition. 1962. ff.[ii].v.pp.vi–xx.273. [3903.]*

PERIODICAL publications in the Patent office library. List of current titles. 1947. pp.v.187. [5500.]*

— Second edition. 1958. pp.v.258. [6500.]*

Hungary

[ZOLTÁNNÉ SEBŐ], Folyóiratok katalógusa 1945–1959. Országos müszaki könyvtár: Budapest 1960. pp.259.273. [4000.]

PÁLNÉ VINCE, Hazai és külföldi időszakos müszaki dokumentációs kiadványok. Országos műszaki könyvtár: Időszerü müszaki dokumentációs kérdések (no.1): Budapest 1960. pp.78.

India

STANLEY KEMP, Catalogue of the scientific serial

publications in the principal libraries of Calcutta. Asiatic society of Bengal: Calcutta 1918. pp.xii. 292. [1607.]

[MANOHAR BHASKAR ARTE], List of scientific periodicals in the Bombay presidency. Bombay 1931. pp.[vi].ii.107.

Indonesia

UNION catalogue of periodical holdings in the main science libraries of Indonesia. Pilot edition. Unesco science co-operation office for south east Asia: Djakarta 1953. pp.[v].302. [3000.]*

Ireland

[D. FERGUSON], List of scientific and technical periodicals in Dublin libraries. National library of Ireland: Dublin 1929. pp.viii.147. [2750.]

Israel

[KAETHE LEWY and E. PFEFFERMANN], Union list of serials in Israel libraries. Pure and applied sciences excluding medicine. Israel national commission of Unesco: Sub-committee for documentation in the natural sciences: Jerusalem 1955. pp.vi.123. [2500.]

SERIAL HOLDINGS of the Weizmann institute of

science. Rehovot 1959. pp.[v].48. [1500.]*

— [another edition]. 1960. pp.[v].62. [1750.]*

Italy

ELENCO delle opere periodiche che trovansi nelle stanze di letteratura del Reale istituto veneto di scienze, lettere ed arti. [Venezia] [printed] [1867]. pp.22. [323.]

ELENCO dei periodici scientifici in corso posseduti dalla biblioteca e dagli instituti delle facoltà universitarie di Pavia. Biblioteca universitaria di Pavia: Pubblicazioni: Pavia 1904. pp.33.

ELENCO dei periodici di scienze fisico-chemiche, naturali e mediche posseduti dagli istituti della R. università e dalla R. biblioteca palatina. Parma 1907. pp.17.

CATALOGO delle pubblicazioni periodiche scientifico-tecniche esistenti nelle biblioteche di Milano e della R. università di Pavia. Comitato nazionale scientifico tecnico per lo sviluppo e l'incremento dell'industria italiana: Milano 1921. pp.48. [1600.]

— Secunda edizione. 1927. pp.92. [3000.]

GIOVANNI MAGRINI, Periodici stranieri che si trovano nelle biblioteche degli istituti scientifici italiani. Consiglio nazionale delle ricerche: Roma 1930. pp.[v].556. [10,000.]

Science

CATALOGO dei periodici scientifici posseduti dalle biblioteche ed istituti di Padova. Padova 1931. pp.iii–xi.90. [2500.]

CATALOGO delle pubblicazioni periodiche scientifiche e tecniche a disposizione del Centro di fotodocumentazione politecnico. Milano 1949. pp. [iii].211. [3000.]*

Japan

CATALOGUE of scientific periodicals in the College of agriculture, Kyoto imperial university. Kyoto 1935. pp.[vi].140. [1394.]

CATALOGUE of the scientific periodicals and allied serials in the Osaka imperial university. Osaka imperial university: Library: Bibliographical series (no.1): Osaka [1937]. pp.[iv].115. [2300.]

CATALOGUE of foreign scientific serial publications in the various Institutions in Japan. Third edition. National research council: Tokyo 1938. pp.[iii].viii.25.827. [12,500.]

Morocco

CATALOGUE des périodiques. Institut scientifique chérifien: [Rabat 1946]. ff.[35]. [520.]*

Netherlands

[W. VROLIK], Catalogus der boekerij, van de

Koninklijke akademie van wetenschappen. . . . Eerste aflevering. Amsterdam 1855. pp.viii.83. [500.]

INVENTARIS van natuurwetenschappelijke, medische en technische periodieken in nederlandsche bibliotheken. Nederlandsch instituut voor documentatie en registratuur: Publicatie (no.197): 's-Gravenhage 1941. pp.[ii].xii.469. [17,200.]

New Zealand

GILBERT ARCHEY, Reference list of the scientific periodicals in the libraries of New Zealand. [Auckland] 1927. pp.46.

Pakistan

A. R. GHANI, A guide to the current scientific journals received in various libraries of west Pakistan. Pakistan association for the advancement of science: Lahore 1950. pp.[ii].iii.26. [750.]

FAZAL ELAHI and AKHTAR H. SIDDIQUI, Union catalogue of periodicals in social sciences held by the libraries in Pakistan. Pakistan bibliographical working group: Publication (no.5): Karachi 1961. pp.v.92. [1000.]*

Poland

KATALOG der bibliothek der Naturforschenden

gesellschaft in Danzig. 3. heft. E. Zeitschriften und gesellschaftsschriften. Danzig 1914. pp.vi.58. [800.]

[HANS JESSEN], Verzeichnis der von der Staats- und universitäts-bibliothek und den instituten der universität gehaltenen zeitschriften aus den gebieten der medizin und naturwissenschaften. Breslau 1931. pp.vii.125. [2000.]

WYKAZ czasopism zagranicznych otrzymywanych w r. 1932 przez 11-cie polskich bibljotek naukowich. Kraków 1932. pp.44. [1724.]

ZOFIA MLECKO and ALOJZY PREISSNER, Katalog wydawnictw ciągłych matematyczno-przyrodniczych znajdujących się w bibliotece Polskiej akademii nauk. Kraków 1962. pp.686. [3251.]*

Portugal

PUBLICAÇÕES periódicas estrangeiras inventariadas nas bibliotecas portuguesas. Matemática, astronomia, física, guimica, engenharia, industria. Instituto de alta cultura: Centro de documentação científica: Lisboa 1955. pp.391. [3023.]

Romania

A[NTON] VALENTINY, Catalogus revistelor ştiinţifice şi medicale din Cluj. Universitatea: Cluj 1926. pp.xxiv.455. [3500.]

Russia

[M.E. GRYUNVALD], Каталогъ періодическихъ исданій библіотеки Кіевскаго политехническаго института императора Александра II. Кіевъ 1909. pp.[iii].28. [700.]

ЗВЕДЕНИЙ каталог періодичної закордонної наукової літератури по библіотеках м. Харкова в 1928 та в 1929 р.р. Харківська центральна наукова библіотека: Харків 1929. pp.64. [1375.]

Scandinavia

SAMKATALOG over utenlandske teknisk-naturvitenskapelige tidsskrifter i norske biblioteker. Norges teknisk-naturvitenskapelige forskningsrad: Trondheim 1953. pp.[vii].249. [3500.]*

Spain

JOSÉ LUIS FONTAN CANDELA, Catálogo de revistas científico-industriales en bibliotecas de Madrid. Madrid 1953. pp.[317]. [3000.]

CATÁLOGO de revistas de las bibliotecas del Patronato e institutos. Consejo superior de investigaciones científicas: Patronato "Juan de la Cierva" de investigación técnica: Madrid 1962. pp.xv.330. [4000.]*

Sweden

HILDA S[OFIA] LINDSTEDT, Katalog över periodiska publikationer tillgängliga i bibliotek i Stockholm och Uppsala och berörande matematik, geodesi, meteorologi, hydrografi, jordmagnetismus, fysik, kemi, geologi och mineralogi samt de tekniska vetenskaperna och arkitektur. Ingeniörsvetenskapsakademien: Stockholm 1927–1931. pp.xvi.301. [6000.]

CARL BJÖRKBOM, Samkatalog över ryska tidskrifter och seriepublikationer inom ren och tillämpad naturvetenskap 1939–1954 i svenska bibliotek. Stockholm 1955. pp.24. [400.]

Switzerland

ALPHONSE MAURICE BOUBIER, Catalogue des périodiques scientifiques et médicaux qui se trouvent aux bibliothèques de Genève. Deuxième édition. Genève 1912. pp.141. [2437.]

VERZEICHNIS der in der bibliothek der ETH laufenden zeitschriften. Eidgenössische technische hochschule: Bibliothek: Schriftenreihe (no.5): Zürich 1960. ff.[ii].242. [5000.]*

Science

United States

CATALOGUE of publications of societies and of other periodical works in the library of the Smithsonian institution. . . . Foreign works. Smithsonian institution: Miscellaneous collections (vol.iii): Washington 1859. pp.[iii].259. [1000.]

CATALOGUE of publications of societies and of periodical works belonging to the Smithsonian institution. Smithsonian institution: Miscellaneous collections (no.179): Washington 1866. pp.v.591. [5000.]

SAMUEL H[UBBARD] SCUDDER, Catalogue of scientific serials of all countries, including the transactions of learned societies in the natural, physical and mathematical sciences, 1633–1876. Harvard university: Library: Special publications (vol.i): Cambridge [Mass.] 1879. pp.xii.358. [4500.]

A LIST of the publications of scientific societies and the periodicals on pure and applied science, in the reference department, of the Carnegie library. Pittsburgh 1900. pp.19. [300.]

FRANCIS K[EESE] W[YNKOOP] DRURY, Technical and scientific serials in the libraries of Providence. Providence, R.I. 1920. pp.ix.63. [2500.]

Science

JUDITH FARRAR, *ed.* Scientific journals in the library of Duke university. Durham, N.C. 1935. pp.[iii].81. [2000.]*

MARJORIE GRAY WYNNE, *ed.* A checklist of all scientific periodicals and of selected serials in the libraries of Duke university. Durham, N.C. 1939. pp.110. [3000.]

BERNARD GESSINESS, Union list of scientific and technical periodicals in the libraries of greater Cincinnati. American chemical society: Cincinnati section: [Cincinnati] 1948. pp.vii.120. [3200.]*

UNION list of scientific and technical serials in the university of Michigan library. Ann Arbor 1954.

— Third edition. Edited by Herbert Sholla. 1964. pp.vii.406.

WIXIE PARKER, A checklist of scientific periodicals and of selected serials in the libraries of Duke university, North Carolina state college, the university of North Carolina, and the Women's college of the university of North Carolina. 1955. [6000.]

CHINESE scientific and technical serial publications in the collections of the Library of Congress: Washington 1955. pp.vii.55. [550.]*

— Revised edition. 1961. pp.v.107. [950.]*

UNION list of scientific and technical periodicals in the libraries of greater Cincinnati and vicinity. Special libraries association: Cincinnati chapter: [Cincinnati] 1958. pp.vi.144.*

UNION list of scientific and technical serials in the university of Michigan library. University of Michigan: Library: Committee on scientific and technical resources: Ann Arbor 1959. pp.[ii].v.171. [7000.]*

A LIST of scientific and technical serials currently received by the Library of Congress: Washington 1960. pp.v.186. [1750.]*

THE HOUSTON list of scientific and technical serial publications. Houston 1960 &c.

the first instalment was entitled Scientific and technical serial publications in the libraries of Rice university and university of Houston.

PERIODICALS in the scientific library, U. S. Patent office. Washington 1962. pp.[iii].53. [4500.]*

CURRENT serials and journals in the M. I. T. libraries. Fifth edition. Cambridge, Mass. [1962]. pp.[ii].289. [4000.]*

RARELY held scientific serials in the Midwest

inter-library center. Chicago 1963. pp.vi.197. [3500.]*

Uruguay

INVENTARIO de las revistas científicas, técnicas y económicas existentes en las bibliotecas del Uruguay. Misión de asistencia técnica de la Unesco: Centro de documentación científica, técnica y económica: Montevideo.*

i. Biblioteca nacional. Por Zulma Pucurull de Perez Gomar. 1953. ff.[22]. [242.]

ii. Biblioteca Artigas-Washington. Por Ermelinda Acerenza. 1953. ff.[9]. [86.]

iii. Biblioteca del Instituto de investigación de ciencias biológicas. Por Elsa Trinkle y Eliana Sánchez Rosas. 1954. ff.[15]. [168.]

iv. Biblioteca del Servicio cooperativo interamericano de salud pública, por Margarita Montedónico, y Biblioteca central de servicios médicos del Banco de seguros del estado, por Nelly Rezzani y Elena Araújo. 1954. ff.[12]. [63.]

v. Biblioteca del Centro de cooperación científica de la Unesco para América latina. 1954. ff.[39]. [432.]

vi. Biblioteca de la Facultad de humanidades y ciencias. Por Esther Dosil de Ramírez,

Matilde Servetti de Torrano y Leonor Pessaj. 1954. ff.[19]. [206.]

vii. Biblioteca de la Dirección de agronomía, por Sofía Naville de Morassi, biblioteca de la Facultad de agronomía, por Esther Cabarco, biblioteca del Servicio de información y prensa del Ministerio de ganaderia y agricultura, por J. F. Tajes. 1954. ff.[32]. [352.]

viii. Biblioteca de la Asociación odontologica. Por Marta Nogueira. 1954. ff.[12]. [120.]

ix. Biblioteca de la Facultad de veterinaria. Por Marta Nogueira. 1954. ff.[34]. [389.]

x. Biblioteca de la Sociedad de arquitectos del Uruguay, por Zulma Pucurull de Pérez Gomar y Marta Nogueira, biblioteca de la Facultad de arquitectura, por Maria S. de Toledo y Juanita Blasco Moll de Sarasqueta. 1954. ff.[20]. [210.]

xi. Biblioteca de la Asociación de quimica y farmacia, por María Teresa Castilla, biblioteca de la Asociación de quimicos industriales y biblioteca del Instituto sodamericano del petroleo, por Esther Cabarco. 1954. ff.[14]. [136.]

xii. Biblioteca del Departamento de biofisica

de la Facultad de medicina, por Zulma Pucurull de Pérez Gomar y Marta Nogueira, biblioteca del Centro de estudios laringeos y biblioteca del Instituto de endocrinología, por María Teresa Castilla. 1954. ff.[8]. [36.]

xiii. Biblioteca del hospital Pedro Visca, por Zulma Pucurull de Pérez Gomar, biblioteca de la Sociedad uruguaya de pediatría y biblioteca del Instituto de clínica pediátrica e higiene infantil dr. Luis Morquio, por Sara E. Scoseria. 1954. ff.[28]. [303.]

xiv. Biblioteca del Instituto de tisiología, por Zulma Pucurull de Pérez Gomar y Adelina Iriarte Platero. 1954. ff.[36]. [415.]

xv. Biblioteca de la Facultad de odontología. Por Nelly Rezzani y Elena Araújo. 1954. ff.[10]. [97.]

xvi. Biblioteca de la Facultad de ingeneria. Por N. Ziegler de Cabrera y Etelvina Rossi Magliano. 1954. ff.[32]. [350.]

xvii. Biblioteca del Instituto de matemáticas y estadística de la Facultad de ingenieria. Por María Teresa Carballal. 1954. ff.[10]. [132.]

xviii. Biblioteca de la Facultad de ciencias económicas y administración, biblioteca del Instituto de estadística, biblioteca del Instituto de finanzas, biblioteca del Instituto de teoría y política económicas. Por Zulma Pucurull. 1954. ff.[19]. [183.]

xix. Bibliotecas de los Institutos de anatomía patológica y parasitología, bacteriología, fisiología, industria animal y de zootecnia de la Facultad de veterinaria, por Zulma Pucurull, biblioteca del Contralor de la leche, por María Luisa Di Vita de Arzadun. 1954. ff.[14]. [98.]

xx. Biblioteca del Instituto de higiene. Por Martha Nogueira Abella, Nelly Huertas y Zulma Pucurull. 1955. ff.[36]. [409.]

Venezuela

[BJORN V. TELL], Catálogo de revistas científicas. Universidad central de Venezuela: Carácas 1959. pp.xiv.112. [3500.]

4. *History*

i. *Periodicals*

LIST of current periodicals on the history of science. Science library: Bibliographical series (no.721): 1953. ff.3. [70.]*

Science

ii. *General*

H[ORST] BRETSCHNEIDER, Wissenschaftliches verzeichniss der in der Herzogl. bibliothek zu Gotha enthaltenen ausgaben, uebersetzungen und erläuterungsschriften medicinisch-physikalischer werke der griechischen, arabischen und der lateinischen literatur bis zum dreizehnten jahrhundert. Gotha 1851. pp.40. [400.]

FERRUCCIO RIZZATTI, Prime linee d'una bibliografia per servire alla storia delle scienze. Parma 1904. pp.115. [1749.]

AKSEL G[USTAV] S[ALOMON] JOSEPHSON, A list of books on the history of science. John Crerar library: Chicago 1911. pp.10.297. [2000.]

— — Supplement. 1917. pp.6.139. [900.]

— — Second supplement. Prepared by Reginald B[rodrick] Gordon. Part I. 1942. pp.12. [120.]

no more published.

ENCYCLOPÉDIE sur fiches. Bibliographie critique extraite de la revue "Isis". [Edited by George Sarton]. Fasc.I. Wondelgem-lez-Gand 1914.

no more published.

ALDO MIELI, La storia della scienza in Italia. Saggio di bibliografia di storia della scienza. Firenze 1916. pp.vii.131. [500.]

CATALOGUE of early scientific literature exhibited in connection with the visit of the British association for the advancement of science. Public libraries: Cardiff 1920. pp.51. [250.]

[W. W. FRANCIS, R. H. HILL and ARCHIBALD MALLOCK], Bibliotheca osleriana. A catalogue of books illustrating the history of medicine and science collected, arranged, and annotated by sir William Osler, bt. and bequeathed to McGill university. Oxford 1929. pp.xxxvi.786. [7950.]

IL CARD. Gregorio Barbarigo e gli studi scientifici del seminario di Padova nel secolo XVIII. Catalogo di opere scientifiche stampate nella tipografia del seminario dal 1684 ad oggi. In occasione della 1ª esposizione nazionale di storia della scienza, Firenze 1929. Padova 1929. pp.68. [300.]

AVENIR TCHEMERZINE, Bibliographie d'ouvrages sur les sciences et les arts édités aux XVᵉ et XVIᵉ siècles. Courbevoie 1933. pp.160. [160.]
A–Compost only; no more published.

[HERBERT MCLEAN EVANS], Exhibition of first editions of epochal achievements in the history of science . . . on display at the university library. University of California: Berkeley 1934. pp.48. [114.]

I[NNA] I[VANOVNA] LYUBIMENKO, Ученая корреспонденция Академии наук XVIII века. Научное описание... 1766–1782. Академия наук: Труды архива (vol.2): Москва 1937. pp.607. [2116.]

NOTES bibliographiques. Comité belge d'histoire des sciences: Bruxelles 1948 &c.*
in progress.

O. A. STAROSELSKAYA-NIKITINA, O. V. KRASNOUKHOVA and YU. D. KATSENELSON, История естествознания. Академия наук СССР: Институт истории естествознания: Фундаментальная библиотека общественных наук: Москва &c.
1917–1947. 1949. pp.519. [10,000.]
1948–1950. 1955. pp.396. [6130.]

ROBERT JAMES FORBES, Bibliographia antiqua. Philosophia naturalis. Nederlandsch instituut voor het nabije oosten: Leiden 1940 [*sic*, 1949]–1950. [10,751.]
— — Supplement I, 1949–1950. 1952. [2489.]

[JEAN OLIVIA LINDSAY], The early history of science. A short handlist, Historical association: Helps for students of history (no.52): 1950. pp.64. [550.]

THE E. DE GOLYER collection in the history of science and technology. A check list of books and other materials. Second edition. University of Oklahoma: Norman 1953. pp.[ii].52. [1000.]*

— [third edition]. A checklist of the E[verette Lee] De Golyer [&c.]. . . . Compiled by Arthur McAnally and Duane H[enry] D[u Bosc] Roller. 1954. pp.vi.127. [3500.]

F[RANÇOIS] RUSSO, Histoire des sciences et des techniques. Bibliographie. Actualités scientifiques et industrielles (no.1204): 1954. pp.187. [2500.]

— — Supplément. 1955. ff.19. [250.]*

WHITFIELD J. BELL, Early american science: needs and opportunities for study. Institute of early american history and culture: Williamsburg 1955. pp.ix.85. [400.]

HANS [FREDRIK] SALLANDER, Bibliotheca walleriana. The books illustrating the history of medicine and science collected by dr Erik Waller and bequeathed to the library of the Royal university of Uppsala. A catalogue. Acta bibliothecae R. universitatis upsaliensis (vols.viii–ix): Stockholm 1955. pp.x.475+[vii].497. [20,446.]

BERN DIBNER, Heralds of science as represented by . . . epochal books and pamphlets selected from the Burndy library. Burndy library: Publication (no.12): Norwalk 1955. pp.96. [200.]

A CLASSIFIED catalogue of old books on science and technology published . . . from the end of 16th century to the middle of 19th century . . . from the [baron Kichizaemon] Sumitomo collection of the Osaka prefectural library. [Osaka] 1957. pp.20. [200.]

A[LBERT] GLODEN, Liste des travaux d'histoire des sciences et de la technique dus à des luxembourgeois de 1839 à 1959. Luxembourg 1959. pp.19. [200.]

THE HERITAGE from Newton and Linnæus. Scientific links between England and Sweden in bygone times. An english guide to the exhibition in the Museum of national antiquities. Stockholm 1962. pp.20. [224.]

WILLIAM PERSEHOUSE DELISLE WICHTMAN, Science and the renaissance. Vol.II. An annotated bibliography of the sixteenth-century books relating to the sciences in the library of the university of Aberdeen. Aberdeen university studies (no.144): Aberdeen 1962. pp.xix.293. [780.]

5. *Manuscripts*

CATALOGUES of the miscellaneous manuscripts

and of the manuscript letters in the possession of the Royal society. 1840. pp.[ii].38.iv.180. [4500.]

the catalogue of manuscripts by J. O. Halliwell, and that of letters, by W. E. Shuckard, were also issued separately.

J[AMES] O[RCHARD] HALLIWELL [-PHILLIPPS], A catalogue of scientific manuscripts in the possession of J. O. Halliwell, esq. [1841?]. pp.8. [200.]

ENRICO NARDUCCI, Catalogo di manoscritti ora posseduti da d. Baldassare Boncompagni. Roma 1862. pp.xxii.220. [2000.]

—— Seconda edizione. 1892. pp.viii.520. [5000.]

supplements by the same authors appear in Speculum *(1939), xiv.93–105, (1942), xvii.342–366.*

[RALPH J. BEEVOR], Inventory of [Stephen Peter] Rigaud papers in the Bodleian library. Oxford [1905]. pp.229–240. [68.]

M[ARJORIE] P[HYLLIS] CRIGHTON, A catalogue of the medical books and manuscripts, including a selection of the scientific works, in Exeter cathedral library. [*s.l.*] 1934. pp.38.[vi]. [200.]

LYNN THORNDIKE and PEARL KIBRE, A catalogue of incipits of mediaeval scientific writings in latin. Mediaeval academy of America: Publications (no.29): Cambridge, Mass. 1937. pp.xvii.coll.926. [15,000.]

supplements appear in Speculum *(1942), xvii.342–366; (1951), xxvi.673–695.*

— — Revised . . . edition. 1963. pp.xxii.coll. 1938. [30,000.]

THEODORE SILVERSTEIN, Medieval latin scientific writings in the Barberini collection [Vatican]. A provisional catalogue. [Chicago 1957]. pp.vii.147. [400.]

S. A. J. MOORAT, Catalogue of western manuscripts on medicine and science in the Wellcome historical medical library.

i. Mss written before 1650 a. d. 1962. pp.vii. 679. [800.]

in progress.

6. *General*

CORNELIUS À BEUGHEM, Bibliographia medica & physica novissima: perpetua, continuanda, sive conspectus primus catalogi librorum medicorum . . . physicorum, &c. quotquot . . . ab . . . 1651. (inclusive) . . . in quavis lingua . . . typis prodierunt. Amstelædami 1681. pp.[viii].503. [4000.]

no more published.

CORNELIUS À BEUGHEM, Syllabus recens exploratorum in re medica, physica & chymica prout in miscellaneis medico-physicis naturæ curiosorum Germaniæ, Galliæ, Daniæ & Belgii sparsum extant, in ordinem redactus. Amstelodami 1696. pp.[xliv].316. [2000.]

SEBASTIAN JOHANN LUDWIG, Critisches repertorium der auf in- und ausländischen höhern lehranstalten vom jahre 1781 bis 1800 herausgekommenen probe- und einladungsschriften aus dem gebiete der arzneygelahrtheit und naturkunde. Erste abtheilung, enthaltend das verzeichniss der schriften von 1781 bis 1790. Herborn 1803. pp.xvi.412. [science: 262.]

WILHELM TRAUGOTT KRUG, Enzyklopädisch-szientifische literatur. Fünftes heft, die enzyklopädisch-anthropologische literatur enthaltend. Leipzig &c. 1805. pp.xvi.54. [500.]

RETROSPECT of philosophical, mechanical, chemical, and agricultural discoveries: being an abridgment of the periodical and other publications, english and foreign.

i. [1805]. pp.xi.404.[iv]. [300.]
ii. [1806]. pp.xii.568.[viii]. [400.]
iii. [1807]. pp.xii.548.[viii]. [400.]
iv. [1808]. pp.xii.488.[viii]. [400.]
v. [1809]. pp.xvi.562.[ix]. [500.]
vi. [1810]. pp.xiv.514.[ix]. [500.]

vii. [1811]. pp.xii.360.[x]. [400.]
viii. [1812–1813]. pp.xv.466. [500.]
no more published.

[M. VAN MARUM, *ed.*], Catalogus der bibliotheek van Teyler's stichting. Haarlem 1826. pp.[i].xxii. 127. [1200.]

CATALOGUE of the works in medicine and natural history contained in the Radcliffe library. [By John Kidd]. Oxford 1835. pp.viii.330. [6000.]

— Catalogue of books on natural science in the Radcliffe library . . . up to December, 1872. With an appendix containing a list of works on India, having reference to comparative national health. Oxford 1877. pp.[iv].566. [14,000.]

— — Catalogue of books added to the Radcliffe library, Oxford university museum, during the year.

1873. pp.15. [300.]
1874. pp.18. [400.]
1875. pp.19. [400.]
1876. pp.20. [400.]
1877. pp.30. [600.]
1878. pp.28. [600.]
1879. pp.34. [750.]
1880. pp.36. [750.]
1881. pp.34. [750.]

1882. [By — Bailey and Joseph Walker]. pp.44. [1000.]
1883. pp.
1884. pp.
1885. [By H. R. Hall and J. Walker]. pp.38. [750.]
1886. pp.32. [750.]
1887. pp.32. [750.]
1888. pp.36. [750.]
1889. pp.36. [750.]
1890. pp.31. [600.]
1891. pp.39. [750.]
1892. pp.48. [1000.]
1893. pp.40. [750.]
1894. pp.36. [750.]
1895. [By J. Walker]. pp.42. [1000.]
1896. pp.52. [1250.]
1897. pp.44. [1000.]
1898. pp.39. [750.]
1899. pp.46. [1000.]
1900. pp.32. [600.]
1901. pp.35. [750.]
1902. pp.26. [500.]
1903. pp.28. [500.]
1904. pp.40. [750.]
1905. pp.39. [750.]
1906. pp.38. [750.]

1907. pp.42. [1000.]
1908. pp.53. [1250.]
1909. pp.41. [1000.]
1910. pp.31. [750.]
1911. pp.32. [750.]
1912. [By James Ford]. pp.27. [600.]
1913. pp.38. [750.]
1914. pp.24. [500.]
1915. pp.25. [500.]
1916. pp.29. [600.]
1917. pp.22. [500.]
1918. pp.24. [500.]
1919. pp.20. [500.]
1920. pp.24. [500.]
1921. pp.34. [750.]
1922. pp.25. [500.]
1923. pp.26. [500.]
1924. pp.19. [400.]
1925. pp.20. [500.]
1926. pp.31. [750.]
1927. pp.31. [750.]

discontinued on the amalgamation of the Radcliffe library with the Bodleian library.

CATALOGUE of the library of the Royal Dublin society. Dublin 1839. pp.vii.279. [5000.]

CATALOGUE of the scientific books in the library

of the Royal society. 1839. pp.[iii].776. [15,000.]
— [another edition]. 1883. pp.[iii].1199. [20,000.]

VERZEICHNIS der in der bibliothek der Schweizerischen naturforschenden gesellschaft vorhandenen bücher. Bern 1850. pp.[iii].172. [2500.]
— [another edition]. Verzeichnis der bibliothek der Schweizerischen naturforschenden gesellschaft. 1864. pp.x.188. [5000.]

BIBLIOTHECA historico-naturalis et physico-chemica [vols.ii–xxxvi: et mathematica], oder systematisch geordnete uebersicht der in Deutschland und dem auslande . . . neu erschienenen bücher. Göttingen.

i. 1851. Edited by Ernst A. Zuchold. pp.[ii]. 120. [1500.]
ii. 1852. pp.[ii].202. [2500.]
iii. 1853. pp.[ii].176. [2000.]
iv. 1854. pp.224. [2000.]
v. 1855. pp.246. [2000.]
vi. 1856. pp.232. [2000.]
vii. 1857. pp.252. [2000.]
viii. 1858. pp.[ii].254. [2000.]
ix. 1859. pp.[ii].254. [2000.]
x. 1860. pp.[ii].206. [1500.]
xi. 1861. pp.[ii].218. [1500.]

xii. 1862. pp.[ii].202. [1500.]
xiii. 1863. pp.[ii].206. [1500.]
xiv. 1864. pp.[ii].214. [1500.]
xv. 1865. pp.[ii].188. [1500.]
xvi. 1866. pp.[ii].238. [2500.]
xvii. 1867. pp.[ii].186. [2500.]
xviii. 1868. Edited by H. Guthe. pp.[ii].210. [3000.]
xix. 1869. pp.[ii].220. [3000.]
xx. 1870. pp.[ii].176. [2500.]
xxi. 1871. pp.[ii].176. [2500.]
xxii. 1872. Edited by A. Metzger. pp.[ii].186. [2000.]
xxiii. 1873. pp.[ii].198. [3000.]
xxiv. 1874. pp.[ii].232. [3500.]
xxv. 1875. pp.[ii].288. [4500.]
xxvi. 1876. pp.[ii].272. [4000.]
xxvii. 1877. pp.[ii].272. [4000.]
xxviii. 1878. pp.[ii].296. [4500.]
xxix. 1879. pp.[ii].280. [4500.]
xxx. 1880. Edited by F. Frenkel. pp.[ii].334. [5500.]
xxxi. 1881. pp.[ii].302. [5000.]
xxxii. 1882. Edited by K. Fricke. pp.[ii].266. [4000.]
xxxiii. 1883. Edited by R[einhard] von Hanstein. pp.[ii].368. [6000.]

xxxiv. 1884. pp.[ii].332. [5500.]
xxxv. 1885. pp.[ii].324. [5500.]
xxxvi. 1886. pp.[ii].338. [5500.]
xxxvii. 1887. pp.[ii].380. [6000.]
no more published.

CATALOGUE of books in Queen's college library, Belfast. Department of mathematics and natural philosophy. Belfast 1856. pp.18. [350.]

J[OHANN] C[HRISTIAN] POGGENDORFF, Biographisch-literarisches handwörterbuch zur geschichte der exacten wissenschaften, enthaltend nachweisungen über lebensverhältnisse und leistungen. Leipzig [1858–]1863. pp.viii.4.coll.1584 +pp.[ii].coll.1468. [60,000.]

— — Dritter band (1858 bis 1883). Herausgegeben von B. W. Feddersen und A. J. von Oettingen. [1896–]1898. pp.xi.846+[ii].847–1496. [60,000.]

— — Vierter band (die jahre 1883 bis zur gegenwart umfassend). Herausgegeben von Arthur von Oettingen. [1902–]1904. pp.xiii.930+[ii].931–1718. [80,000.]

— — Band v: 1904 bis 1922. . . . Redigiert von P. Weinmeister. Sächsische akademie der wissenschaften: [1925–]1926. pp.vii.695+[ii].697–1423. [70,000.]

— — Band VI: 1923 bis 1931. . . . Redigiert von

Hans Stobbe. 1936–1940. pp.lxxii.696+[iv].697–1438+xl.1439–2256+vii.2257–2975. [75,000.]
in progress; the title varies.

CATALOGUE of the American philosophical society library. Philadelphia 1863–1884. pp.1390. [30,000.]

JAHRBUCH der erfindungen und fortschritte auf den gebieten der physik und chemie, der technologie und mechanik, der astronomie und meteorologie. Leipzig.

i. Edited by H[einrich] Hirzel and H[einrich] Gretschel. 1865. pp.viii.396. [750.]
ii. 1866. pp.viii.373. [750.]
iii. 1867. pp.viii.440. [1000.]
iv. 1868. pp.viii.414. [750.]
v. 1869. pp.viii.414. [750.]
— Alphabetisches sach- und namen-register zu dem I. bis v. jahrgang. 1869. pp.28.
vi. 1870. pp.viii.472. [1000.]
vii. 1871. pp.vii–xv.390. [750.]
viii. 1872. pp.v–xv.396. [750.]
ix. 1873. pp.viii.368. [750.]
x. 1874. pp.viii.414. [750.]
xi. Edited by H. Gretschel and G. Wunder. 1875. pp.vii.416. [750.]

xii. 1876. pp.vii.469. [1000.]
xiii. 1877. pp.vi.460. [1000.]
xiv. 1878. pp.vi.452. [1000.]
xv. 1879. pp.viii.460. [1000.]
xvi. 1880. pp.vi.409. [750.]
xvii. 1881. pp.vi.442. [750.]
xviii. 1882. pp.vi.452. [750.]
xix. Edited by H. Gretschel and G[eorg] Bornemann. 1883. pp.viii.439. [750.]
xx. 1884. pp.viii.416. [750.]
xxi. 1885. pp.vi.408. [750.]
xxii. 1886. pp.vi.409. [750.]
xxiii. 1887. pp.vi.404. [750.]
xxiv. 1888. pp.vi.408. [750.]
xxv. 1889. pp.vi.436. [750.]
xxvi. 1890. pp.vi.384. [750.]
xxvii. 1891. pp.vi.392. [750.]
xxviii. Edited by H. Gretschel [*and others*]. 1892. pp.viii.386. [750.]
xxix. Edited by G. Bornemann, Otto Müller and A. Berberich. 1893. pp.vi.388. [750.]
xxx. 1894. pp.vi.392. [750.]
xxxi. 1895. pp.vi.379. [750.]
xxxii. 1896. pp.vi.380. [750.]
xxxiii. 1897. pp.vi.384. [750.]
xxxiv. 1898. pp.vi.384. [750.]
xxxv. 1899. pp.vi.388. [750.]

xxxvi. 1900. pp.iii[*sic*, viii].392. [750.]
xxxvii. 1901. pp.vi.400. [750.]
no more published.

INDEX to foreign scientific periodicals contained in the Patent office library.

i. June–December 1866. pp.[iv].226. [6000.]
ii. 1867. pp.[iv].402. [10,000.]
iii. 1868. pp.[iv].coll.424.pp.cxcviii. [12,500.]
iv. 1869. pp.[iv].coll.440.pp.cxcvi. [12,500.]
v. 1870. pp.[iv].coll.400.pp.clxxxviii. [12,500.]
vi. 1871. pp.[iv].coll.368.pp.clxxxiv. [10,000.]
vii. 1872. pp.[iv].coll.726.pp.cccxxiii. [20,000.]
no more published.

CATALOGUE of scientific papers. Royal society. London [vols.xiii–xix: Cambridge].

i–vi. 1800–1863. 1867–1872. pp.lxxix.960+iv.1012+v.1002+iv.1006+iv.1000+xi.764. [172,500.]
vii–viii. 1864–1873. 1877–1879. pp.xxxi.1048+[ii].1310. [70,000.]
ix–xi. 1874–1883. 1891–1896. pp.xxxii.1016+[iv].1048+[iv].902. [90,000.]
xii. 1800–1883. Supplementary volume.

1902. pp.xxxii.807. [25,000.]
xiii–xix. 1884–1900. 1914–1925. pp.c.951+ [v].1024 + vi.1012 + vi.1054 + [v].1053 +v.1067+[vi].879. [384.478.]
— 1800–1900. Subject index. [Edited by] Herbert McLeod. Cambridge 1908–1914. pp.lviii.667 + lxxiii.356 + c.550.vii + xv. 551–927.vii.

the index volumes have been published only for pure mathematics, mechanics and physics.

CATALOGUE of the library in connection with the chair of natural science in the New college, Edinburgh. [Edinburgh 1868]. pp.12. [250.]

REPERTORIUM der technischen, mathematischen und naturwissenschaftlichen journal-literatur. Mit genehmigung des Königl. preuss. ministeriums für handel, gewerbe und öffentliche arbeiten, nach amtlichen materialen herausgegeben von F. Schotte. Berlin [Leipzig].

i. 1869. pp.654. [20,000.]
ii. 1870. pp.vi.vii.558.lxii. [17,500.]
iii. 1871. pp.xiv.vi.588. [17,500.]

no more published.

ANNUAL record of science and industry. . . . Edited by Spencer F. Baird. New York.

1871. pp.xxxii.634. [1000.]
1872. pp.lxvii.651. [1000.]
1873. pp.cxxxii.714. [1250.]
1874. pp.iv.cciv.665. [1250.]
1875. pp.v.iii–ccxc.656. [1250.]
1876. pp.[iv].iii.ccxxxvi.609. [1250.]
1877. pp.xiv.480. [1500.]
1878. pp.xiii.715. [2500.]
no more published.

[SIR] GEORGE C[HRISTOPHER] T[OUT] BARTLEY, A catalogue of modern works on science and technology. 1872. pp.[vii].77. [2250.]

[—] — Forty-second edition. 1914. pp.[vi].160. [4000.]

THE STUDENTS' library. A selection from the catalogue of the Radcliffe library at the Oxford museum. For the use of students in the reading-room. Third edition. Oxford 1877. pp.viii.60. [1000.]

THE SCIENCE index. A monthly guide to the contents of the scientific periodicals. Edited by A. Hildebrandt. Vol.I, no.1[–2]. Manchester 1879. pp.64.60. [5000.]
no more published.

NATURAE novitates. Bibliographie neuer erscheinungen aller länder auf dem gebiete der naturgeschichte und der exacten wissenschaften. Herausgegeben von R. Friedländer & sohn. Berlin 1879–1943.

details of this work are entered under Natural history, above.

KATALOG öfver Finska vetenskaps-societetens bibliothek. Helsingfors 1881. pp.[iv].179. [1800.]

CATALOGUE des livres de la bibliothèque de l'Académie royale. . . . Sciences. Bruxelles 1883. pp.xcviii.516. [5871.]

A. HARTLEBEN, Verzeichniss der neuesten erscheinungen auf dem gebiete der elektricität, elektrotechnik, physik, chemie und mechanik. Wien 1883. pp.96. [1750.]

[CAREL PHILIP SLUITER], Catalogus der bibliotheek van de Koninklijke natuurkundige vereeniging in Nederlandsch-Indië. Batavia 1884. pp.vi. 393.

— — Supplement-catalogus. 1895. pp.lxxvi.19.

SCIENCE. Public libraries: Class list (no.3): Nottingham 1884. pp.44. [1000.]

— Supplement. 1888. pp.17. [500.]

— 2nd Supplement. 1890. pp.12. [400.]

— 3rd Supplement. 1896. pp.16. [500.]

C. EKAMA [vol.iii: G. C. W. BOHNENSIEG], Catalogue de la bibliothèque. . . . Tome I[III]. Sciences exactes et naturelles. Fondation Teyler: Harlem 1885–1888, 1904. pp.vii.827+vii.1168.2. [13,500.]
vol.ii is entered under other headings.

[— ESTABEL and JULES LEPREUX], Catalogue méthodique des imprimés de la Bibliothèque publique de Douai. Sciences. Douai 1886. pp. [vii.]285. [4000.]

NATURWISSENSCHAFTLICHE rundschau. Wöchentliche berichte über die fortschritte auf dem gesammtgebiete der naturwissenschaften. Braunschweig.

i. Herausgegeben von Friedrich Vieweg und Sohn 1886. pp.xiii.496. [3000.]
ii. Herausgegeben von W. Sklarek. 1887. pp.xv.512.lxxx. [3500.]
iii. 1888. pp.xv.672.lxxii. [3000.]
iv. 1889. pp.xv.672.lxiv. [2500.]
v. 1890. pp.xii.672.lxxii. [3000.]
vi. 1891. pp.xii.680.lxxxviii. [3500.]
vii. 1892. pp.xii.676.lxviii. [2500.]
viii. 1893. pp.xii.672.lxiv. [2500.]
ix. 1894. pp.xv.672.lxxvi. [3000.]
x. 1895. pp.xvi.672.lx. [2500.]
xi. 1896. pp.xvi.672.lxiv. [2500.]

xii. 1897. pp.xvi.672.lxxii. [3000.]
xiii. 1898. pp.xvii.696.lxiv. [2500.]
xiv. 1899. pp.xvii.672.lxiv. [2500.]
xv. 1900. pp.xviii.672.lxiv. [2500.]
xvi. 1901. pp.xvii.672.lxiv. [2500.]
xvii. 1902. pp.xix.672.lxviii. [2500.]
xviii. 1903. pp.xviii.684.lxviii. [2500.]
xix. 1904. pp.xviii.672.lxxx. [3500.]
xx. 1905. pp.xviii.672.lxxx. [3500.]
xxi. 1906. pp.xviii.694.lxviii. [2500.]
xxii. 1907. pp.xx.672.lxxvi. [3000.]
xxiii. 1908. pp.xxii.672.lxxii. [3000.]
xxiv. 1909. pp.xxi.672.lxxvi. [3000.]
xxv. 1910. pp.xx.672.lxxvi. [3000.]
xxvi. 1911. pp.xx.672.lxxx. [3500.]
xxvii. 1912. pp.xix.672.lxxvi. [3000.]

transformed into a general journal entitled Die Naturwissenschaften.

SOCIETATUM litterae. Verzeichniss der in den publikationen der akademieen und vereine aller länder erscheinenden einzelarbeiten auf dem gebiete der naturwissenschaften. Berlin.

[i]. 1887. Herausgegeben von Ernst Huth. pp.viii.108. [3000.]
[ii]. 1888. pp.viii.192. [4000.]
[iii]. 1889. pp.x.168. [4000.]

[iv]. 1890. pp.viii.164. [4000.]
v. 1891. Herausgegeben von E. Huth und Arthur Hering. pp.vii.176. [4000.]
vi. 1892. pp.vii.184. [4000.]
vii. 1893. Herausgegeben von E. Huth und Max Klittke. pp.viii.192. [5000.]
viii. 1894. Herausgegeben von M. Klittke. pp.viii.192. [5000.]
ix. 1895. pp.viii.192. [5000.]
x. 1896. pp.viii.192. [5000.]
xi. 1897. pp.viii.192. [5000.]
xii. 1898. pp.viii.248. [5000.]
xiii. 1899. pp.viii.240. [5000.]
xiv. 1900. pp.viii.256. [5000.]

no more published.

CATALOGUE of the scientific books in the library of the Royal society of New South Wales. Part I. General catalogue. Sydney 1889. pp.110. [1250.]

[JOSÉ MARIA VIGIL], Catálogos de la Biblioteca nacional de México. Quinta división. Ciencias matemáticas, físicas y naturales. México 1890. pp.103. [2000.]

PANTOBIBLION. International bibliographical review of the world's scientific literature. Editor: A. Kersha. [No.1–2]. St.-Petersburg &c. [1891].

pp.iv.coll.506. [10,000.]
no more published.

CATALOGUE of the Science library in the South Kensington museum. 1891. pp.viii.504. [15,000.]
— Supplement. 1895. pp.iv.118. [3500.]

OSCAR [ANTON] GRULICH, Katalog der bibliothek der Kaiserlichen leopoldinisch-carolinischen deutschen akademie der naturforscher. Halle 1891–1905. pp.xxxii.732 + l.1435 + viii.253. [22,500.]
incomplete; no more published.

J[OHN] POTTER BRISCOE, Class-list of science. Public lending library: Nottingham 1895. pp.26. [2000.]

WILLIAM SWAN SONNENSCHEIN, A bibliography of science. Being the sections relating to that subject in The best books and The reader's guide. 1897. pp.[ii].487–536.375–426. [2000.]

SCIENCE WORK: a monthly review of scientific literature. Edited by Walter Jeffs. Vol.i, no.1[–3]. Manchester 1898–1899. pp.36. [750.]
no more published.

CATALOGUS van de bibliotheek van het Bataafsch genootschap der proefondervindelijke wijsbegeerte te Rotterdam. Rotterdam 1899. pp.[ii].216. [2500.]

[JOHN HEY], Catalogue of books in the library of the Yorkshire philosophical society. York 1900. pp.[iv].103. [2500.]

SIR JAMES HECTOR, Catalogue of the Colonial museum library. (Second edition). Wellington [printed] 1900. pp.[v].156. [3000.]

SUBJECT list of works on general science, physics, sound, music, light, microscopy, and philosophical instruments, in the library of the Patent office. Patent office library series (no.11 = Bibliographical series, no.8): 1903. pp.184. [1849.]

[PETER JOHN ANDERSON], Subject catalogue of the science library in Marischal college. University of Aberdeen: University studies (no.23): Aberdeen 1906. pp.319. [4000.]

[J. H. SHAW], Classified catalogue . . . of books on physical science (including music) in the lending library. Public library: Bury 1912. pp.60. [science: 750.]

CATÁLOGO de la biblioteca de la Real academia de ciencias y artes. Barcelona 1913. pp.760. [20,000.]

CATALOGUE of books in the Presidency college library. Part II. Science. Calcutta 1915. pp.[ii]. ix.231.7. [3500.]

THE ATHENÆUM subject index to periodicals, 1915. . . . Science and technology. 1916. pp.79. [2500.]

— 1916. Science and technology, including hygiene and sport. 1918. pp.162. [5000.]

[*continued as:*]

The subject index to periodicals . . . K. Science and technology. Library association.

1917–1919. 1922. pp.[ii].coll.556. [11,000.]
1920. 1923. pp.[ii].coll.232. [5000.]
1921. 1924. pp.[ii].coll.252. [5000.]
1922. 1926. pp.[ii].coll.272. [5500.]

subsequent issues were not published in sections.

[HELEN PATERSON], Cruickshank science library subject catalogue. University of Aberdeen: University studies (no.85): Aberdeen 1921. pp. 337. [5000.]

CATÁLOGO de las obras existentes en la biblioteca. Universidad nacional de Buenos Aires: Facultad de ciencias exactas, físicas y naturales: Buenos Aires 1921. pp.226. [6000.]

— [another edition]. Catálogo de la biblioteca. Universidad: Facultad de ciencias exactas, físicas y naturales: Publicación (ser. A, no.6): 1930–1931. pp.xxi.1359. [20,000.]

MITTEILUNGEN aus den gebieten der naturwissenschaften und der technik herausgegeben von Friedr. Vieweg & sohn. [Heft 1–28]. Braunschweig 1922–1932. [10,000.]

LISTE des acquisitions des bibliothèques scientifiques de Belgique. Bruxelles.

i. 1930. pp.1292. [16,000.]
ii. 1931. pp.1498. [20,000.]
iii. 1932. pp.799. [10,000.]

AN ANNOTATED bibliography of theses accepted for the master of science degree. State teachers college: Studies in education (no.17 &c.): Topeka 1930 &c.

in progress.

BIBLIOGRAPHICAL series. No.1 &c. Science library: 1930 &c.

in progress; these bibliographies have been entered individually under their subjects; a list and index of nos.1–555, by Samuel Baig, appears in the Bulletin of the New York public libraiy *(1942), xlvi.707–731.*

WEEKLY list of accessions to the library. No.1 [–254]. Science library 1931–1935.*

[*continued as:*]

Weekly bibliography of pure and applied science. No.255 [&c.]. 1935 &c.*

in progress; each issue sets out about a hundred titles.

SCIENCE and its applications. A select list of books. Reference and lending libraries: Leicester 1934. pp.vi.65. [1750.]

M[ARJORIE] P[HYLLIS] CRIGHTON, A catalogue of the medical books and manuscripts, including a selection of the scientific works, in Exeter cathedral library. [*s.l.*] 1934. pp.38.[vi]. [200.]

CLASSIFIED catalogue of books and periodicals in the Shanghai science institute. [Shanghai] 1937. pp.[815]. [4000.]

— Supplement I. [1939]. pp.[118]. [500.]

BULLETIN analytique [signalétique]. Centre national de la recherche scientifique 1940 &c.

in progress.

SCIENZE fisiche, matematiche ed agrarie. Istituto nazionale per le relazioni culturali con l'estero: Roma 1941. pp.xi.211. [2100.]

CATÁLOGO sistemático de los libros de la biblioteca. Universidad de Córdoba: Facultad de ciencias exactas, físicas y naturales: Córdoba 1942. pp.xii.3–194. [2500.]

PURE science. National book council: Book list (no.198): 1943. pp.12. [500.]

LIBRARY-CATALOGUE. Royal institute of science: Bombay 1946. pp.iv.567. [10,000.]

C.S.I.R. information. South african council for scientific and industrial research: Pretoria May 1946 &c.*

in progress.

DOCUMENTS digest. Defence research board: Ottawa 1946 &c.*

in progress.

SCIENCE and technology. A record of literature recently added to the Carnegie library. Pittsburgh.

i. 1946.
ii. 1947.
iii. 1948. pp.[iv].165. [1500.]
iv. 1949. pp.[iv].225. [2000.]
v. 1950. pp.[iv].227. [2000.]
vi. 1951. pp.[iv].188. [1750.]

no more published.

ЕСТЕСТВОЗНАНИЕ. Рекомендательный указатель литературы для бесед и лекций. Государственная... библиотека СССР: Москва 1947. pp.90. [399.]

новая литература о науке и научно-исследовательской работе за рубежом. Академия наук СССР: Фундаментальная библиотека общественных наук: Москва 1947 &c.

новые книги за рубежом. Москва 1948 &c. *in progress.*

занимательное чтение. Рекомендательный указатель занимательной литературы. Государственная публичная библиотека: Москва.

i. [By] М. И. Бурдина, К. П. Кипарисова. 1948. pp.62. [100.]

no more published?

abstracts from the scientific and technical press.

additions to the library. Department of scientific and industrial research.*

April–December 1949. pp.[111]. [2000.]

1950. pp.[184]. [3000.]

December 1950–January 1951. pp.xv. [200.]

[*continued as:*]

D.S.I.R. libraries.

1951.

свободный бюллетень новых иностранных книг, поступивших в крупнейшие биб-

лиотеки гор. Москвы [В библиотеки СССР.] Серия А. Естественные науки, медицина, сельское хозяйство, техника. Всесоюзная государственная библиотека иностранной литературы: Москва 1949 &c.
in progress.

INDICE di periodici scientifici e tecnici. Consiglio nazionale delle ricerche: [Roma] 1949–1950.

BOLETÍN de información documental . . . (Sección de ciencias). Consejo superior de investigaciones científicas: Madrid 1950 &c.
in progress.

THOMAS P[AUL] FLEMING, Guide to the literature of science for use in courses on science literature. Columbia university: School of library service: New York 1952. ff.[ii].46. [451.]*
— — Second edition. [1957]. ff.[ii].69. [596.]*

BOLETÍN del Centro de documentación científica y tecnica. México.
i. 1952. pp.2731. [62,089.]
ii. 1953. pp.2592. [67,518.]
iii. 1954. pp.2375. [66,428.]
iv. 1955. pp.2423. [66,108.]
in progress.

ФИЗИКА, механика, математика, астрономия. Систематический указатель статей в иностранных журналах. Всесоюзная государственная библиотека иностранной литературы: Москва.

1952. pp.73+60+64+60+52+56+72+96 +88. [12,706.]

1953. pp.64+84+72+80+65+60+68+68 68+65+52. [15,000.]

1954. pp.89+96+96+93+84+72+65+68. [15,000.]

1955. pp.73+76+68+72+80+87+69+72 +80+88. [15,000.]

1956. pp.96+ +93+80+81+73+65+77. [12,500.]

1957.

1958. pp.136+112+116+112+109+123+ 106+124+136+93. [25,000.]

1959. pp.100+92+101+95+92.

no more published?; from 1955 astronomy was dropped from the title.

NEUERWERBUNGEN der Öffentlichen wissenschaftlichen bibliothek zu Berlin. Naturwissenschaftlich-technische literatur. Berlin.

i. 1952. pp.51.31.20.24.24.24.24.28.28.36. [3500.]

ii. 1953. pp.158.70.71.71.70.68.74.84.86.79. 92.88. [9000.]

iii. 1954. pp.101.110.106.98.75.81.72.78.82. 77.83.74. [10,000.]

[*continued as:*]

Das neue wissenschaftliche buch. Neuerwerbungen der wissenschaftlichen bibliotheken der DDR: Reihe A: Naturwissenschaften — technik.

iv. 1955. pp.76.76.75.80.69.73.66.63.68.74.70. 77. [8000.]

v. 1956. pp.71.71.72.72.76.102.92.98.87. [7000.]

vi. 1957. pp.71.71.64.74.74.72.86.84.80. [6000.]

vii. 1958. pp.79.80.69.68.67.64.64.64.64.68. 64. [6000.]

viii. 1959. pp.68.68.112.96.103.116.120. [6000.]

CATALOGUE mensuel des traductions effectuées dans les services et centres de documentation français. Centre national de la recherche scientifique: Centre de documentation 1952 &c.*

S.L.A. list of translations. Special libraries association: Translation committee: New York [1953]. ff.72. [850.]*

— Supplement. [1954]. ff.62.*

TRANSLATION title list and reference guide. Atomic energy commission: Technical information service: Oak Ridge, Tenn. 1953. pp.iii.204. [3200.]

CATALOGUE mensuel des traductions effectuées dans les services et centres français de documentation. Centre national de la recherche scientifique: 1953 &c.*

in progress?

INFORMATION. Being a list of scientific articles.

1953. pp.16.16.16.20. [1250.]
1954. pp.24.24.24.24. [1500.]
1955. pp.24.24.24.24. [2500.]
1956. pp.29.36. [1500.]

in progress; limited to articles appearing in the periodicals published by Taylor & Francis ltd.

BIULETYN osrodka bibliografii i dokumentacji naukowej PAN. Polske akademia nauk: Zakład imienia Ossolińskich: Wrocław &c.

i. 1953.
ii. 1954.
iii. 1955. pp.[575]. [18,081.]
iv. 1956. pp.[895]. [31,051.]
v. 1957. pp.[943]. [29,476.]

[*continued as:*]

Biuletyn zagranicznych nabytkow bibliotek PAN.

vi. 1958. pp.[911]. [21,773].

vii. 1959. pp.[1111]. [28,177.]
in progress.

FRANCES BRIGGS JENKINS, Science reference sources. A selected list of titles for use in courses on science literature . . . in the university of Illinois library school. Champaign [1954]. ff.[iii].90. [1350.]*

— — Second edition. [1958]. ff.[iii].93. [974.]*

BOLETÍN de información documental. Publicación trimestral. Sección de ciencias. Consejo superior de investigaciones científicas: Biblioteca general: Madrid.

v. 1954. pp.686. [14,000.]
vi. 1955. pp.796. [16,000.]
vii. 1956. pp.996. [20,000.]
viii. 1957. pp.997. [20,000.]
ix. 1958. pp.937. [19,000.]
x. 1959. pp.855. [17,000.]
xi. 1960. pp.1097. [20,000.]
xii. 1961. pp.874. [18,000.]
xiii. 1962. pp.873. [18,000.]

in progress; the earlier volumes are entered under Periodical publications: Indexes, above.

SCIENCE reference notes. Columbia university: Science libraries: New York 1954–1958. [941.]*

[VLADIMÍR BERNÁŠEK], Vesmír – příroda – člověk. Výběrová bibliografie české populárně vedecké literatury. Universita: Knihovna: Čteme a studujeme (1955, no.3): Praze [1955]. pp.[ii].25. [200.]

ПЕРЕЧЕНЬ переводов выполненных Отделом переводов Всесоюзной торговой палаты за 1950–1954 гг. Москва 1955 &c.

TRANSLATION monthly. Special libraries association: New York.*

i. 1955.
ii. 1956.
iii. 1957. pp.453.[lvi]. [9000.]
iv. 1958. pp.530.[lxxiii]. [10,000.]

[*continued as:*]

Technical translations. Department of commerce: Office of technical services: Washington 1959 &c.

in progress.

IRMA JOHNSON, Selected books and journals in science and engineering. Massachusetts institute of technology: Technology monographies: Library series (no.1): Cambridge.

— — Second edition. [1959]. pp.[vi].63. [900.]

AVDNA T. CLUM, General science bibliography.

Bureau of elementary curriculum development: Albany 1956.

[—] — [another edition]. Science bibliography . . . for general science, grades 7, 8 and 9. 1960. pp.20.

HILARY J. DEASON, An inexpensive science library. A selected list of paper-bound science books. American association for the advancement of science [&c.]: Washington 1957. pp.16. [200.]

— — Fourth edition. By H. J. Deason and Robert W. Lyon. 1960. pp.[ii].70. [500.]

INTERNATIONAL technical titles. . . . A monthly compilation of tables of contents of journals in pure and applied aeronautics, physics, chemistry, engineering, mathematics and electronics. Los Angeles 1957–1960.*

no more published?

HELENA WINKLEROVÁ, Matematika – fysika – chemie – geologie. Universita: Knihovna: Čteme a studujeme (1958, no.4): Praze [1958]. pp.48. [400.]

IRMA JOHNSON [*and others*], Selected books and journals in science and engineering. Massachusetts institute of technology: Technological monographs: Library series (no.1): Cambridge, Mass. [1958]. pp.[vi].58. [900.]*

— — Second edition. [1959]. pp.[iv].63. [900.]*

ИНФОРМАЦИОННЫЙ бюллетень новых иностранных книг, поступивших в библиотеку имени В. И. Ленина. Серия 1. Физико-математические, химические и геолого-географические науки. Москва 1958 &c.

in progress.

HILARY J. DEASON, The AAAS science book list. American association for the advancement of science: Washington 1959. pp.[ii].140. [1000.]*

SYLVIA GOLDMAN, Guide to the literature of engineering, mathematics and the physical sciences. Johns Hopkins university: Applied physics laborators: Silver Spring, Md. 1959. ff.52.*

AUREL AVRAMESCU and VIRGIL CÂNDEA, Introducere în documentaşea ştiinţifica. [Bucureşti] 1960. pp.520. [500.]

GHID de documentare în ştiinţele naturii şi ştiinţele aplicate. Biblioteca Academiei Republicii Populare Romîne: Secţia de documentare ştiinţifica: [Bucharest] 1960. pp.437. [3036.]

LIBRARY information and accessions. South afri-

can council for scientific and industrial research: Pretoria 1960 &c.*
in progress?

RECENT additions on science and technology. Public libraries: Gateshead [1961]. pp.38. [500.]

SCIENCE citation index. An international interdisciplinary index to the literature of science. Institute for scientific information: Philadelphia.*
1961. pp.xxx.496+[ii].497–1040+[ii].1041–1584 + [ii].1585–2128 + [ii].2129–2672. [1,400,000.]
in progress?

NORMA OLIN IRELAND, Index to scientists of the world from ancient to modern times: biographies and portraits. Useful reference series (no.90): Boston 1962. pp.xliii.662. [15,000.]

YVONNE RUYSSEN, Choix d'ouvrages et de périodiques proposé pour la constitution d'un fonds scientifique de base. 1ère partie. Généralités, sciences mathématiques, physiques et chimiques. Direction des bibliothèques de France: Bibliothèques des collèges scientifiques universitaires: 1962. ff.vi.168. [1500.]*

7. *Research and methodology*

[*see also* **Research**]

LIST of references on scientific research. Library of Congress: Washington 1917. ff.8. [107.]*

— Additional references. 1918. ff.7. [82.]*

LISTS of scientific institutions, associations and research laboratories, 1939–1949. Science library: Bibliographical series (no.693): 1950. ff.5. [75.]*

ORGANISATION scientifique. Répertoire bibliographique 1944–1951. Comité national belge de l'organisation scientifique: [Bruxelles *c.*1952]. pp.32. [600.]

КНИГИ о природе и её законах. Указатель научно-популярной литературы по естествознанию. Библиотека СССР имени В. И. Ленина: Москва 1953–1955. pp.68+36+44+143+48. [500.]

ROBERT BLANCHE, Logique et philosophie des sciences. Association pour la diffusion de la pensée française: Bibliographie française établie à l'intention des lecteurs étrangers: 1959. pp.54. [241.]

limited to works published since 1945; an english edition was published, New York 1959.

[THEODORE SURANYI-UNGER and ELIZABETH

HARRIS], Current projects on the economic and other impacts of scientific research and development, 1959. National science foundation: Office of special studies: Washington [1959]. pp.vi.59. [150.]

MILDRED CATHERINE BENTON, Creativity in research and invention in the physical sciences. An annotated bibliobraphy [*sic*]. Naval research laboratory: Bibliography (no.19): Washington 1961. pp.ix.151. [1359.]*

PAUL C. JANASKE, *ed.* Information handling and science information. A selected bibliography 1957–1961. American institute of biological sciences: Biological sciences communication project: Washington 1962. pp.[256]. [1121.]*

[A. ARNOŠT *and others*], Seznam vědecké a technické literatury. Výběr zahraniční a domácí literatury, která došla do státních vědeckých knihovna ČSSR v letech 1950–1960. Státní technická knihovna: Praha 1962–1963. pp.616+[v].615–1016+[iii].1083. [40,000.]

HILARY J. DEASON and WILLIAM BLACKLOW, A guide to science reading: Signet science library (no.P2283): [New York 1963]. pp.220. [900.]

MELVIN BENNETT, Science and technology. A

purchase guide for branch and small public libraries. Carnegie library: Pittsburgh 1963. pp.64.

OUT-OF-PRINT books reproduced by xerography. Science. Ann Arbor 1963. pp.[iv].26. [750.]*

8. *Science teaching*

BIBLIOGRAPHY of science teaching. Compiled by a committee of the American federation of teachers of the mathematical and the natural sciences. Bureau of education: Bulletin (no.446): Washington 1911. pp.27. [328.]

A LIST of science books suitable for school libraries. Science masters' association: 1921. pp.31. [500.]

— Third edition. Science books for a school library. 1949. pp.xii.82. [1600.]

EARL R. GLENN, Bibliography of science teaching in secondary schools. Bureau of education: Bulletin (1925, no.13): Washington 1925. pp.x.161. [4000.]

A[VRIL] S[YLVESTER] BARR and MABEL RUDISILL, An annotated bibliography on the methodology of scientific research as applied to education.

University of Wisconsin: Bureau of educational research: Bulletin (no.13): Madison 1931. pp.[ii]. 129. [1000.]

RUTH BUDD, *ed.* Science books for the elementary schools. Reading for background (no.5): New York 1937. pp.23. [150.]

REFERENCE books for elementary school science. University of Iowa extension bulletin (no.584): Iowa City 1945. pp.15. [350.]

L'ENSEIGNEMENT des sciences à l'école primaire. Unesco: Revue analytique de l'éducation (vol.vii, no.7): 1955. pp.24. [80.]

JOHN S[ANFORD] RICHARDSON [*and others*], Resource literature for science teachers. Ohio state university: College of education: Columbus 1957. ff.v.65. [600.]*

— — Second edition. 1961. pp.iv.103. [800.]*

GEORGE GREISEN MALLINSON and JACQUELINE V. BUCK, A bibliography of reference books for elementary science. National science teachers association: Washington [1958]. ff.[iv].35. [600.]*

SCIENCE books for a school library. Science masters' association: [1959]. pp.54. [750.]

— Supplement no.1. 1960. pp.16. [200.]

LA FORMATION de professeurs de sciences pour les écoles secondaires. Unesco: Revue analytique de l'éducation (vol.xiii, no.4): 1961. pp.40. [200.]

JOHN H. WOODBURN, *ed.*, Keys to careers in science and technology. A bibliography. National science teachers association: Washington [1964]. pp.ii.53. [250.]

[GEORGE GREISEN MALLINSON and JACQUELINE BUCK MALLINSON], A bibliography of text books for courses in science for the elementary, junior high and senior high school levels. National science teachers association: Washington 1964. pp.[iii].23. [250.]*

9. *Popular science*

J. B. CLARK, Popular science. A selection of titles. [National book council:] Bibliography (no.107): 1929. pp.[4]. [150.]

— — Popular science. Second edition. National book council: Book list (no.107): 1938. pp.[4]. [150.]

HOW to understand the sciences. A graduated list of books. Public libraries: Middlesbrough 1935. pp.28. [150.]

Science

BOOKS on popular science. Science library: Bibliographical series (no.274): 1936. ff.7. [158.]*
— Revision . . . (no.361): 1937. ff.7. [144.]*

V. V. NEIMAN, Пропаганда естественно-научной литературы в массовых библиотеках. Сборник материалов. Государственная публичная библиотека: Научно-методический кабинет библиотековедения: Москва 1951. pp.88. [400.]
— — [another edition]. [Edited by V. V. Neiman and E. N. Fomina]. 1957. pp.128. [250.]

E. V. SEGLIN and A. F. SADOFEV, *edd*. Пропаганда естественно-научной литературы в массовых библиотеках. Государственная... публичная библиотека СССР имени В. И. Ленина: Научно-методический кабинет библиотековедения: Москва 1953. pp.64. [400.]

M[IKHALINA] P[ETROVNA] BRONSHTEIN and V[ERA] V[LADIMIROVNA] GNUCHEVA, Библиография естественнонаучной литературы. Управление учебных заведении: Москва 1956. pp.192. [1000.]

NATUR und technik. Eine auswahl populärwissenschaftlicher bücher. Stadt- und bezirksbibliothek [&c.]: Leipzig [1957]. pp.79. [600.]

Science

SCIENCE for all. An annotated reading list for the non-specialist. National book league: Cambridge 1958. pp.viii.178. [800.]

[G. S. MULTANOVSKAYA and V. V. NEIMAN], Научно-технические знания — в массы. Сборник материалов в помощь массовым библиотекам. Государственная . . . библиотека СССР имени В. И. Ленина: Научно-методический кабинет библиотековедения: Москва 1958. pp.199. [600.]

THE WORLD of science. Books for boys and girls. Bureau of elementary curriculum development: [Albany 1958]. pp.12. [200.]

— [another edition]. [1960]. pp.17. [250.]

ESTER K. NELSON, A bibliography of science books for elementary school children. California state department of education: Bulletin (vol.xxviii, no.5): [Sacramento 1959]. pp.[vii].115. [1000.]

HILARY J[OHN] DEASON, The science book list for children. American association for the advancement of science: Washington 1960. pp.xii.139. [1100.]*

TIBOR PELEJTEI, Szép könyvek a természetről. Budapest 1961. pp.[116]. [55.]

Science

10. *Countries*

Africa

JEAN LEYDER, L'enseignement supérieur et la recherche scientifique en Afrique intertropicale. Centre de documentation économique et sociale africaine: Enquêtes bibliographiques (no.ii): Bruxelles 1959–1960. pp.67+[ii].220. [1325.]

America, latin

LISTA de artículos científicos publicados en América latina. United nations educational, scientific and cultural organization: Centro de cooperación científica para América latina: Montevideo.

i. 1948. pp.[iii].296. [6000.]*
ii. 1949. pp.[iii].xlvi.506. [10,000.]*
iii. 1950. pp.324. [13,000.]

no more published.

Arabic

[HEINRICH] FERDINAND WUSTENFELD, Geschichte der arabischen aerzte und naturforscher. Göttingen 1840. pp.xvi.168.[xvi]. [2000.]

the title is misleading; this is a bio-bibliography.

Asia

LISTE de [des] travaux scientifiques publiés au

moyen-orient. Unesco: Centre de coopération scientifique — moyen-orient: Le Caire.

i–ii. 1948. ff.37+65. [1200.]*
iii–iv. 1949. pp.59+72. [1400.]
v–vi. 1950. pp.78+76. [2336.]
vii. 1951. pp.77. [1063.]
viii. 1952. pp.86. [1046.]
ix. 1953. pp.64. [930.]
x. 1954. pp.74. [1145.]
xi. 1955. pp.84. [1110.]
xii. 1955. pp.130. [1268.]

[*continued as:*]

Résumés analytiques des travaux scientifiques et techniques publiés en Égypte et travaux reçus de l'Afghanistan, Chypre, Iran, Irak, Liban, Pakistan, Soudan et Syrie. National research council of Egypt: Bulletin of the scientific and technical documentation centre (part 2): Cairo.

i. 1955. pp.[59]. [705.]
ii. 1956. pp.150. [1755.]
iii. 1957. pp.158. [1839.]
iv. 1958. pp.149. [1743.]
v. 1959. pp.144. [1167.]
vi. 1960.
vii. 1961. pp.61. [495.]

in progress.

BIBLIOGRAPHY of scientific publications of south

Asia (India, Burma, Ceylon). Unesco: Science co-operation office for south Asia [iv–xi: South Asia science co-operation office; xii: Indian national scientific documentation centre]: New Delhi &c.

i–ii. 1949. pp.[vi].107+[ii].88. [4305.]
iii–iv. 1950. pp.[iv].xiii.84+[v].86. [4245.]
— Name index to . . . 1–4. [1952]. pp.[ii].48.
v–vi. 1951. pp.[iii].ii.97.v+[v].viii.ii.124. [5321.]
vii–viii. 1952. pp.[vii].178+[v].157. [6679.]
ix–x. 1953. pp.[v].174+[v].158. [6912.]
xi–xii. 1954. pp.[v].212+[v].129. [7796.]

[*continued as:*]

Bibliography of scientific publications of south & south east Asia.

Australia

JOHN SHIRLEY, International catalogue of scientific literature. Queensland volume. Brisbane 1899. pp.[iv].vii.154. [850.]

AUSTRALIAN science abstracts. Australian national research council: Sydney.

i. 1922. Editor: A[rthur] B[ache] Walkom. pp.32. [245.]
ii. 1923. pp.47. [379.]
iii. 1924. pp.42. [299.]
iv. 1925. pp.31. [225.]

v. 1926. pp.34. [287.]
vi. 1927. pp.36. [308.]
vii. 1928. pp.35. [350.]
viii. 1929. pp.35. [312.]
ix. 1930. Editor: G[ustavus] A[thol] Waterhouse. pp.51. [476.]
x. 1931. pp.53. [465.]
xi. 1932. pp.60. [498.]
xii. 1933. pp.69. [701.]
xiii. 1934. pp.67. [696.]
xiv. 1935. pp.12.16.14.15. [799.]
xv. 1936. pp.14.12.15.17. [655.]

in progress.

LIST of publications. Commonwealth scientific and industrial research organization: Melbourne 1950. pp.28. [600.]

AUSTRALIAN science index. Commonwealth scientific and industrial research organization: Melbourne 1957 &c.*

Belgium

[AUGUSTE VINCENT], Histoire des sciences en Belgique jusqu'à la fin du XVIIIe siècle. Exposition. Bibliothèque royale de Belgique: Bruxelles 1938. pp.3–160. [441.]

Cuba

CARLOS M. TRELLES, Biblioteca científica cubana.

Matanzas 1918–1919. pp.[iii].vii.473+[v].x.505. [8000.]

200 (vol.ii: 250) copies printed.

Czechoslovakia

REVUE des travaux scientifiques tchécoslovaques. Section deuxième. Mathématiques, physique, chimie, sciences biologiques . . . techniques . . . médicales. [Česká akademie věd a umění: Prague.]

i. 1919. 1924. pp.264. [500.]
ii. 1920–1921.
iii. 1922–1924. 1931. pp.[xi].873. [3000.]

Denmark

CHR[ISTIAN] V. BRUUN, Bibliotheca danica. Systematisk fortegnelse over den danske literatur fra 1482 til 1830. . . . Fjerde hefte. . . . De mathematiske, physiske, naturhistoriske, økonomiske og techniske videnskaber. Kjøbenhavn 1879. coll.lvi.382. [5750.]

FORTEGNELSE over bøger of danske forfattere udkommet paa fremmede sprog indenfor fagene matematik, naturvidenskab og medicin. [*s.l.* 1939]. ff.14. [200.]*

East Indies

H[ENRI] C[ONSTANTINOS] A[LEXANDROS] MULLER,

Science

Catalogue of manuscripts, old curious and more recent and richly illustrated books concerning the study of science in the Dutch East Indies exhibited in the Technical high school at Bandoeng. [Bandoeng 1929]. pp.38. [136.]

Ecuador

CARLOS MANUEL LARREA, Bibliografía científica del Ecuador. Quito.

i. Geografía, descripciones generales, viajes, datos estadísticos. 1948. pp.240. [1900.]

ii. Geología, petrografía, mineralogía, paleontología, vulcanología. 1952. pp.319. [2500.]

iii. Antropología, etnografía, arqueología, linguistica, folklore, prehistoria, historia antigua. 1952. pp.[v].325–559. [1800.]

iv. Historia general, historia de la cultura, historia eclesiastica, cuestiones políticas, etc. 1953. pp.[v].565–1027. [3600.]

v. Apendice e indices. 1953. pp.[ii].1037–1197. [1000.]

Egypt

CLASSIFIED list of egyptian scientific papers. Fouad I national research council [ii–iii: Al-Maʻhad al-qawmī lil-buḥūth]: Cairo.

i. 1951. pp.55. [500.]
ii. 1952.
iii. 1953.
no more published.

France

[LOUIS] ATH[ÉNAÏS] MOURIER, Notice sur le doctorat ès sciences, suivie du catalogue des thèses admises par les facultés des sciences depuis 1810. 1856. pp.79. [300.]

REVUE des travaux scientifiques. Comité des travaux historiques et scientifiques.
[i]. 1881. pp.vi.1146. [2000.]
[ii]. 1882. pp.[iii].1002. [2000.]
[iii]. 1883. pp.xv.967. [2000.]
[iv]. 1884. pp.[iii].969. [2000.]
[v]. 1885. pp.[iii].880. [2000.]
[vi]. 1886. pp.[iii].828. [2000.]
[vii]. 1887. pp.[iii].847. [2000.]
[viii]. 1888. pp.[iii].952. [2500.]
[ix]. 1888 [*sic*]. pp.[iii].1240. [2500.]
[x]. 1889 [*sic*]. pp.[iii].1200. [2500.]
[xi]. 1890 [*sic*]. pp.[iii].1064. [2500.]
[xii]. 1892. pp.[iii].1272. [3000.]
[xiii]. 1893. pp.[iii].1204. [3000.]
[xiv]. 1894. pp.[iii].1203. [3000.]

xv. 1895. pp.[iii].1305. [3000.]
xvi. 1896. pp.[iii].1280. [3000.]
xvii. 1897. pp.[iii].1240. [3000.]
xviii. 1898. pp.[iii].1308. [3000.]

NOTICE sur les travaux personnels des professeurs des facultés des sciences durant l'année 1881, avec rappel des années antérieures. Tome premier. 1882. pp.[v].190. [600.]

ALBERT MAIRE, Catalogue des thèses de sciences soutenues en France de 1810 à 1890 inclusivement. Bibliographie des thèses (no.2): 1892. pp.xi.224. [1000.]

— — Revue décennale des thèses présentées à la Faculté des sciences de Paris en vue du grade de docteur ès sciences du 1er janvier 1891 au 31 décembre 1900, avec l'indication des périodiques contenant la plupart de ces mémoires ou leurs analyses. Par E[ugène Pierre] Estanave. Arcis-sur-Aube 1901. pp.115. [347.]

J[OSEPH] DENIKER, Bibliographie des travaux scientifiques (sciences mathématiques, physiques et naturelles) publiés par les sociétés savantes de la France depuis l'origine jusqu'en 1888. . . . Tome premier [-II, 1re livraison]. Ain-Orne[-Sarthe]. [1895-]1916-1922. pp.[iii].iii.607 + vii.235. [23,244.]

no more published.

Science

BIBLIOGRAPHIE scientifique française. Bureau français du Catalogue international de la littérature scientifique [vol.xix &c.: Commission du répertoire de bibliographie scientifique].

i. 1902. pp.vi.374. [4905.]
ii. 1903. pp.[iii].140+[iii].236. [5366.]
iii. 1904–1905. pp.[iii].144+[iii].232. [5010.]
iv. 1905–1906. pp.[iii].140+[iii].236. [5470.]
v. 1906–1907. pp.152+232. [6262.]
vi. 1907–1908. pp.[iii].136+[iii].240. [6982.]
vii. 1908–1909. pp.132+244. [7260.]
viii. 1909–1910. pp.[iii].132+[iii].244. [7491.]
ix. 1910–1911. pp.[iii].132+[iii].244. [7086.]
x. 1911–1912. pp.[iii].136+[iii].240. [6666.]
ix [*sic*, xi]. 1912–1913. pp.[iii].136+[iii].240. [6720.]
xii. 1913–1914. pp.[iii].140+[iii].240. [6720.]
xii [*sic*, xiii]. 1915–1916. pp.[iii].92+[iii].176. [4721.]
— Tables des tomes I à XIII. . . . E. Astronomie (K. Paléontologie; L. Biologie]. 1919–1921. pp.48+22+42.
xiv. 1916–1917. pp.[iii].56+[iii].128. [3300.]
xv. 1918. pp.[v].83+[iv].95. [3510.]
xvi. 1919. pp.[iv].32+[iv].64. [2534.]
xvii. 1920. pp.[iv].84+104. [4039.]
xviii. 1921. pp.87+144. [4622.]

xix. 1922. pp.[iv].98+[iv].186. [5691.]
xx. 1923. pp.[iv].66+[iv].11.128. [4507.]
xxi. 1924. pp.[iv].100+[iv].250. [7638.]
xxii. 1925. pp.[iv].160.16 + [iv].288.24. [9329.]
xxiii. 1926. pp.[iv].96+[iv].452. [12,107.]
xxiv. 1927. pp.[iv].96.16 + [iv].256.24. [8144.]
xxv. 1928. pp.[iv].96.12 + [iv].288.30. [8705.]
xxvi. 1929. pp.[iv].128.16 + [iv].288.28. [11,446.]
xxvii. 1930. pp.[iv].96.16 + [iv].288.32. [8546.]
xxviii. 1931. pp.[iv].96.16 + [iv].288.31. [8647.]
xxix. 1932. pp.[iv].96.14 + [iv].224.24. [7136.]
xxx. 1933. pp.[iv].128+[iv].192.24. [7045.]
xxxi. 1934. pp.[iv].64+[iv].192.22. [5634.]
xxxii. 1935. pp.[iv].128.18+[iv].192. [7044.]
xxxiii. 1936. pp.[iv].160.23 + [iv].160.23. [7113.]
xxxiv. 1937. pp.[iv].96.15 + [iv].224.24. [7040.]
xxxv. 1938. pp.[iv].96.14 + [iv].224.24. [7224.]

xxxvi. 1939. pp.[iv].160. [3000.]
xxxvii–xxxviii. 1940. pp.192.19 + 608. [16,000.]
xxxix. 1941. pp.128. [3100.]
xl–xli. 1942. pp.192+256. [9000.]
no more published.

BIBLIOGRAPHIE des livres français de médecine & de science . . . 1900–1912. Publiée par la Section de médecine du Syndicat des éditeurs. [1913]. pp.144. [3500.]

BIBLIOGRAPHIE des livres français de médecine et de sciences publiée par les maisons J.-B. Baillière et fils, G. Doin et c^ie^, N. Maloine, Masson et c^ie^, Vigot f^res^, 1919–1928. 1928. pp.[viii].xlviii.167. [2250.]

FRENCH bibliographical digest. Science. French embassy: Cultural division: New York 1949 &c.
a fragmentary list of recently published french books.

RÉPERTOIRE des travaux publiés par les chercheurs. Centre national de la recherche scientifique.
[1952–1953]. pp.575. [5000.]
[*continued as:*]

Répertoire des travaux publiés avec le concours du C.N.R.S.

1954. pp.xxv.560. [5000.]

DOUZE mois d'édition scientifique et technique. Département étranger Hachette.

1959. pp.272. [2000.]
1960. pp.288. [2000.]
1961. pp.355. [2500.]
1962. pp.331. [2500.]
in progress.

CATALOGUE collectif des livres français de sciences et techniques, 1950–1960. Cercle de la librairie: [1960]. pp.viii.243. [3000.]

— 1960–1962. [1963]. pp.146. [1500.]

Germany

BIBLIOGRAPHIE der deutschen naturwissenschaftlichen litteratur. Deutsches bureau der internationalen bibliographie [Berlin]: Jena [vols.v–xii, xiv–xviii: Berlin; xiii: Burg b. M.]

i–ii. 1901–1902. pp.[vi].iv.1280+[ii].1280. [23,227.]
iii. 1903. pp.[ii].1246. [20,000.]
iv–v. 1904. pp.[ii].1280+[ii].1240. [40,000.]
vi–vii. 1905. pp.[ii].1300+[ii].1280. [50,000.]
viii. 1906. pp.[ii].922. [17,500.]

ix. 1907. pp.[ii].977. [17,500.]
x–xi. 1908. pp.[ii].1018+[ii].929. [35,000.]
xii–xiii. 1909. pp.[ii].918+[ii].752. [30,000.]
xiv. 1910. pp.[ii].760.xxxiv. [25,000.]
xv. 1911. pp.[ii].776.36. [25,000.]
xvi. 1912. pp.[ii].766.36. [25,000.]
xvii. 1913. pp.[ii].766.38. [25,000.]
xviii. 1914. pp.[ii].760.40. [25,000.]
no more published.

BOOK republication program. Titles suggested for republication. Current foreign imprints, 1940–1943. Office of alien property custodian: [Washington] 1944. pp.vii.42.

Great Britain

A LIST of periodicals and bulletins containing abstracts published in Great Britain. With an appendix giving partial list of journals containing abstracts published in the british commonwealth. Royal society: 1949. pp.64. [135.]

— Second edition. 1950. pp.80. [134.]
limited to science and technology.

[P. PASSENGER], A catalogue of british scientific and technical books. British science guild: 1921. pp.xviii.376. [7500.]

— — Third edition . . . by Daphne Shaw. 1930. pp.[iv].xxi.754. [15,000.]

A CLASSIFIED list of new & forthcoming books on natural science. [National book council:] Reader's guides (no. 5).

1928. pp.12+12. [125.]
1929. pp.8+8+8+8. [150.]
1930. pp.8+8+8+8. [100.]
1931. pp.8+8+12+8. [150.]
1932. pp.8+8+12. [100.]
1933. pp.8. [25.]
1934. pp.8. [25.]

no more published.

TECHNICAL books, including books on chemistry, physics, engineering, electricity, mechanics, and technical handbooks, published during 1936. pp.28. [500.]

limited to works published in Great Britain.

ASLIB book-list. Quarterly recommendations of recently published scientific and technical books. Association of special libraries and information bureaux.

i. 1935–1936. pp.152. [1000.]
ii. 1937. pp.142. [1070.]
iii. 1938. pp.142. [1050.]
iv. 1939. pp.144. [1031.]
v. 1940. pp.116. [747.]

vi. 1941. pp.104. [596.]
vii. 1942. pp.108. [764.]
viii. 1943. pp.118. [810.]
ix. 1944. pp.100. [925.]

— Author and subject index to vols.I–IX. [Compiled by Alfred Loewenberg, under the direction of Theodore Deodatus Nathaniel Besterman]. 1945. pp.98.

x. 1945. pp.viii.107. [1018.]
xi. 1946. pp.viii.95. [997.]
xii. 1947. pp.100. [1051.]
xiii. 1948. pp.viii.[112]. [802.]
xiv. 1949. pp.98. [978.]
xv. 1950. pp.96.19. [931.]
xvi. 1951. pp.102.18. [991.]
xvii. 1952. pp.98.18. [953.]
xviii. 1953. pp.96.19. [991.]
xix. 1954. pp.98. [1010.]
xx. 1955. pp.92.19. [1010.]
xxi. 1956. pp.[159]. [752.]
xxii. 1957. pp.[166]. [792.]
xxiii. 1958. pp.[183]. [745.]
xxiv. 1959. pp.[160]. [751.]
xxv. 1960. pp.[166]. [744.]
xxvi. 1961. pp.[166]. [773.]
xxvii. 1962. pp.[170]. [848.]

xxviii. 1963. pp.[164]. [739.]

in progress; limited to works in english.

SELECT list of standard british scientific and technical books. Association of special libraries and information bureaux: 1937. pp.36. [321.]

— Third edition. [General editor: Theodore Deodatus Nathaniel Besterman]. 1946. pp.64. [900.]

— Fourth edition. Edited by E. R. McColvin. 1952. pp.72. [1000.]

— Fifth edition. Edited by L[aurance] J. Anthony. 1957. pp.88. [1600.]

LIST of documents received from British Commonwealth sources. British Commonwealth scientific office: Washington 1946–1953.*

A LIST of british scientific publications reporting original work or critical reviews. Royal society: Information services committee: [1950]. pp.95. [212.]

limited to periodical publications.

BRITISH scientific and technical books. A select list of recommended books published in Great Britain and the Commonwealth. Aslib.

1935–1952. 1956. pp.xi.364. [5250.]

1953–1957. Edited by L. J. Anthony. 1960. pp.ix.251. [4000.]

Science

India

INDIAN science abstracts. (Being an annotated bibliography of science in India). National institute of sciences of India: Calcutta.

[i]. 1935. pp.[ii].5.10.48.81.37.72.77.32. [2000.]
[ii]. 1936. pp.18.59.33.66.34.61.67.35.45. [2000.]
[iii]. 1937. pp.18.38.46.63.39.43.61.69.64. [2000.]
[iv]. 1938. pp.22.50.73.76.69. [1500.]
[v]. 1939. pp.21.46.48.59.92.59. [2000.]

in progress?

INDIAN scientific & technical publications exhibition. Council of scientific & industrial research: New Delhi 1960. pp.198.195.

Indonesia

INDONESIAN abstracts. Abstracts on current scientific indonesian literature. Madjelis ilmu pengetahuan Indonesia: Djakarta.

i. 1958–1959. pp.129. [400.]
ii. 1960. pp.74. [200.]
iii. 1961. pp.62. [150.]
iv. 1962.
v. 1963.

in progress.

Science

Ireland

SELECT references on irish science. [Science library: Bibliographical series (no.42): 1932]. single sheet. [13.]*

Israel

[HANNAH FARKAS, *ed.*], Catalogue des publications scientifiques d'Israël, 1948–1955. Paris 1955. pp.[viii].166. [3235.]

an exhibition catalogue.

CATALOGUE of scientific publications by Israel scientists 1948–1955. Research council: [Jerusalem 1956]. pp.259. [3235.]

also issued in a french edition.

Italy

IGNAZIO CANTÙ, L'Italia scientifica contemporanea, notizie sugli italiani ascritti ai cinque primi congressi. Milano 1844. pp.[ix].320.173. [2500.]

ALDO MIELI, La storia della scienza in Italia. Saggio di bibliografia di storia della scienza. Firenze 1916. pp.vii.131. [500.]

ALDO MIELI, Gli scienziati italiani dall'inizio del medio evo ai nostri giorni. Repertorio bibliografico. Roma 1921–1923. pp.viii.236+[ii].235–466. [2500.]

Science

BIBLIOGRAFIA scientifico-tecnica italiana, 1928 [&c.]. Consiglio nazionale delle ricerche: Bologna &c. 1928 &c.

in progress; details of this work are entered under Italian literature.

[GIULIO PROVENZAL], Prolusioni di argomento scientifico lette nelle università e negli istituti superiori d'Italia per l'inaugurazione dell'anno scolastico dal 1860 al 1930. Consiglio nazionale delle ricerche: Roma 1932. pp.viii.150. [800.]

SANDRO PIANTANIDA, LAMBERTO DIOTALLEVI and GIANCARLO LIVRAGHI, Autori italiani del seicento. Catalogo bibliografico. II. Le scienze. Milano 1949. pp.xiii.281. [1198.]

Mexico

MANUEL DE OLAGUÍBEL, Memoria para una bibliografía científica de México en el siglo XIX. México 1899. pp.99. [1350.]

MANUEL DE OLAGUÍBEL and ENRIQUE IGLESIAS, Bibliografía científica del estado de México. Toluca 1899. pp.30. [200.]

ÍNDICE bibliográfico del Centro de investigación y de estudios avanzados. Sección 1ª. Matemáticas, astronomia y astrofísica, física, geología, geofísica, geodesia. Instituto politécnico nacional: México 1962 &c.

in progress.

Netherlands

CATALOGUS van proefschriften vervaardigd ter verkrijging van de graad van doctor in één der natuurwetenschappen aan universiteiten of hoogscholen in Nederland in de jaren 1900–1930. [*s.l.*] 1931. pp.206.

NATUURWETENSCHAPPELIJK onderzoek in Nederland. Een overzicht van hetgeen in de laatste vijf jaren in Nederland verricht is op het gebied der natuurwetenschappen, der medische en der technische wetenschappen. Werkgemeenschap van wetenschappelijke organisaties in Nederland: Amsterdam 1942. pp.261. [275.]

NEDERLANDSE technisch-wetenschappelijke literatuur. Bibliografie van in Nederland verschenen publicaties. Nederlands instituut voor documentatie en registratuur: 's-Gravenhage.

i. 1953.
ii. 1954.
iii. 1955.
iv. 1956.
v. 1957. pp.96. [2435.]

vi. 1958. pp.96. [2460.]
vii. 1959. pp.96. [2424.]
viii. 1960. pp.96. [2190.]
ix. 1961. pp.104. [2335.]
x. 1962. pp.104. [2376.]

in progress; an abridged translation is also issued under the title of Technical digest.

Pakistan

PAKISTAN scientific literature. Current bibliography. Pakistan national scientific and technical documentation centre: Karachi.*

i. 1961. pp.[iii].ii.34+[iii].iii.50 + [iii].91. [1432.]

in progress.

PAKISTAN science abstracts. Pakistan national scientific & technical documentation centre: Karachi 1961 &c.*

in progress.

Poland

HENRYK ARCTOWSKI, Materyały do bibliografii prac naukowych polskich . . . dla Komitetu bibliograficznego Towarzystwa królewskiego londyńskiego. Warszawa 1897. pp.[ii].78. [700.]

KATALOG literatury naukowej [vol.xv &c.: literatury matematyczno-przyrodniczej] polskiej.

Matematyczno-przyrodniczego Akademii umiejętności [vol.xv &c.: Polska akademja umiejętności]: Kraków.

i. 1901. 1901. pp.[iii].106. [757.]
ii. 1902. 1902. pp.[iii].98. [670.]
iii. 1903. 1904. pp.[iii].102. [687.]
iv. 1904. 1905. pp.[v].86. [627.]
v. 1905. 1906. pp.[ii].79. [570.]
vi. 1906. 1907. pp.[iii].86. [617.]
vii. 1907. 1908. pp.[iii].76. [597.]
viii. 1908. 1909. pp.[iii].77. [578.]
ix. 1909. 1910. pp.[iii].88. [621.]
x. 1910. 1910–1911. pp.43+45–94. [720.]
xi. 1911. 1912. pp.[iii].91. [700.]
xii. 1912. 1913. pp.[iii].76. [650.]
xiii–xiv. 1913–1914. 1924. pp.84. [650.]
xv. 1915–1918. 1928. pp.[iv].130. [1000.]
xvi. 1919–1922. 1930. pp.[iv].239. [2000.]
xvii. 1923–1926. 1934. pp.[v].487. [5000.]
xviii. 1927–1930. 1936. pp.[v].507. [5000.]
xix. 1931–1934. 1939. pp.[vi].616. [6000.]
xx. 1935–1939. [*not yet published.*]
xxi. 1945–1949. 1952. pp.[vii].441. [9000.]

in progress.

QUARTERLY review of scientific publications of the Polish academy of sciences, the Ossolineum and the polish scientific publishers. Series C. Pure and technical sciences. Polish academy of sciences: Distribution centre for scientific publications: Warsaw.

1958. pp.30.33.33.36. [289.]
1959. pp.44.36.40.34. [338.]
1960. pp.52.68.8. [325.]
1961. pp.42.39.44.43.10. [347.]
1962.
1963. pp.115.106.108.124. [1249.]

in progress?

Russia

i. *Bibliographies and history*

M[INHALINA] P[ETROVNA] BRONSHTEIN and V[ERA] V[LADIMIROVNA] GNUCHEVA, Библиография естественно-научной литературы. Москва 1956. pp.192. [250.]

Z. V. GURKO and O. V. STULOVA, Развитие советской науки за 40 лет. Указатель юбилейной литературы 1957–1958 гг. Академия наук СССР: Фундаментальная библиотека общественных наук: Москва 1960. pp.88. [1579.]

ii. *General*

. R[UDOLF] KREBEL, Russlands naturhistorische und medicinische literatur. Schriften und abhandlungen in nichtrussischer sprache. Jena 1847. pp. vi.220. [science: 1250.]

the scope is wider than is suggested by the title.

УКАЗАТЕЛЬ русской литературы по математикѣ, чистымъ и прикладнымъ естественнымъ наукамъ, медицинѣ и ветеринаріи. Общество естествоиспытателей: Кіевъ.

i. 1872.

ii. 1873.

iii. 1874.

iv. 1875.

v. 1876. Составлена Л. Л. Лундомъ [and К. О. Румшевичемъ]. pp.viii.186+[ii]. ii.254. [10,000.]

vi. 1877. Составленъ В. К. Совинскимъ и Ф. Ф· Рындовскимъ. pp.viii.338. [7500.]

[*continued as:*]

Указатель русской литературы по математикѣ, чистымъ и прикладнымъ естественнымъ наукамъ.

vii. 1878. Составленъ В. К. Совинскимъ, подъ редакціею проф. Н. А. Бунге. pp.viii.181. [4000.]

viii. 1879. pp.viii.148. [3000.]

ix. 1880. pp.ix.224. [5000.]

x. 1881. pp.xviii.427. [7500.]

xi. 1882. pp.[ii].v.289. [6000.]

xii. 1883. pp.[iv].iv.257. [5000.]
xiii. 1884. pp.[iv].vi.265. [5000.]
xiv. 1885. pp.x.349. [7500.]
xv. 1886. pp.xi.389. [8000.]
xvi. 1887. pp.xi.331. [7500.]
xvii. 1888. pp.xii.379. [8000.]
xviii. 1889. pp.xii.433. [9000.]
xix. 1890. pp.xii.421. [9000.]
xx. 1891. pp.xxxvii.852. [12,500.]
no more published.

РУССКАЯ библіографія по естествознанію и математикѣ. Императорская академія наукъ. С.-Петербургское бюро международной библіографіи: Санктпетербургъ.
i. 1901. 1904. pp.[ii].xv.207. [3500.]
ii. 1901–1902. 1906. pp.[ii].vi.246. [4500.]
iii. 1903–1904. 1908. pp.[ii].vi.390. [7600.]
iv. 1905. 1910. pp.[ii].vi.223. [4295.]
v. 1906. 1912. pp.[ii].vi.196. [3654.]
vi. 1907. 1914. pp.[ii].vi.212. [3973.]
vii. 1908–1909. 1914. pp.[ii].viii.413. [8418.]
viii. 1910–1911. 1915. pp.x.432. [9155.]
ix. 1912–1913. 1917. pp.viii.426. [9000.]
no more published.

B[ORIS] V[ARNAVOVICH] IGNATEV, Что читать советскому учителю по естествознанию. Москва 1924. pp.48. [146.]

НАУЧНАЯ литература СССР. Систематический указатель книг и журнальных статей. Естествознание. Москва 1928 &c.
in progress.

G. M. AKIMOVA, Роль советской науки в борьбе за дальнейший расцвет нашей родины. Краткий рекомендательный указатель литературы. Государственная ордена Ленина [&c.]: Москва 1947. pp.32. [100.]

INHALTSVERZEICHNISSE sowjetischer fachzeitschriften. Reihe III. Naturwissenschaften, mathematik. Zentralstelle für wissenschaftliche literatur: Berlin.

i. 1952. pp.557. [2500.]
ii. 1953. pp.383. [1600.]
iii. 1954. pp.846. [3500.]
iv. Januar-Juni 1955. pp.493. [2000.]

continued in two parts, one of which is entered under Chemistry, the other below.

— Reihe III A. Mathematik, physik, astronomie, geophysik, geologie, geographie.*

v. 1956. pp.1195. [5000.]
vi. 1957. pp.1790.147. [6000.]
vii. 1958. pp.2056.103. [6500.]
viii. 1959. pp.2163.421. [6500.]
ix. 1960. pp.2046. [6500.]

in progress.

JAN HANÁK [*and others*], Sovětská matematická, astronomická, fysikalní a chemická literatura z let 1950–1953 v státních studijních knihovnách. Universita: Knihovna: Výběrový seznam (no.4): Brně 1953. pp.[114].xii. [753.]*

BIBLIOGRAPHY of translations from russian scientific and technical literature. List no.1[–39]. Library of Congress: Scientific translations center: Washington 1953–1956. pp.[576]. [7500.]*

— Author index.

1–15. [1955]. pp.[ii].30.
16–27. [1956]. pp.[ii].26.
28–39. [1957]. pp.[ii].29.

LIST of translations from russian, yugoslav, ukrainian, polish. Defence research board: Defence scientific information service: Ottawa [1955]. pp.[29]. [250.]*

— [another edition]. List of translations from russian, yugoslav, ukrainian, polish, japanese and chinese. [1959]. pp.[44]. [500.]*

GÜNTHER REICHARDT, Sowjetische literatur zur naturwissenschaft und technik. Bibliographischer wegweiser. Deutsche forschungsgemeinschaft: Wiesbaden 1957. pp.[iv].181. [2000.]

— — Zweite . . . auflage. 1959. pp.306. [3000.]

SOVIET science and technology. English table of contents of . . . all soviet journal [*sic*] being translated into english on a continuing basis. New York.*

i. 1957. pp.[ii].498. [10,000.]

ii. 1958. pp.66.82.68.88.56.32.38.46.38.144. [12,500.]

iii. 1959–1960. pp.86.62.58.70.62.66.106.54. 78.82.65.86. [15,000.]

[*continued as:*]

SST. Soviet science in translation.

LIST of books received from the USSR [and translated books]. Department of scientific and industrial research: Lending library unit.*

[1]. 1957. pp.[ii].39. [550.]

2–5. 1958. pp.iv.24+[ii].72+iii.72+iii.33. [2500.]

6–16. 1959. pp.[249]. [3000.]

17–28. 1960. pp.[248]. [3000.]

29–40. 1961. pp.[257]. [3000.]

41–52. 1962.

in progress.

[A. L. VAINSHTEIN *and others*], Документы ЦГИА СССР в Ленинграде в работах советских исследователей. Библиографический указатель. Статьи и публикации на русском языке в периодических и продолжающихся изданиах 1917–1957 гг. Центральный государственный исторический архив СССР: Ленинград 1960. pp.284. [1655.]

A GUIDE to the scientific and technical literature of eastern Europe. National science foundation: [Washington] 1962 [1963]. pp.[v].94. [440.]*

Z. V. GURKO and O. V. STULOVA, Развитие советской науки за 40 лет. Указатель юбилейной литературы 1957–1958 гг. Академия наук СССР: Фундаментальная библиотека Общественных наук: Москва 1960. pp.88. [1579.]

SOVIET science and technology. A bibliography on the state of the art 1955–1961. Library of Congress: Science and technology division: Washington 1962. pp.[ii].209. [2000.]*

Scotland

SUMMARIES of theses approved for higher degrees in the faculty of science during the academical year. University of Glasgow: Glasgow 1949–1950 [&c.].

details of this work are entered under Glasgow, university, of, above.

Slovakia

DUŠAN FRANCŮ and MAGDA HORSÁKOVÁ, Bibliografický prehl'ad pôvodnej slovenskej technickej a prírodovednej literatúry. Bratislava 1953. pp. 127. [1500.]

Spain

FELIPE PICATOSTE Y RODRÍGUEZ, Apuntes para una biblioteca científica española del siglo XVI. Estudios biográficos y bibliográficos de ciencias exactas, físicas y naturales y sus inmediatas aplicaciones en dicho siglo. Madrid 1891. pp.ix.419. [1007.]

Switzerland

BIBLIOGRAPHIE der schweizerischen naturwissenschaftlichen [vol.xvi &c.: und geographischen] literatur. Schweizerische landesbibliothek: Bern.

i. 1925. ff.xvi.152. [1750.]
ii. 1926. pp.xii.169. [1750.]
iii–iv. 1927–1928. pp.xvi.321. [3500.]
v. 1929. pp.xii.176. [2000.]
vi–vii. 1930–1931. pp.xii.262. [3000.]
viii. 1932. pp.xii.191. [2000.]
ix. 1933. pp.xii.206. [2000.]
x. 1934. pp.xii.251. [2500.]
xi. 1935. pp.xii.221. [2250.]

xii. 1936. pp.xii.243. [2500.]
xiii. 1937. pp.xii.196. [2000.]
xiv. 1938. pp.xii.210. [2000.]
xv. 1939. pp.xii.215. [2000.]
xvi. 1940. pp.xiii.211. [2000.]
xvii. 1941. pp.xiii.219. [2000.]
xviii. 1942. pp.xii.237. [2500.]
xix. 1943. pp.xii.256. [2500.]
xx. 1944. pp.xii.236. [2500.]
xxi. 1945. pp.xi.260. [2750.]
xxii. 1946. pp.xi.329. [3545.]
xxiii. 1947. Redaktion: Walter Rytz. 1949. pp.xi.339. [4016.]

[*continued as:*]

Bibliographia scientiae naturalis helvetica.

xxiv. 1948. pp.xi.435. [5062.]
xxv. 1949. pp.xi.457. [6300.]
xxvi. 1950. pp.xi.278. [3439.]
xxxvii. 1961. pp.xxiv.365. [4042.]
xxxviii. 1962. pp.xxiv.391. [4163.]

in progress.

United States

THE ANNUAL of scientific discovery . . . exhibiting the most important discoveries and improvements . . . together with a list of recent scientific publications; a classified list of patents; . . . an index of important papers in scientific journals, reports, etc. Boston.

1850. Edited by David A. Wells . . . and George Bliss. [Second edition]. pp.5–392. [500.]

1851. pp.xix.428. [600.]

1852. Edited by D. A. Wells. pp.xxiv.408. [500.]

1853. pp.411. [400.]

1854. pp.398. [400.]

1855. pp.394. [400.]

1856. pp.398. [400.]

1857. pp.406. [400.]

1858. pp.419. [400.]

1859. pp.410. [400.]

1860. pp.430. [400.]

1861. pp.424. [400.]

1862. pp.415. [400.]

1863. pp.343. [300.]

1864. pp.351. [300.]

1865. pp.355. [300.]

1866–1867. Edited by Samuel Kneeland. pp. 370. [300.]

1868. pp.331. [300.]

1869. pp.377. [300.]

1870. Edited by John Trowbridge. pp.354. [300.]

1871. pp.349. [300.]

the volumes of 1852–1871 do not contain the index of papers and are limited to the publications of the U.S.A.

LIST of references on the development of science in the United States. Library of Congress: Washington 1914. ff.5. [63.]*

DOCTORATES conferred in the sciences by american universities. National research Council: Reprint and circular series (no.12 &c.): Washington.

1919–1920. Compiled by Callie Hull. . . . (no.12). pp.19. [330.]
1920–1921. Compiled by C. Hull and Clarence J[ay] West . . . (no.26). pp.20. [336.]
1921–1922 . . . (no.42). pp.42. [455.]
1922–1925. [*not published*].
1925–1926 . . . (no.75). pp.34. [748.]
1926–1927 . . . (no.80). pp.36. [796.]
1927–1928 . . . (no.86). pp.38. [842.]
1928–1929 . . . (no.91). pp.[ii].46. [1025.]
1929–1930 . . . (no.95). pp.49. [1074.]
1930–1931 . . . (no.101). pp.55. [1147.]
1931–1932 . . . (no.104). pp.59. [1241.]
1932–1933 . . . (no.105). pp.63. [1343.]

the issue for 1921–1922 is entitled Doctorates conferred in the arts and sciences *&c.*

[THEODORE HORNBERGER], Science and the new world (1526–1800). An exhibition to illustrate the scientific contributions of the new world and the spread of scientific ideas in America. Henry E. Huntington library and art gallery: San Marino 1937. pp.18. [39.]

REGINALD ROBERT HAWKINS, *ed.* Scientific, medical, and technical books published in the United States of America, 1930–1944. National academy of sciences: National research council: Committee on bibliography of american scientific and technical books: Washington 1946. pp.xv.1114. [5193.]

—— Second edition. Books published to December 1956. National academy of sciences: National research council: Committee on bibliography of american scientific and technical books: Washington 1958. pp.xii.1491. [7954.]

BARTON BLEDSOE, Masters' theses in science, 1952. Washington [1954]. pp.252. [5588.]

PHYLLIS B. STECKLER, *ed.* American scientific books, 1960–1962. A basic selection. New York [1962]. pp.[v].474. [7500.]

FROM the United States. Books on science. Department of state: [Washington 1948]. pp. [iii].102. [330.]

BIBLIOGRAPHY of technical reports & articles of Manpower branch sponsored research classified by title of task. Office of naval research: Human resources division: Manpower branch: Washington 1952. ff.[ii].7. [70.]*

11. *Miscellaneous*

ADAM VETULANI, Bibliographie des études scientifiques publiées en Suisse par les internés militaires polonais. Brougg 1944. pp.13. [99.]

[MORRIS CECIL LEIKIND], The social impact of science: a select bibliography. With a section on atomic power. Senate: Committee on military affairs: Sub-committee on war mobilization: Monograph (no.3): Washington 1945. pp.vii.51. [525.]

D.S.I.R. consolidated list of translations. Department of scientific and industrial research.*

1940–1949. pp.[ii].78. [640.]

MABEL H. ELLER and JACK WEINER, Scientific personnel: a bibliography. . . . Edited by Barton

Bledsoe. Library of Congress: Navy research section: Washington 1950. pp.vii.164. [1746.]*

SCIENTIFIC manpower: supply, demand and utilization, 1940–1950. A selected and annotated bibliography. Research and development board: Washington [1951]. pp.iii.27. [242.]

— Supplement, 1950–1951. 1952. pp.iii.40. [219.]

[ELISABETH SAUVENIER-GOFFIN], Maîtres liégeois de l'illustration scientifique. Catalogue de l'exposition. Bibliotheca universitatis leodiensis: Publications (no.7): Liège 1955. pp.26. [52.]

SOCIOLOGY of science. Unesco: Current sociology (vol.v, no.2): Paris 1956. pp.[ii].91–153. [572.]

E[FIM SEMENOVICH] LIKHTENSHTEIN, Редактирование научной книги, некоторые вопросы издательской культуры. Москва 1957. pp.272. [180.]

CURRENT projects on the economic and other impacts of scientific research and development. National science foundation: [Washington].

1959. [By Theodore Suranyi–Unger and Elizabeth Harris]. pp.vi.53. [235.]

1960. pp.v.124. [300.]

1961. [By Th. Suranyi-Unger]. pp.vii.116. [262.]

in progress; the title varies.

BIBLIOGRAPHY of publications designed to raise the standard of scientific literature. Unesco: Documentation and terminology of science: [Paris 1963]. pp.83. [359.]

DIRECTORIES in science and technology. A provisional checklist. Library of Congress: Science and technology division: Washington 1963. pp. vi.65. [304.]*

Astronomy.

1. *Bibliographies*

EDWARD S[INGLETON] HOLDEN, Catalogue of the United States naval observatory, Washington. Part I.—Astronomical bibliography. Washington 1879. pp.10. [125.]

includes works not in the observatory.

N. B. LAVROVA, Библиография астрономических библиографий. Академия наук СССР: Астрономический совет [&c.]: Москва 1962. pp.[ii].vi.110. [248.]*

2. *Manuscripts and history*

[J. M. F.] GUÉRIN, Catalogue de manuscrits

orientaux en langue sanscrite sur l'astronomie ... recueillis dans les Indes orientales. 1855. pp.20. [63.]

JONAS GURLAND, Kurze beschreibung der mathematischen, astronomischen und astrologischen hebräischen handschriften der Firkowitsch'schen sammlung in der Kayserlichen öffentlichen bibliothek zu St. Petersburg. Neue denkmäler der jüdischen literatur (no.2): St. Petersburg 1866. pp.[iv].57. [100.]

HṚISHÍKEŚA ŚÁSTRÍ and ŚIVA CHANDRA GUI, A descriptive catalogue of sanskrit manuscripts in the library of the Calcutta sanskrit college. Volume ix. Astrology and astronomy manuscripts. Calcutta [1904–] 1906. pp.[ii].180.ii. [193.]

L[ÉON] BULTINGAIRE, Catalogue des incunables de la bibliothèque de l'Observatoire de Paris. 1913. pp.[ii].16. [38.]

A. DE POORTER and M. ALLIAUME, Catalogue des manuscrits mathématiques et astronomiques de la Bibliothèque de Bruges. Bruges 1922. pp.13–50. [100.]

E. ZINNER, Verzeichnis der astronomischen handschriften des deutschen kulturgebietes. Mün-

chen 1925. pp.544. [12,750.]
reproduced from handwriting.

ERNST ZINNER, Geschichte und bibliographie der astronomischen literatur in Deutschland zur zeit der renaissance. Leipzig 1941. pp.[vii].452. [5236.]

FRANCIS J[AMES] CARMODY, Arabic astronomical and astrological sciences in latin translation. A critical bibliography. Berkeley &c. 1956. pp.vi. 193. [2000.]

M[ARCEL] DESTOMBES, Globes célestes et catalogues d'étoiles orientaux du moyen-âge. Congrès international d'histoire des sciences: Florence [1956]. pp.11. [catalogues: 27.]

3. *Periodicals*

LIJST van tijdschriften in het astronomisch observatorium (sterrenwacht) te Utrecht. Rijksuniversiteit: Bibliotheek: Utrecht 1934. pp.41. [350.]

MAX REICHEL, Die fortlaufenden astronomischen veröffentlichungen in ihrer geschichtlichen entwicklung, mit einer gesamtbibliographie. Bibliothekar-lehrinstitut des landes Nordrhein-Westfalen (no.12): Köln 1957. pp.124. [1250.]*

4. *General*

JO[HANN] FRIEDRICH WEIDLER, Bibliographia astronomica. Wittenbergae 1755. pp.[vii].126. [x].44. [1250.]

[JOHANN EPHRAIM SCHEIBEL], Astronomische chronologische bibliographie. Einleitung zur mathematischen bücherkenntniss (vol.iii = parts 13–20): Breslau 1784–1798. pp.[xxv].155+[vi]. 182+[iii].197–254+[iii].232+233–380. [2000.]

limited to works printed up to 1650; also published independently under the title of Astronomische bibliographie, *with the author's name on the title-page, and with shorter preliminary matter.*

[JOSEPH] JÉRÔME [LE FRANÇAIS] DE LA LANDE, Bibliographie astronomique. [1803]. pp.[ii].viii. 966. [5000.]

J[EREMIAS] D[AVID] REUSS, Repertorium commentationum a societatibus litterariis editarum. . . . Tom. v. Astronomia. Gottingae 1804. pp.viii.548. [6000.]

ÅRSBERÄTTELSE i astronomien. Af S. A. Cronstrand. [Kongliga svenska vetenskaps academi]: Stockholm.

1826. pp.[iii].94. [75.]
1827. pp.[ii].66. [50.]

1828. pp.[iv].67. [50.]
1829. pp.
1830. pp.[iv].111. [50.]
1831. pp.iv.62. [50.]
1832. pp.
1833. pp.[ii].ii.98. [75.]
1834. pp.iv.31. [50.]
1835. pp.[ii].ii.46. [50.]
1836. pp.[iv].56. [75.]

[*continued as:*]

Berättelse om astronomiens framsteg. Af N. H. Selander.

1837–1841. pp.[ii].iii.137. [100.]
1842.

earlier issues formed part of the Academy's Årsberättelse; *the dates shown are those of publication.*

F[RIEDRICH] G[EORG] W[ILHELM] S[TRUVE], Librorum in bibliotheca Speculae pulcovensis contentorum catalogus systematicus. Petropoli 1845. pp.[ii].xlviii.458. [5411.]

— — [another edition]. Edendum curavit . . . Otto [Wilhelm] Struve. 1860. pp.[ii].xxx.970. [20,000.]

— — — [supplement]. Pars secunda, ab Eduardo Lindemanno elaborata. Edendum curavit . . . O. Struve. 1880. pp.xiii.640. [15,000.]

CATALOGUE des livres de la bibliothèque de l'observatoire royal. Bruxelles 1847. pp.[vii].80. [800.]

CATALOGUE of the library of the Royal astronomical society. 1850. pp.[iii].140. [2500.]

— [another edition]. [By W. H. Wesley]. 1886. pp.iv.408. [6000.]

— — Supplementary catalogue. 1900. pp.[iii]. 175. [3000.]

— — — [By — Williams]. 1926. pp.[iii].176. [3000.]

[MORITZ LUDWIG GEORG WICHMANN], Systematisches verzeichniss der in der bibliothek der Königl. universitätssternwarte zu Königsberg enthaltenen bücher. 1852. pp.102. [1738.]

A[ugust] L[udwig] Busch, whose name appears on the titlepage, is not the author; see the 'vorbemerkung'.

ADOLPH BÜCHTING, Bibliotheca astronomica et meteorologica, oder verzeichniss der . . . in den letzten x jahren . . . im deutschen buchhandel erschienenen bücher und zeitschriften. Nordhausen 1872. pp.55. [900.]

[ALBERT LANCASTER], Catalogue des ouvrages d'astronomie et de météorologie qui se trouvent dans les principales bibliothèques de la Belgique. Observatoire royal: Bruxelles 1878. pp.[iii].xxiii. 645. [Astronomy: 3171.]

[KARL CHRISTIAN BRAHMS], Catalog der bibliothek der Astronomischen gesellschaft. [Vierteljahrsschrift der Astronomischen gesellschaft (vol. xiv, supplementheft):] Leipzig 1880. pp.vii.75. [1000.]

H. G. V. D. [HENDRIK WILLEM VAN DE] SANDE BAKHUYZEN, Catalogus van de boeken op 1 Januari 1877 aanwezig in de bibliotheek der Sterrenwacht te Leiden. Annalen der Sterrenwacht (bijvoegsel): 's-Gravenhage.

— Eerste supplement. 1881. pp.[ii].74. [1000.]

— Tweede supplement. 1882. pp.[iv].98. [1250.]

— Deerde supplement. 1893. pp.[v].198. [2250.]

— Vierde supplement. 1902. pp.vii.152. [2000.]

an index to the main work and the first three supplements is included in the third of these.

J[EAN] C[HARLES] HOUZEAU, Vade-mecum de l'astronome. Observatoire royal de Belgique: Annales: (appendice): Bruxelles 1882. pp.xxviii. 1144. [3447.]

J[EAN] C[HARLES] HOUZEAU and A[LBERT] LANCASTER, Bibliographie générale de l'astronomie. Bruxelles 1882–1889. pp.vii.858+viii.cxx.

859–1623+[iii].lxxxix.87.coll.2225. [42,500.]
the last volume was published first.

CATALOGUE of Nautical almanach library. Bureau of navigation: Washington 1883. pp.110. [1000.]
many pages are blank.

[A. R. KREISBERG and M. I. KUDRYASHEV], Catalogus alphabeticus librorum qui in bibliotheca speculae Imperialis literarum universitatis petropolitanae asservantur. Протоколы Засѣданій совѣта Императорскаго С.-Петербургскаго университета (no.37, Приложеніе): Petropoli 1888. pp.[iv].74. [1250.]

CATALOGUE of the library. Astronomical society: Liverpool 1888. pp.20. [370.]

CATALOGO de la biblioteca del Instituto y observatorio de marina de San Fernando. San Fernando 1889. pp.viii.360. [3375.]

WILLIAM C[RAWFORD] WINLOCK, Bibliography of astronomy: for the year 1887. Smithsonian institution: Miscellaneous collections (no.664): Washington 1888. pp.63. [2500.]

CATALOGUE of the Crawford library of the Royal observatory. Edinburgh 1890. pp.viii.499. [10,000.]

JAHRBUCH der astronomie und geophysik. Enthaltend die wichtigsten fortschritte auf den gebieten der astrophysik, meteorologie und physikalischen erdkunde. Leipzig.

i. 1890. Herausgegeben von Hermann J. Klein. pp.xvi.356. [300.]
ii. 1891. pp.xi.400. [250.]
iii. 1892. pp.x.349. [250.]
iv. 1893. pp.viii.360. [200.]
v. 1894. pp.x.351. [250.]
vi. 1895. pp.ix.376. [250.]
vii. 1896. pp.x.400. [250.]
viii. 1897. pp.viii.370. [250.]
ix. 1898. pp.viii.384. [250.]
x. 1899. pp.x.392. [300.]
xi. 1900. pp.viii.379. [250.]
xii. 1901. pp.viii.416. [250.]
xiii. 1902. pp.viii.366. [250.]
xiv. 1903. pp.viii.368. [250.]
xv. 1904. pp.viii.344. [250.]
xvi. 1905. pp.viii.368. [250.]
xvii. 1906. pp.viii.402. [250.]
xviii. 1907. pp.viii.372. [250.]
xix. 1908. pp.viii.384. [250.]
xx. 1909. pp.viii.416. [250.]
xxi. 1910. pp.viii.352. [200.]
xxii. 1911. pp.viii.392. [250.]

xxiii. 1912. pp.viii.368. [250.]
[*continued as:*]
Kleins jahrbuch der astronomie [&c.].
xxiv. 1913. Herausgegeben von Theodor Arldt. pp.xii.[iii].384. [3000.]
no more published.

EDWARD S[INGLETON] HOLDEN, Catalogue of the library of the Lick observatory [at Mount Hamilton] of the university of California. Part 1 — to July 1, 1890. Sacramento 1891. pp.122. [1750.]
no more published.

BIBLIOGRAPHIA hopkinsiensis, 1876–1893. Parts IV, V, and VI. Physics, astronomy, mathematics. Johns Hopkins university: Baltimore 1894. pp.vii. 57. [1500.]

BIBLIOGRAPHIA astronomica. Office international de bibliographie: Bibliographia universalis: Bruxelles 1897. pp.15.ff.16. [150.]

ASTRONOMISCHER jahresbericht. Berlin.
i. 1899. Herausgegeben von Walter F. Wislicenus. pp.xxiii.537. [1768.]
ii. 1900. pp.xxvi.633. [2320.]
iii. 1901. pp.xxxii.673. [2513.]
iv. 1902. pp.xxxiii.650. [2411.]
v. 1903. pp.xxxv.663. [2582.]

vi. 1904. pp.xxxviii.615. [2280.]
vii. 1905. Herausgegeben von A[dolf] Berberich. pp.xxxviii.647. [2336.]
viii. 1906. pp.xxxvi.672. [1961.]
ix. 1907. pp.xxxvi.654. [1854.]
x. 1908. pp.xxxviii.708. [1876.]
xi. 1909. pp.xxxvi.727. [1988.]
xii. 1910. Bearbeitet im Kgl. astronomischen rechen-institut. pp.xxviii.602. [2000.]
xiii. 1911. pp.xxvi.492. [1500.]
xiv. 1912. pp.xxv.564. [2000.]
xv. 1913. pp.xxvii.545. [2000.]
xvi. 1914. pp.xxiv.391. [1500.]
xvii. 1915. pp.xvi.299. [1000.]
xviii. 1916. pp.xvi.494. [2000.]
xix. 1917. pp.xiii.344. [1250.]
xx. 1918. pp.xii.323. [1250.]
xxi. 1919. pp.xii.397. [1250.]
xxii. 1920. pp.xii.280. [1000.]
xxiii. 1921. pp.xii.347. [1500.]
xxiv. 1922. pp.xviii.230. [2000.]
xxv. 1923. pp.xviii.198. [2000.]
xxvi. 1924. pp.xx.300. [3000.]
xxvii. 1925. pp.xiv.250. [2500.]
xxviii. 1926. pp.xxii.240. [2500.]
xxix. 1927. pp.xxv.270. [2500.]
xxx. 1928. pp.xxvii.276. [2500.]

xxxi. 1929. pp.xxvi.274. [2500.]
xxxii. 1930. pp.xxviii.295. [3000.]
xxxiii. 1931. pp.xxix.261. [2500.]
xxxiv. 1932. pp.xxx.261. [2500.]
xxxv. 1933. pp.xxxii.267. [2500.]
xxxvi. 1934. pp.xxxii.287. [3000.]
xxxvii. 1935. pp.xxxvi.341. [3500.]
xxxviii. 1936. pp.xxxvi.401. [4000.]
xxxix. 1937. pp.xxxvii.329. [3500.]
xl. 1938. pp.xl.359. [4000.]
xli. 1939. pp.xxxiii.301. [3500.]
xlii. 1940. pp.xxxiv.258. [3000.]
xliii. 1941. pp.xxvi.235. [3000.]
xliv. 1942. 1947. pp.xix.151. [2000.]
xlv–xlvi. 1943–1946. 1948–1951. pp.xlii.333 +lv.268. [7500.]
xlvii. 1947. pp.xxvii.208. [3000.]
xlviii. 1948. pp.xliii.364. [5000.]
xlix. 1949. 1952. pp.l.420. [6000.]
l. 1950. 1953. pp.lxiii.445. [7000.]
li. 1951. 1954. pp.lviii.468. [5000.]
lii. 1952. 1955. pp.lviii.454. [5000.]
liii. 1953. 1955. pp.lxii.468. [5000.]
liv. 1954. Bearbeitet von K. Heinemann, F. Henn, W. Lohmann. 1956. pp.lxiv.517. [5000.]
lv. 1955. 1957. pp.xxxiii.542. [5000.]

lvi. 1956. 1958. pp.xxxvii.503. [5000.]
lvii. 1957. 1959. pp.xxxviii.525. [5000.]
lviii. 1958. 1960. pp.xliii.567. [5000.]
lix. 1959. Bearbeitet von W. Lohmann, F. Henn. 1961. pp.xv.517. [5000.]
lx. 1960. 1962. pp.xvi.656. [6000.]
lxi. 1961. Bearbeitet von W. Lohmann, F. Henn, U. Güntzel-Lingner. 1963. pp. xvi.668. [6000.]

in progress.

PUBLICATIONS faites à l'observatoire des missionnaires de la Compagnie de Jésus, à Zi-ka-wei, près Chang-hai, Chine. [*c.*1901]. pp.4. [43.]

ASTRONOMY. International catalogue of scientific literature (section E): Royal society.

i. 1902. pp.xiii.303. [2096.]
ii. 1903. pp.viii.223. [1223.]
iii. 1904. pp.viii.304. [1909.]
iv. 1905. pp.viii.246. [1391.]
v. 1906. pp.viii.298. [2036.]
vi. 1907. pp.lv.232. [1946.]
vii. 1909. pp.viii.246. [1750.]
viii. 1910. pp.viii.226. [1749.]
ix. 1911. pp.viii.239. [1961.]
x. 1912. pp.viii.276. [2418.]
xi. 1913. pp.viii.234. [2013.]

xii. 1914. pp.viii.230. [1656.]
xiii. 1915. pp.viii.204. [1582.]
xiv. 1918. pp.viii.175. [1310.]
no more published.

V[INCENZO] MESSERI, Catalogo della biblioteca dell' Osservatorio astronomico di Arcetri. R. Istituto di studi superiori pratici e di perfezionamento: Sezione di scienze fisiche e naturali: Pubblicazioni (no.27): Firenze 1909. pp.203. [2000.]

A[UGUSTE] COLLARD, Catalogue alphabétique des livres, brochures et cartes. Bibliothèque de l'Observatoire royal de Belgique, à Uccle: Bruxelles 1910–1912 [1913]. pp.xxiv.698+716. [13,447.]

— — Accroissements de 1910–1912. 1913. pp.131. [1105.]

— — — 1913–1922. 1927. pp.432.xlviii. [4199.]

W[ILLIA]M D. HORIGAN, List of publications issued by the United States naval observatory, 1845–1908. Washington 1911. pp.[ii].36. [350.]

ASTRONOMICAL papers . . . for sale. Superintendent of documents: Price list (no.57): Washington 1912. pp.20. [150.]

——5th edition. Astronomy. 1917. pp.8. [150.]

no more published in this form.

LITERARISCHES beiblatt zu den Astronomischen nachrichten. Herausgegeben von H[ermann] Kobold. Kiel.

i. 1912–1913. pp.[iii].coll.x.176. [500.]
ii. 1913–1914. pp.[ii].coll.xii.192. [500.]

no more published.

PHYLLIS A. CASEY, Stories of the constellations. A list of books. Public library: St. Louis 1920. pp.8. [200.]

AUGUSTE COLLARD, L'astronomie et les astronomes. Répertoires des ouvrages à consulter: Bruxelles 1921. pp.iii–viii.3–119. [758.]

CATALOGUE of the [Edmond Herbert] Grove-Hills library. Royal astronomical society: 1924. pp.iii.48. [500.]

limited to works published before 1700.

ARTHUR JOHN HAWKES, The solar eclipse of 1927. Catalogue of books on astronomy in the Wigan public libraries. Wigan 1927. pp.14. [400.]

BULLETIN bibliographique. . . . Tome 1. Fasc. 1[–3]. Association française d'observateurs

d'étoiles variables: [Lyon] 1932. pp.88. [1000.]
no more published.

BIBLIOGRAPHIE mensuelle de l'astronomie. Société astronomique de France.
i. 1933. pp.236. [4000.]
ii. 1934. pp.230. [4000.]
iii. 1935. pp.264. [5000.]
iv. 1936. pp.266. [5000.]
v. 1937. pp.254. [5000.]
vi. 1938. pp.266. [5000.]
vii. 1939. pp.207. [3500.]
viii. nos.1–11. pp.204. [3500.]
ix. nos.1–9. pp.96. [1500.]
x. nos.1–9. pp.60. [1000.]
xi. nos.1–3. pp.32. [400.]
no more published?

CATALOGUE of the library of the British astronomical association. 1936. pp.94. [2000.]

READERS' guide to books on astronomy. Library association: County libraries section [no.15]: 1938. pp.[i].7. [150.]

OTTO SEYDL, *ed.* Knihovna astronoma Antonína Strnada, ředitele Pražské hvězdárny (1746–1799). Podle rukopisu knihovny královské kanonie Premonstrátů na Strahově v Praze [&c.]. Publikace Pražské hvězdárny (vol.xiii): Praha 1939. pp.80. [895.]

K[ONSTANTIN] L[VOVICH] BAEV, Астрономия. Государственная библиотека СССР им. В. И. Ленина: Книга о лучших книгах: [Moscow] 1941. pp.128. [100.]

D. TEN BRUGGENCATE [*and others*], Astronomy, astrophysics and cosmogony. Office of military government for Germany: Field information agencies technical: Fiat review of german science, 1939–1946: Wiesbaden 1948. pp.[xi].441. [1500.]
the text is in german.

V[ITALY] A[LEKSEEVICH] SHISHAKOV, Строение вселенной. Краткий обзор дитературы. Библиотека СССР имени З. И. Ленина: Москва 1950. pp.48 [25.]

РЕФЕРАТИВНЫЙ журнал. Астрономия [from 1954: Астрономия, геодезия]. Академия наук СССР: Институт научной информации: Москва.

1953. coll.208.pp.iv + coll.208.pp.iv + coll. 228.pp.v. [1468.]

1954. pp.100.iv + 89.vi + 94.vi + 84.iv + 78.v + 76.iv + 76.iv + 72.iv + 83.iv + 80.iv+99.iv+92. [5900.]

Astronomy

— Предметный указатель 1953–1954. pp.71.

1955. pp.80.v + 84.iv + 108.v + 106.vi + 88.iv + 103.iv + 108.v + 117.vi + 105.vi + 124.v+98.vi+73+69. [5125.]

1956. pp.176.iv + 146.v + 140.iv + 113 + 72 + 80 + 81 + 81+76+84+93+81. [7110.]

1957. pp.114 + 136 + 108 + 119 + 137 + 136 + 114 + 129 + 112 + 116 + 112 + 101. [10,000.]

1958. pp.119 + 96 + 100 + 108 + 85 + 93 + 85 + 100 + 105 + 104 + 129 + 96 + 84 + 81. [7902.]

1959. pp.114 + 124 + 119 + 116 + 120 + 132 + 139 + 114 + 127 + 156 + 162 + 116. [9700.]

1960. pp.138 + 128 + 120 + 120 + 132 + 136 + 168 + 164 + 156 + 159 + 132 + 135 + 120 + 174 + 100. [12,850.]

1961. pp.80.40 + 80.40 + 80.39 + 96.48 + 79.40 + 80.40 + 80.40 + 96.36 + 88.40 + 104.48+96.48. [12,000.]

1962. pp.96.32 + 80.40 + 96.44 + 89.43 + 84.47 + 88.44 + 108.36 + 108.36 + 88.40 +92.48 + 108.40 + 92.44 + 181 + 96. [15,000.]

[*continued as:*]

Реферативный журнал. Отдельный выпуск. 51. Астрономия.

1963. pp.103+92+88+96+92+100+96+83+92+92+92+96. [15,000.]

in progress.

MIROSLAV PETRTÝL, Astronomie. Výběrový seznam populárně vědecké literatury. Městska lidová knihovna: Praha 1955. ff.[i].13. [60.]*

[EUGENIUSZ RYBKA *and others*], Polska bibliografia analityczna. Astronomia-geodezja-geofizyka. Polska akademia nauk: Ośrodek bibliografii i dokumentacji naukowej: Warszawa.

i. 1955. pp.   . [246.]
ii. 1956. pp.48.88. [328.]
iii. 1957. pp.159. [356.]

no more published.

ASTRONOMIE. Association pour la diffusion de la pensée française: Bibliographie française établie à l'intention des lecteurs étrangers: 1956. pp.100. [400.]

limited to works published since 1949.

S. A. SHORVIGIN, Что читать по физике и астрономии. Аннотированный указатель. Москва 1961. pp.224. [2500.]

5. *Cartography*

[J. H. HEUSER], Katalog von kartographischen werken . . . und veranschaulichungsmitteln für den unterricht in der astronomischen geographie. Neuwied &c. 1877. pp.[viii].xxxvi.283. [astronomy: 35.]

BASIL [J. W.] BROWN, Astronomical atlases, maps & charts. An historical & general guide. 1932. pp.200. [1000.]

6. *Radio astronomy*

F[RANK] R[OYLE] TAYLOR, Radio astronomy. Library association: Special subject list (no.26): 1958. pp.15. [204.]*

N. G. NEISHILD and B. N. PANOVKIN, Радиоастрономия. Аннотированный библиографический указатель отечественной и иностранной литературы 1932–1958 гг. Библиотека физического института им. П.Н. Лебедева: Москва 1960. pp.216. [2000.]

RADIO astronomy antennas and processing of radio astronomy data. Annotated bibliography of soviet literature. Library of Congress: Aerospace information division: AID report (no.B–64–1): [Washington] 1964. pp.iii.ff.42. [97.]*

7. *Miscellaneous*

[VICTOR MAUVAIS], Notice sur les travaux astronomiques de m. Victor Mauvais. [1843]. pp.6. [12.]

[URBAIN JEAN JOSEPH LE VERRIER], Exposé sommaire des recherches astronomiques de m. U.-J. Le Verrier. [1834]. pp.22. [11.]
—— [another edition]. [1845]. pp.8. [17.]

EDWARD S[INGLETON] HOLDEN, Index catalogue of books and memoirs relating to nebulæ and clusters, &c. Smithsonian institution: Smithsonian miscellaneous collections (no.311): Washington 1877. pp.[iii].ix.109. [1500.]

[GUSTAVE LEVEAU], Notice sur les travaux astronomiques de m. Gustave Leveau. [1884]. pp.8. [20.]

F[RIEDRICH WILHELM] RISTENPART, Verzeichniss von 336 sterncatalogen. Breslau 1901. pp.41. [336.]

PAUL W[ILLARD] MERRILL and CORA G[ERTRUDE] BURWELL, Catalogue and bibliography of stars of classes B and A whose spectra have bright hydrogen lines. Carnegie institution of Washington: Mount Wilson observatory: Contributions (no.471): [Chicago 1933]. pp.54. [363.]
—— Supplement. . . . (no.682): [1943]. pp.32. [407.]

R[ICHARD] PRAGER, Geschichte und literatur des lichtwechsels der veränderlichen sterne. Zweite ausgabe, enthaltend die literatur der jahre 1916–1933. Veröffentlichung der Universitätssternwarte Berlin-Babelsberg: Berlin 1934–1957. pp.xv.318 +viii.354+xv.360+xii.549. [20,000.]

— — History and bibliography of the light variations of variable stars. Second edition, supplementary volume . . . 1931–1938. Harvard college: Astronomical observatory: Annals (vol.cxi): Cambridge, Mass. 1941. pp.251. [4000.]

ERNST ZINNER, Geschichte und bibliographie der astronomischen literatur in Deutschland zur zeit der renaissance. Leipzig 1941. pp.[vii].452. [6000.]

WILLIAM P. BIDELMAN, Catalogue and bibliography of emission-line stars of types later than B. Astrophysical journal: Supplement series (no.7): Chicago 1954. pp.[ii].175–267. [903.]*

S. N. KORUITNIKOV, M. I. LAVROV and D[MITRY] YA[KOVLEVICH] MARTUINOV, Библиография спектрально-двойных звезд. Академия наук СССР: Астрономический совет: Москва 1961 &c.

in progress.

METEOROLOGICAL & geoastrophysical titles. American meteorological society: [Washington] 1961 &c.

in progress.

SOVIET infrared stellar spectroscopy. Annotated bibliography of the period 1947–1962. Library of Congress: Aerospace information division: AID report (no.62–153): [Washington] 1962. ff.ii.46. [127.]*

Atom, atomic energy. [*see also* **Nuclear physics**].

1. *Bibliographies*

PAUL E. POSTELL and HUGH E. VORESS, Unclassified bibliographies of interest to the atomic energy program. Atomic energy commission: Technical information service: Oak Ridge, Tenn. 1953. pp.vi.56. [500.]

2. *Periodicals*

M. INAGAKI-TROUW, Catalogue des périodiques de la bibliothèque du CERN. Organisation européenne pour la recherche nucléaire: CERN bibliography (no.5): Genève 1960. pp.91.*

3. *General*

WALTER GERLACH, Materie, elektrizität, energie. Die entwickelung der atomistik in den letzten zehn jahren. Wissenschaftliche forschungsberichte: Naturwissenschaftliche reihe (vol.iii): Dresden &c. 1923. pp.viii.195. [1000.]

—— Zweite . . . auflage. 1926. pp.xii.296. [1500.]

PRELIMINARY cumulative list of documents in unrestricted series issued by the General assembly, Security council, Atomic energy commission, Economic and social council and its commissions, International health conference, to 31 July 1946. [Lake Success] 1946. pp.[iii].83. [1500.]*

ATOMIC energy: a selected list of references. Library of Congress: Washington 1946. ff.21. [228.]*

ATOMIC energy: a brief list of books and pamphlets. Library of Congress: Washington 1946. ff.2. [16.]*

GUIDE to published research on atomic energy. United States atomic energy commission: Technical information on division: [Oak Ridge, Tenn.]*

i.

ii.

iii. 1948. pp.569. [1623.]

[*continued as:*]
Nuclear science abstracts.
in progress.

ATOMIC energy. House of commons: Library: Bibliography (no.30): [1947]. ff.10. [60.]*

ATOMIC energy: significant references. Library of Congress: Legislative reference service: Washington.*

i. 1946–1947. [Compiled by Janie Ellis Mason]. ff.(pp.)[106]. [750.]
ii. 1947–1948. ff.(pp.)[320]. [2131.]
iii. 1949. ff.[ii].193. [1378.]
iv. 1950. ff.207. [1383.]
v. January–June 1951. ff.143. [864.]

• *no more published.*

ISRAEL LIGHT, Annotated bibliography on atomic energy . . . for schools and discussion groups. Columbia university: Teachers college: New York 1947. pp.29. [257.]*

ABSTRACTS of declassified documents. United States atomic energy commission: Technical information service: Oak Ridge.*

i. 1947. pp. [ii].813. [1506.]
ii. 1948. pp.[ii].379. [717.]

SELECTED bibliography on atomic energy. [Second edition]. Atomic energy commission: Technical information division: Washington 1948. pp.[iii].15. [150.]

— [another edition]. 1954. pp.[iii].21. [150.]

AN INTERNATIONAL bibliography on atomic energy. United nations: Atomic energy commission: Lake Success.

i. Political, economic and social aspects. 1949. pp.ix.45. [1000.]

— Supplement no. 1. 1950. pp.v.22. [294.]

— Supplement no.2. 1953. pp.v.31. [394.]

ii. Scientific aspects. 1951. pp.xiii.[864]. [24,282.]*

— Supplement no.1. 1952. pp.vii.[300]. [8231.]*

— Supplement no.2. 1953. pp.vii.[300]. [7997.]*

NUMERICAL index of AEC reports, with indications of their availability. Atomic energy commission: Technical information service: Oak Ridge, Tenn. 1950. pp.1187–1226. [2000.]

— Fifth edition. Public availability of reports in Nuclear science abstracts. 1960. pp.ii.323+[i]. 324–618. [37,500.]*

each edition has a different title.

SELECTED readings on atomic energy. Atomic energy commission: Washington 1951. pp.iii.23. [175.]

RÉPERTOIRE analytique à l'usage des physiciens nucléaires. Commission à l'énergie atomique: Service de physique nucléaire: Fontenay-aux-Roses 1950–1951 &c.

INDEX to documents, United nations: Atomic energy commission: Official records: New York [1952]. pp.[ii].71. [3000.]

GEORGE P. PERROS, Preliminary inventory of the records of the Special committee of the Senate on atomic energy, 1945–46. National archives: Preliminary inventories (no.62): Washington 1953. pp.v.8. [large number.]*

CHECKLIST of United Nations documents, part 3, Atomic energy commission, 1946–1952. United nations: New York 1953. pp.vii.46. [400.]

V[ÁCLAV] ČEJCHAN, *ed.* Atomová energie—zbran míru ne války. Vyběrový seznam literatury knižní a časopisecké. Universitní knihovna: Bibliografické přiručky (no.3): Praze 1954. pp.[iv].38. [129.]*

SELECTED readings on atomic energy. United States atomic energy commission: Washington 1954. pp.iii.21. [250.]

— [another edition]. 1955. pp.[iii].53. [350.]

JANE E. BOSWELL, A bibliography of current materials dealing with atomic power and related atomic energy subjects for non-specialists. University of Michigan and Fund for peaceful atomic development: [Ann Arbor] 1955. pp.[viii].236. [2500.]*

ПЕРЕЧЕНЬ экспонируемой литературы. [International conference on the peaceful uses of atomic energy. Geneva 1955]. pp.89. [531].

ARTICLES published in technical journals and papers presented at scientific meetings during 1954 by ORNL staff members. National laboratory: Oak Ridge, Tenn. 1955. pp.[ii].26. [470.]*

LIST of publications. Reports on atomic energy published by Atomic energy of Canada limited. Chalk River, Ont. [1955]. ff.15. [140.]*

UNCLASSIFIED papers appearing in technical journals 1949–1954. University of California: Los Alamos scientific laboratory: Division of chemistry and metallurgy: Los Alamos [1955]. pp.16. [119.]

[BORIS NIKOLAEVICH RZHONSNITSKY and EVA ISAEVNA GENDELMAN], Атомная энергия и ее применение в мирных целях. Рекомендательный указатель литературы. Государственная ордена Трудового Красного Знамени публичная библиотека им. М. Е. Салтыкова-Щедрина: Ленинград 1955. pp.63. [100.]

LIST of translations issued by the library. Atomic energy research establishment: Harwell 1955. pp.5. [40.]*

LIST of publications available to the public. Atomic energy research establishment: Library: Harwell 1955 &c.*

in progress; issued monthly, with irregular cumulations and supplements, and also in a different form as A. E. R. E. unclassified reports.

[HELMUT VOGT], Bücher zum studium der atomphysik. Deutsche staatsbibliothek: Bibliographische mitteilungen (no.10): Berlin 1956. pp.[v].47. [275.]*

L. I. SOROKINA, Атомная энергия и ее

использование. Рекомендательный указатель литературы. Издание второе. Государственная ... библиотека СССР имени В. И. Ленина: Москва 1956. pp.40. [100.]

PEACEFUL uses of atomic energy. United States information agency: Information center service: Subject bibliography (no.14): [Washington] 1956. ff.[i].14. [65.]*

NONNA NIKOLAEVNA TIMOFEEVA, Мирное применение атомной энергии. Рекомендательный список литературы для учащихся 7–10 классов. Государственная . . . публичная библиотека имени М. Е. Салтыкова-Щедрина: Ленинград 1956. pp.28. [50.]

A[LEKSEI] N[IKOLAEVICH] BUCHENKOV, Меченые атомы и их применение в народном хозяйстве. Рекомендательный обзор литературы. Государственная . . . библиотека СССР имени В. И. Ленина [&c.]: Москва 1956. pp.14. [23.]

CATÁLOGO de Biblioteca de energia atomica. Comissão de energia atomica: Rio de Janeiro.

i. 1956. pp.207. [1571.]
ii. 1957. pp.587. [4589.]

L'ÉNERGIE nucléaire. Organisation européenne

de coopération économique: Presse-documentation: Bibliographies spéciales (no.31): 1956. pp.96. [700.]*

— Supplément. 1957. pp.81. [600.]*

ÉNERGIE nucléaire 1946–1955. Haute autorité du charbon et de l'acier: Bibliographies (no.1): [Luxembourg 1956]. ff.[i].9. [215.]*

— [another edition]. Énergie nucléaire 1946–1957. . . . (no.13): [1958]. ff.iv.82. [800.]*

— — Supplément. 1958. pp.iv.33. [250.]

BIBLIOGRAPHIE scientifique hebdomadaire. Commissariat à l'énergie atomique: Service de documentation: Saclay 1956 &c.*

in progress; consists of very numerous and voluminous parts.

UNCLASSIFIED reports of limited value. United States atomic energy commission: Technical information service: Oak Ridge, Tenn. 1957. pp.iii.104. [2000.]*

UNITED KINGDOM atomic energy authority unclassified reports, 1946–December 1956. [1957]. pp.[30].[650.]*

C. G. ALEXANDRIDES, Atomic energy as a means of economic development. A selected bibliography. New York 1957. pp.iii.40. [600.]*

LIST of publications. Joint establishment for nuclear energy research: Kjeller 1957. ff.[6]. [65.]*

A[LEKSEI] N[IKOLAEVICH] BUCHENKOV, Атомная энергетика в шестой пятилетке. Реко. мендательный обзор литературы. Государственная . . . библиотека СССР имени В. И, Ленина: Новости техники (no.13): Москва 1957. pp.20. [20.]

MEMBERSHIP, list of atomic energy legislation- and committee publications. Joint committee on atomic energy: Washington.

1957. pp.ii.12. [150.]
1958. pp.iii.15. [250.]

EURATOM. Bibliographie. Europäisches parlament: [Luxemburg].*

i. [1955–1957]. pp.[ii].iv.105. [757.]
ii. [1957–1958]. pp.[ii].iv.81. [531.]
iii. [1958]. pp.[ii].80. [650.]

HUGH E. VORESS and NAOMI K. SMELCER, Bibliographies of interest to the atomic energy program. United States atomic energy commission: Technical information service: [Oak Ridge, Tenn. 1958]. pp.xiv.149. [1556.]

LIST of U. S. technical papers presented at

Second United nations international conference on the peaceful uses of atomic energy, Geneva. Washington [1958]. pp.16. [750.]

ATÓMOVÉ nebezpečenstvá a ochrona proti nim. Odporúčajúai zoznam literatúry. Ústredná technická knižnica: Publikácie (no.51): Bratislava 1958. pp.[ii].41. [150.]*

ATOMIC energy. Surrey county library: Book list (no.12): [Esher 1958]. pp.[iv].20. [125.]

N. [B]. SKEATS, The United Kingdom atomic power programme. United Kingdom atomic energy authority: Research group: A. E. R. E. reading list (no.2): Harwell 1958. pp.[ii].9. [100.]*

•N[ATALYA] I[VANOVNA] KRASTING, Крылатые и атомные суда. Рекомендательный обзор литературы. Государственная... библиотека СССР имени В. И. Ленина: Новости техники (no.15): Москва 1958. pp.24. [25.]

ATOMKERNENERGIE-DOKUMENTATION beim Gmelin institut. Max-Planck-gesellschaft zur förderung der wissenschaften: Gmelin-institut für anorganische chemie und grenzgebiete: Frankfurt (Main) 1958 &c.*

in progress; for 1958–1961 consisted of two parts: A. Berichte zur atomkernenergie *and* B. Informationen zur atomkernenergie.

J. ROLAND SMITH, *ed.* Guide to U. K. A. E. A. documents. United Kingdom atomic energy authority: 1958. pp.20. [100.]*
— — Second edition. 1960. pp.32. [150.]

N. [B]. KEATS, A list of british books and periodicals on atomic energy. United Kingdom atomic energy authority: Research group: A. E. R. E. reading list (no.3): Harwell 1958. pp.[ii].21. [125.]*
— — 1st revision. 1961. pp.[ii].24. [207.]*

TECHNICAL books. Atomic energy commission: Technical information service extension: Oak Ridge, Tenn. 1959. pp.38.

ENERGÍA atómica. Congreso: Biblioteca: Serie bibliografias (no.7): Santiago de Chile 1959. ff.7. [125.]*

PROCEEDINGS of technical meetings. United States atomic energy commission: Technical information service: [Washington] 1959. pp.[ii].29. [125.]*

MARCO DE ORELLANA SEGOVIA, El atomo. Cien fichas sobre... (no.xxii): Madrid 1959. ff.6. [100.]

ELISE P. FISHBEIN and ELIZABETH C. WESTCOTT, An introduction to research tools for AEC and government scientific reports. Pennsylvania state university: Library: [University Park 1959]. ff.[ii].15. [65.]*

LIST of references on nuclear energy [. . . received by the Agency]. International atomic energy agency: Vienna.*

i. 1959. pp.[v].23.25.22.35.19.76. [2011.]
ii. 1960. pp.[v].29.14.45.41.14.18. [1726.]
iii. 1961. pp.[1180]. [10,354.]
— Index 1959–1961. 1963. pp.450.
iv. 1962. pp.[1225]. [10,910.]
v. 1963. pp.[1330]. [10,964.]

in progress.

ATOOMKERNTECHNIEK en reactorkunde. Literatuuroverzicht. Technisch documentatie- en informatie-centrum voor de krijgsmacht: 's-Gravenhage.

i. 1959.
ii. 1960. pp.161. [608.]
iii. 1961. pp.156. [589.]

in progress.

SEGNALAZIONI bibliografiche. Comitato nazionale per l'energia nucleare: [Rome] 1959 &c.*

ANTHONY J. WOOD, A guide to information on atomic energy in Britain. A selected list. Central office of information: 1960. pp.63. [400.]

L[EV] S[ERGEEVICH] RUBENAU, Атомная энергия на службе мира. Материалы для книжно-иллюстративных выставек. Государственная . . . публичная библиотека имени М. Е. Салтыкова-Щедрина: Ленинград 1960. pp.45. [176.]

TRANSATOM bulletin. Information on translations covering nuclear literature. European atomic energy community: Brussels [1962– : Amsterdam].*

i. 1960. pp.239.20. [2935.]
ii. 1961. pp.386.20.67. [5102.]

in progress.

NUCLEAR engineering abstracts.

i. 1960. pp.234. [200.]
ii. 1961. pp.344. [300.]
iii. 1962.
iv. 1963. pp.170. [200.]

in progress.

LISTE récapitulative des rapports C. E. A. publiés par le Commissariat à l'énergie atomique, juillet 1948–décembre 1960. Commissariat à l'énergie atomique: Centre d'études nucléaires de Saclay: Service de documentation: Série "Bibliographies" (no.8): Gif-sur-Yvette 1961. ff.88. [1777.]

NUCLEAR reactions. A bibliography. [Library of Congress:] Aerospace information division: AID report (no.61–7): [Washington] 1961. ff.iii.96. [682.]*

[PETER EATON], The Eaton- [sir Owen Willems] Richardson library on the atom. 1960 [1962]. ff.[468]. [17,000.]*

50 copies reproduced from typewriting; the collection now forms part of the Texas university library.

NUCLEAR theory reference book for 1957 and 1958. National research council: Nuclear data project: Washington 1963. pp.v.136. [1276.]*

LISTE des traductions envisagées. Commissariat à l'énergie atomique: Service central de documentation: Groupe traduction: Saclay 1962. ff.[6]. [32.]

4. *Special subjects*

GEORGE F[ERDINAND] BECKER, Atomic weight determinations: a digest of the investigations published since 1814. Smithsonian institution: Smithsonian miscellaneous collections (no.358 = The constants of nature, part iv): Washington 1880. pp.iii.149. [500.]

Atomic Energy

A BIBLIOGRAPHY on the scientific and political aspects of atomic energy. Library of Congress: Washington 1945. ff.39. [234.]*

— Addendum. 1946. ff.40–44. [44.]*

HELEN L[AWRENCE] SCANLON, Security in the atomic age: international control of the bomb. An annotated bibliography. Carnegie endowment for international peace: Library: Select bibliographies (no.15): Washington 1946. pp.20. [225.]*

SELECTED references on the ethical implications of atomic energy. New York committee on atomic information: New York [1948]. ff.7. [75.]*

ATOMIC energy for industry. Available atomic energy commission declassified documents of particular interest to industry. [Oak Ridge, Tenn. 1949]. pp.19. [300.]*

SELECTIVE bibliography on educational effects and implications of atomic energy in the social sciences. New York committee on atomic information: Religious and welfare committee: [New York 1949]. ff.13. [100.]*

BIBLIOGRAPHY of published literature on the

preparation and properties of radioactive neutron sources. Atomic energy commission: Oak Ridge, Tenn. 1950. pp.3–43. [138.]*

CIVIL defense against atomic warfare: a selected reading list. Atomic energy commission: Washington 1950. pp.24. [424.]
— [second edition]. 1953. pp.iv.48. [618.]

A BIBLIOGRAPHY on fission. Atomic energy research establishment: Harwell 1951. pp.32. [333.]*

CIVIL defense and atomic warfare. A selected reading list. Federal civil defense administration: [Washington 1953]. pp.iv.28. [400.]

M. ELAINE TOMS, Bibliography of photonuclear reactions. Naval research laboratory [Bellevue, D.C.]: Bibliography (no.2): Washington 1955. pp.62.
— — [Third edition]. . . . (no.14): 1960. pp.124.

NUCLEAR warfare. A bibliography June 1955–June 1956. Engineeı school: Library: Fort Belvoir, Va. 1956. pp.[v]. 96. [800.]*

ROBERT J. LIST, Annotated bibliography on the transport and deposition of atomic debris. Atomic energy commission (TID-3061): Washington 1956. pp.iv.27. [76.]

A[LEKSEI] N[IKOLAEVICH] BUCHENKOV, Меченые атомы и их применение в народном хозяйстве. Государственная ордена Ленина библиотека СССР имени В. И. Ленина: Центральная политехническая библиотека: Новости техники (no.4): Москва 1956. pp.14. [23.]

A. G. HORAD, M. EISENBUD and J. H. HARLEY, Annotated bibliography on fallout resulting from nuclear explosions. United States atomic energy commission: New York operations office: New York 1956. pp.48. [175.]

— — Supplement 1. Annotated bibliography on long-range effect of fallout from nuclear explosions. 1956. pp.77. [75.]

ECONOMIA dell'energia atomica. Rassegna bibliografica e di documentazione. Banco nazionale del lavoro: Roma 1956.

— III edizione. [1958]. pp.104. [2000.]

DEUTSCHES und ausländisches schrifttum zur frage der abrüstung 1945–1956 unter besonderer berücksichtigung . . . der kernwaffen und der internationalen kontrolle der kernenergie. Europaarchiv: Aktuelle bibliographien (no.13): Frankfurt/Main 1957. pp.44. [803.]

GIFFORD A. YOUNG, Feed materials. A bibliography of unclassified report literature. United States atomic energy commission. Technical information service extension: Oak Ridge, Tenn. 1957. pp.viii.109. [754.]*

HAZARDS and safety measures for nuclear-powered merchant ships. An annotated bibliography. Atomic energy commission: Technical information service extension: Oak Ridge, Tenn. 1957. pp.vi.272. [626.]*

ANNOTATED bibliography on the reduction of urban vulnerability to nuclear attack. Department of commerce: Office of area development: [Washington] 1957. pp.[iii].91. [250.]*

EKONOMICKÉ aspekty mírového využiti atomové energie. Soubor literárnic rešerší. Ústřední technická knihovna ČSR: Praha 1958. pp.vi.246. [1656.]*

HEAT transfer fluids for nuclear power generation. Literature abstracts 1947–1957. North Carolina state college: Department of engineering research: Raleigh 1958. pp.[v].169. [475.]*

R[OSWELL] C[LIFTON] GIBBS and KATHARINE WAY, A directory to nuclear data tabulations. National research council: Nuclear date project: Washington 1958. pp.xiii.185. [300.]

LIST of scientific reports on plasma physics, magnetohydrodynamics and thermonuclear reactions in the CERN library. European organization for nuclear research: Library: CERN bibl[iography] (no.1): Genève [19]58. ff.11. [200.]*

THOMAS W. SCOTT, Controlled thermonuclear processes. A selective bibliography. United States atomic energy commission: [Oak Ridge, Tenn.] 1958. pp.viii.26. [126.]

C. S. SABEL, A selected reading list on controlled thermonuclear reactions. United Kingdom atomic energy authority: Research group: A. E. R. E. reading list (no.5): Hawell 1958. pp.[ii].5. [50.]*

B. J. WILSON, Long range fallout of radioactive materials from nuclear explosions. United Kingdom atomic energy authority: Research group: A. E. R. E. reading list (no.1): Harwell 1958. pp.[ii].6. [30.]*

P. J. JONES and G. L. COOPER, Metallurgical

aspects of atomic energy. United Kingdom atomic energy authority: Research group: A. E. R. E. reading list (no.4): Harwell 1958. pp. [ii]. 14. [150.]*

C. S. SABEL, A selected reading list on controlled thermonuclear reactions. United Kingdom atomic energy authority: Research group: A. E. R. E. reading list (no.5): Harwell 1958. pp.[ii].5. [50.]*

[HERBERT COBLANS], Time reversal: a bibliography. European organization for nuclear research: Cern bibliography (no.4): Genève 1959. pp.15. [122.]*

HENRY H[ERMAN] HAUSNER and HELEN C. FRIEDEMANN, Bibliography on powder metallurgy in nuclear engineering (1956–1960). Reutte/Tyrol 1961. pp.[ii].iv.119. [643.]*

O[LGA] N[IKOLAEVNA] LEVSHINA, Проблема управления термоядерными реакциями. Рекомендательный обзор литературы. Государственная библиотека СССР имени В. И. Ленина: Новости науки и техники (no.27): Москва 1961. pp.17. [50.]

RESEARCH on controlled thermonuclear fusion. International atomic energy agency: Bibliographical series (no.7): Vienna 1962. pp.584. [4957.]*

ABBOTT WAINWRIGHT and JOHN A. BALDWIN, Radiological warfare. Armed services technical information agency: Arlington, Va. 1962. ff.[i].10. pp.125. [750.]*

S[TANLEY] L[EROY] HARRISON, Arms control. Selected bibliography, some recent writings on disarmament, arms control, and the spread of nuclear weapons. Weapons systems evaluation group. [Washington] 1962. pp.iii.28. [450.]*

SEMICONDUCTOR nuclear particle detectors. International atomic energy agency: Bibliographical series (no.8): Vienna 1962. pp.95. [612.]*

PARTICLE accelerators, principles, theory, and design. Bibliography. Library of Congress: Aerospace information division: AID report (no.B-63-9): [Washington] 1963. ff.[iii].41. [422.]*

Chemistry.

1. *Bibliographies*

JULES GARCON, Les sources bibliographiques des sciences chimiques. Société bibliographique: 1899. pp.27. [350.]

SELECT list of bibliographies of chemistry in the

Library of Congress. Library of Congress: Washington 1913. ff.4. [39.]*

CLARENCE J[AY] WEST and D. D. BEROLZHEIMER, Bibliography of bibliographies on chemistry and chemical technology, 1900–1924. National research council: Bulletin (vol.ix, part 3, no.50): [Washington] 1925. pp.308. [8000.]

— — First supplement, 1924–1928. . . . (no.71): 1929. pp.161. [4500.]

— — Second supplement, 1929–1931. . . (no.86): 1932. pp.150. [4000.]

JULIAN ARELL SOHON and WILLIAM L. SCHAAF, A reference list of bibliographies: Chemistry, chemical technology and chemical engineering published since 1900. New York 1924. pp.x.100. [2000.]

BIBLIOGRAPHY of chemical reviews. American chemical society: Columbus.*

i. 1958. pp.[iii].140. [5000.]
ii. 1959. pp.[ii].154. [5000.]

in progress.

2. *History*

J[AMES] CAMPBELL BROWN and ALEXANDER WATT, Catalogue of books illustrating the history of alchemy and early chemistry exhibited at the associated soirée. Liverpool 1890. pp.11. [100.]

LIST of references on the history of the chemical industry. Library of Congress: Washington 1916. ff.2. [11.]*

EDGAR F[AHS] SMITH, Old chemistries. New York 1927. pp.xi.89. [50.]

M[AKS] A[BRAMOVICH] BLOCH, Хронология важнейших событий в области химии и смежных дисциплин и библиография по истории химии. Государственное научно-техническое издательство химической литературы: Ленинград 1940. pp.xii.756. [33,000.]

BIBLIOTHECA alchemica et chemica. An annotated catalogue of printed books on alchemy, chemistry and cognate subjects in the library of Denis I[an] Duveen. 1949. pp.vii.669. [2000.]

PAUL WALDEN, Chronologische übersichtstabellen zur geschichte der chemie von den ältesten zeiten bis zur gegenwart. Berlin &c. 1952. pp.xii. 118. [1500.]

CATALOGUS van de historische boeker der Koninklijke nederlandse chemische vereniging in bruikleen geplaatst in het Rijksmuseum voor de geschiedenis der natuurwetenschappen. [Amsterdam 1954]. pp.43. [617.]

— Aanvulling. [1957]. pp.8. [100.]

CATALOG of the Edgar Fahs Smith memorial collection in the history of chemistry. University of Pennsylvania [Philadelphia]. Boston 1960. pp.[v]. 524. [10,970.]*

3. *Manuscripts*

N. M. RASKIN, Рукописные материалы химиков второй половины XVIII в. в архиве Академии наук СССР. Труды архива (no.15): Москва 1957. pp.214. [1023.]

4. *Periodicals*

H[ENRY] CARRINGTON BOLTON, Short titles of chemical periodicals current in 1887. [*s.l.*] 1887. pp.4. [75.]

LIST of periodicals abstracted by Chemical abstracts. Ohio state university: Columbus [*c.*1914]. pp.xviii. [600.]

— — [another edition]. American chemical society: Columbus 1957. pp.314. [10,000.]

ELIZABETH GATES, Union list of chemical periodicals in the Cincinnati libraries. University of Cincinnati studies (2nd ser., vol.xi, part 2): Cincinnati 1925. pp.58. [900.]

M. S. PAUTUINSKAYA, Каталог выставки мировой химической периодики на VI Менделеевском съезде. Всесоюзное научное инженерно-техническое общество химиков: Харьков 1932. pp.48. [579.]

CHEMISTRY bibliography. A check list of periodical holdings and standard chemical reference tools to be found in the libraries of the chemical plants and in the public libraries of Buffalo and the Niagara frontier. Grosvenor library: Buffalo, N.Y. 1936. ff.[i].52.

MAXIMILIAN PFLÜCKE, Periodica chimica. Verzeichnis der im Chemischen zentralblatt referierten zeitschriften, mit den entsprechenden genormten titelabkürzungen, sowie angaben über den besitz in bibliotheken Grossdeutschlands. Deutsche chemische gesellschaft: Berlin 1940. pp.vii. 208. [3000.]

pages 178 a–b are inserted after p.178.

RELAÇÃO das publicações periodicâs sôbre química e assuntos afins. Títulos abreviados de acordo com ai citações comumente encontradas na literatura científica. Associação química do Brasil: Anais: [Rio de Janeiro *c.*1940]. pp.180. [10,000.]

R[AYMOND] CORNUBERT, Contribution au développement de la recherche chimique. Tome I. La littérature chimique mondiale. 1943. pp.vii.74. [100.]

consists in a large part of a select list of chemical periodicals.

NELSON F[RANCIS] MURPHY and RALPH J. GOLLADAY, Chemical and chemical engineering journals available in southwest Virginia. Virginia polytechnic institute: Blacksburg 1953. ff.[ix].14. [100.]*

JAMES G[OODWIN] HODGSON, Current serials in chemistry. A study of the titles held in Colorado and Wyoming libraries, with a supplement including titles of serials in chemistry and a union list of holdings. Colorado A. & M. college: Library bulletin (no.17): Fort Collins 1945. ff.14. 51. [500.]*

JOURNALS on pure and applied chemistry in languages other than latin, teutonic and russian. Science library: Bibliographical series (no.636): 1947. ff.2. [29.]*

A[UGUSTUS] GORTER, Inventaris van periodieken op chemisch en verwant gebied aanwezig in nederlandse bibliotheken. Nederlandse chemische vereniging: Amsterdam 1951. pp.173. [4500.]

REAL sociedad española de fisica y química. Índice de las revistas existentes en su biblioteca. [Madrid 1951]. pp.14. [300.]

F[REDERIK] BIRKET-SMITH, K[AI] A[RNE] JENSEN and P. K. LYSTBÆK, Fortegnelse over kemiske tidsskrifter og tidsskrifter af kemisk interesse i danske biblioteker. Kemisk forening: København 1953. pp.216. [1954.]

G[ASTON] GOETHALS, Répertoire des périodiques et grands traités de chimie et sciences connexes existant dans les bibliothèques de Belgique. Fédération des industries chimiques de Belgique: Bruxelles [1954]. pp.143. [2000.]

— — Supplément. [1962]. pp.83. [2000.]

SOUPIS některých chemických časopisů v knihovnách ČSR. Třetí vydání. Vyzkumný ústav pro farmacii a biochemii: Praha 1955. ff.[vii].97. [175.]*

PERIODICALS in the Chemical society library. [1960]. pp.48. [1500.]

JAMES VAN LUIK [*and others*], An analysis of 150 chemical and technical journals, and a guide to the pharmacological literature. Purdue university: West Lafayette [1956]. pp.[316]. [150.]*

— — Second edition. Searching the chemical and chemical engineering literature. 1957. pp. [414]. [229.]*

УКАЗАТЕЛЬ иностранных периодических изданий, выписанных Центральная научно-техническая библиотека: Москва 1957. pp.32. [600.]

5. *General*

PERRE BOREL, Bibliotheca chimica. Sev catalogvs librorvm philosophicorvm hermeticorvm. Parisiis 1654. pp.[xii].276. [4000.]

a copy in the Bodleian library contains volumnious ms. notes and additions by Elias Ashmole.

— — [another edition]. Heidelbergæ 1656. pp. [x].254. [4000.]

WILLIAM COOPER, A catalogue of chymicall books . . . written originally, or translated into english. 1675. pp.[139]. [1250.]

OLAUS BORRICHIUS, Conspectus scriptorum chemicorum illustriorum. Havniæ 1687. pp.[xii]. 48. [200.]

a copy in the Bibliothèque nationale contains an index in ms.

[FRIEDRICH ROTH-SCHOLTZ], Bibliotheca chemica, oder catalogus von chymischen-büchern, darinnen man alle die jenigen autores findet, die . . . in der Roth-Scholtzischen bibliotheque verhanden seyn. Nürnberg &c. [1725–]1727[–1729]. pp.14 + 48 + 49–96 + [ii].97–172 + 173–250 + 251–238 [*sic*, 328]. [1000.]

— — Editio secvnda. Fascicvlvs primvs. Norimbergae &c. 1735. pp.80. [200.]

JOHANN WILHELM BAUMER, Bibliotheca chemica. Giessae 1782. pp.[xii].116. [750.]

GEORG FRIEDRICH CHRISTIAN FUCHS, Versuch einer uebersicht der chymischen litteratur und ihrer brangschen. Altenburg 1785. pp.144. [1500.]

J[EREMIAS] D[AVID] REUSS, Repertorium commentationum a societatibus litterariis editarum. . . . Tom III. Chemia et res metallica. Gottingae 1803. pp.[ii].viii.221. [2000.]

[GEORG FRIEDRICH CHRISTIAN FUCHS], Repertorium der chemischen litteratur von 494 vor Christi geburt bis 1806 . . . von den verfassern der systematischen beschreibung aller gesundbrunnen und bäder in und ausser Europa. Jena &c. 1806–1812. pp.viii.302 + vi.303–648 + vii.340 + [ii].341–694. [5000.]

[JÖNS] JACOB BERZELIUS, Jahres-bericht über die

fortschritte der physischen wissenschaften. Tübingen.

i. 1822. Übersetzt von C[hristian] G[ottlob] Gmelin. pp.[iii].163. [150.]
ii. 1823. pp.[ii].134. [100.]
iii. 1824. pp.[ii].225. [250.]
iv. 1825. Übersetzt von F. Wöhler. pp.[ii].270. [250.]
v. 1826. pp.[iii].300. [250.]
vi. 1827. pp.viii.312. [250.]
vii. 1828. pp.x.364. [300.]
viii. 1829. pp.x.334. [300.]
ix. 1830. pp.viii.288. [250.]
x. 1831. pp.viii.270. [250.]
xi. 1832. pp.x.354. [300.]
xii. 1833. pp.x.334. [300.]
xiii. 1834. pp.x.398. [350.]
xiv. 1835. pp.x.394. [350.]
xv. 1836. pp.x.482. [350.]
xvi. 1837. pp.x.410. [350.]
xvii. 1838. pp.[ii].viii.432. [350.]
— Vollständiges sach- und namen-register zu dem ersten bis siebenzehnten jahrgang. 1839. pp.[iii].84.
xviii. 1839. pp.xii.660. [400.]
xix. 1840. pp.xii.760. [400.]
xx. 1841. pp.xiii.604. [450.]

[*continued as:*]

Jahres-bericht über die fortschritte der chemie und mineralogie.

xxi. 1842. pp.xii.578. [400.]
xxii. 1843. pp.x.606. [350.]
xxiii. 1844. pp.xiii.690. [400.]
xxiv. 1845. pp.xiii.711. [450.]
xxv. 1846. pp.xvi.913. [550.]

— Vollständiges sach- und namen-register . . . erster bis fünfundzwanzigster jahrgang. 1847. pp.[iii].180.

xxvi. 1847. pp.xvi.926. [550.]
xxvii. 1848. pp.xiii.690. [450.]

[*continued as:*]

Jahres-bericht über die fortschritte der chemie; nach Berzelius' tode fortgesetzt von L. Svanberg.

xxviii. 1849. pp.xi.557. [400.]
xxix. 1850. pp.xii.596. [450.]

the dates shown are those of publication; the swedish originals appeared in the Årsberättelse *of the Kongliga svenska vetenskaps academien, Stockholm; the german translation has been set out here in preference to the swedish edition because the earlier issues of the latter were not published separately; a french translation was also published for 1837–1847 under the title of* Annuaire des sciences chimiques, *afterwards*

Rapport annuel sur les progrès des sciences physiques et chimiques.

PHARMACEUTISCHES central-blatt. [1897–1945: Deutsche chemische gesellschaft (1919–1923: and Verein deutscher chemiker); 1946– : Deutsche akademie der wissenschaften zu Berlin (&c.)]: Leipzig [1882–1896: Hamburg &c.; 1897–1920; Berlin; 1921–1924: Leipzig &c.; 1925– : Berlin].

i. 1830.
ii. 1831.
iii. 1832. pp.[ii].462+[ii].463–991. [1000.]
iv. 1833. pp.[ii].462+[ii].463–1018. [1000.]
v. 1834. pp.[ii].494+[ii].495–1006. [1000.]
vi. 1935. pp.[ii].446+[ii].447–967. [800.]
vii. 1836. pp.[ii].414+[ii].415–932. [600.]
viii. 1837. pp.[ii].414+[ii].415–936. [600.]
ix. 1838. pp.[ii].462+[ii].463–967. [600.]
x. 1839. [Edited by Christian Albert Weinlig]. pp.[ii].414+[ii].415–938. [700.]
xi. 1840. pp.[ii].400+[ii].401–918. [400.]
xii. 1841. pp.[ii].446+447–964. [500.]
xiii. 1842. pp.[ii].432+[ii].433–944. [500.]
xiv. 1843. pp.[ii].478+479–986. [500.]
xv. 1844. pp.[ii].446+449–958. [800.]
xvi. 1845. [Edited by Rudolf Buchheim]. pp. [ii].446+447–966. [600.]
xvii. 1846. pp.[ii].446+447–992. [700.]

xviii. 1847. [Edited by Johann August Ludwig. Wilhelm Knop]. pp.[ii].462+463–974. [600.]
xix. 1848. pp.[ii].462+463–994. [700.]
xx. 1849. pp.[ii].446+447–952. [700.]

[*continued as:*]

Chemisch-pharmaceutisches central-blatt.

xxi. 1850. pp.[ii].462+463–938. [600.]
xxii. 1851. pp.[ii].464+465–988. [700.]
xxiii. 1852. pp.[ii].460+461–982. [700.]
xxiv. 1853. pp.[ii].965. [600.]
xxv. 1854. pp.[iii].968. [700.]
xxvi. 1855. pp.[ii].950. [700.]

[*continued as:*]

Chemisches central-blatt.

new series.

i. 1856. pp.[ii].480+[ii].481–980. [700.]
ii. 1857. pp.[ii].480+[ii].481–984. [600.]
iii. 1858. pp.[ii].480+[ii].481–980. [500.]
iv. 1859. pp.[ii].96[*sic*]+[ii].497–1012. [500.]
v. 1860. pp.[ii].512+[ii].513–1048. [700.]
vi. 1861. pp.[ii].496+[ii].497–1011. [500.]
vii. 1862. [Edited by Rudolf Friedrich Eugen Arendt]. pp.[ii].496+[ii].497–1034. [800.]
viii. 1863. pp.[ii].544+[ii].545–1119. [800.]
ix. 1864. pp.xvi.560+[ii].561–1157. [1000.]
x. 1865. pp.xxii.592+[ii].593–1216. [1500.]

xi. 1866. pp.[ii].560+[ii].561–1162. [1000.]
xii. 1867. pp.[ii].544+[ii].545–1115. [700.]
xiii. 1868. pp.[ii].560+[ii].561–1179. [600.]
xiv. 1869. pp.[ii].528+[ii].529–1120. [1200.]

third series.

i. 1870. pp.xxxvi.918. [2000.]
ii. 1871. pp.xxxviii.886. [2000.]
iii. 1872. pp.xxxii.882. [2000.]
iv. 1873. pp.xxviii.888. [1750.]
v. 1874. pp.xxvii.898. [2250.]
vi. 1875. pp.xxiv.886. [2000.]
vii. 1876. pp.xxviii.888. [2000.]
viii. 1877. pp.xxxi.896. [2250.]
ix. 1878. pp.xxxii.884. [2250.]
x. 1879. pp.xxxi.916. [2250.]
xi. 1880. pp.xxvi.915. [2500.]
xii. 1881. pp.xxix.896. [2000.]
— General-register . . . III. folge, I.–XII. jahrgang. . . . Bearbeitet von Rudolf Arendt. 1883. pp.[iii].884.
xiii. 1882. pp.xxx.902. [2250.]
xiv. 1883. pp.xxii.886. [1500.]
xv. 1884. pp.xxvii.1036. [2000.]
xvi. 1885. pp.xxiv.1034. [2000.]
xvii. 1886. pp.xxiv.1016. [2000.]
xviii. 1887. Redigiert von Rud. Arendt. pp. xxxviii.1726. [4000.]

xix. 1888. pp.xlvi.1818. [5000.]
lx. 1889. pp.xxvi.940+xxviii.1164. [5000.]
lxi. 1890. pp.xxix.1184+xxvi.1122. [5000.]
lxii. 1891. pp.xxviii.1190+xxiv.1028. [5000.]
lxiii. 1892. pp.xxxii.1104 + xxviii.1168. [5000.]
lxiv. 1893. pp.xxviii.1188+xxx.1204. [5000.]
lxv. 1894. pp.xxviii.1256+xxvi.1152. [5000.]
lxvi. 1895. pp.xxxi.1288+xxxii.1276. [5500.]
lxvii. 1896. pp.xl.1392+xxxv.1236. [5500.]
lxviii. 1897. pp.xxxvii.1380+xxxvii.1346. [5689.]
lxix. 1898. pp.xl.1447+xlii.1432. [6061.]
lxx. 1899. pp.xliii.1448 + xxxviii.1280. [6027.]
lxxi. 1900. pp.xlvi.1462+xliii.1442. [6344.]
lxxii. 1901. pp.xliii.1548+xlii.1532. [5889.]
— General-register . . . 1897–1901. Bearbeitet von Rudolf Arendt. 1902. pp.[ii].1297.
lxxiii. 1902. pp.xliv.1584+xlvii.1695. [6359.]
lxxiv. 1903. Redigiert von Albert [Friedrich] Hesse. pp.lv.1630+lviii.1690. [7030.]
lxxv. 1904. pp.lxx.1882+lviii.1972. [7677.]
lxxvi. 1905. pp.lvi.1970+lviii.2068. [8282.]
lxxvii. 1906. pp.lx.1128+[ii].1129–2155+[ii].lxi.1096+[ii].1097–2140. [8647.]
— Generalregister über die jahrgänge 1902–

1906. . . . Bearbeitet von Albert Hesse und Ignaz Bloch. 1907. pp.iv.1663.

[*continued as:*]

Chemisches zentralblatt.

lxxviii. 1907. pp.[ii].lviii.1004 + [ii].1005–2038 + [ii].lxiv.1140 + [ii].1141–2343. [8655.]

lxxix. 1908. pp.[ii].lxviii.1236 + [ii].1237–2472 + [ii].lxvi.1144 + [ii].1145–2316. [9894.]

lxxx. 1909. pp.[ii].lxx.1132 + [ii].1133–2304 + [ii].lxxvi.1104 + [ii].1105–2506. [10,164.]

lxxxi. 1910. pp.[ii].lxxiv.1200 + [ii].1201–2428 + [ii].lxxiv.1012 + [ii].1013–2240. [10,587.]

lxxxii. 1911. pp.[ii].lxvi.1020 + [ii].1021–2168 + [ii].lxviii.1000 + [ii].1001–2255. [9672.]

— Generalregister über die jahrgänge 1907–1911. . . . Bearbeitet von Ignaz Bloch. 1912. pp.[ii].642 + [iv].645– .

lxxxiii. 1912. pp.[ii].lxx.1072 + [ii].1073–2372 + [ii].lxxviii.1172 + [ii].1173–2446. [10,533.]

lxxxiv. 1913. pp.[ii].lxxx.1252 + [ii].1253–2498 + [ii].lxxx.1188 + [ii].1189–2492. [11,219.]

lxxxv. 1914. pp.[ii].lxxx.1320 + [ii].1321–2524+[ii].lxii.856+[ii].857–1718. [9597.]

lxxxvi. 1915. pp.[ii].lxii.720 + [ii].721–1592 +[ii].lxii.732+[ii].733–1533. [7376.]

lxxxvii. 1916. pp.[ii].lvi.594+[ii].595–1482 +[ii].liv.534+[ii].535–1385. [6518.]

— Generalregister IV über die jahrgänge 1912–1916. . . . Bearbeitet von Ignaz Bloch. 1921. pp.vi.2016.

lxxxviii. 1917. pp.[ii].xxxii.548+[ii].549–1342+[ii].xlii.348+[ii].349–968. [5321.]

lxxxix. 1918. pp.[ii].lvi.504+[ii].505–1397+[ii].l.496+[ii].497–1255. [6232.]

xc. 1919. pp.[iv].xxxix.1050+[iii].l.1278+[iv].xlvi.1084+[ii].lxvi.1519. [14,457.]

xci. 1920. pp.[iv].xl.898+[ii].lx.1130+[iv].lii.940+[ii].lxviii.1160. [14,063.]

xcii. 1921. pp.[iv].liii.1096+[ii].lxxix.1571+[iv].lxxi.1514+[ii].c.1973. [22,453.]

— Generalregister v über die jahrgänge 1917–1921. . . . Bearbeitet von Ignaz Bloch. 1924. pp.vi.882+[ii].883–2404.

xciii. 1922. pp.[iv].lxii.1416+[ii].xcii.1824+[iv].lxi.1388+[ii].lxxxviii.1771. [23,121.]

xciv. 1923. Redigiert von A. Hesse und L[eopold Julius] Spiegel [von Maximilian Pflücke und Ernst (Wilhelm) Behrle]. pp.

[iii].1640 + [ii].1844 + [ii].1654 + [iv]. 1652. [23,410.]

xcv. 1924. pp.[iv].1616+[ii].1617–3204+ [iv].1764+[ii].1765–3415. [25,225.]

— Generalregister VI über die jahrgänge 1922–1924. 1925–1927. pp.iv.1132+vi. 1887.897.104.

xcvi. 1925. pp.[iv].1552+[ii].1553–3064+ [iv].1328+[ii].1329–3206. [24,589.]

xcvii. 1926. pp.[iv].2424+[ii].2425–3977+ [iv].1812 + [ii].1813–3442 + [ii].3443–4362. [33,176.]

xcviii. 1927. pp.[iv].1916+[ii].1917–3550+ [iv].1656 + [ii].1657–3134 + [ii].3135–3832.306. [33,638.]

xcix. 1928. pp.[ii].1740+[ii].1741–3501+ [iv].1416 + [ii].1417–3114 + [ii].3115–3818.328. [36,424.]

c. 1929. Redigiert von M. Pflücke. pp.[iii]. 1652 + [ii].1653–3501 + [iii].1760 + [ii]. 1761–3632+[ii].3633–4416.380. [37,622.]

ci. 1930. pp.[iii].2044+[ii].2045–4286+[v]. 2096 + [ii].2097–4361 + [ii].4363–5306. 394. [48,009.]

cii. 1931. pp.[v].2156+[ii].2157–4220+[v]. 1960 + [ii].1961–4133 + [ii].4135–5018. 352. [45,000.]

ciii. 1932. pp.[v].1864+[ii].1865–3938+[v]. 2004 + [ii].2005–4448 + [iii].4449–5182. 330. [50,000.]

— Generalregister VII über die jahrgänge 1925–1929. 1930–1932. pp.iv.1088+[ii]. 1089–2109.146 + vi.1172 + iv.1184 + [ii].1185–2252.

civ. 1933. pp.[v].2208+[ii].2209–4552+[v]. 2092 + [ii].2093–3948 + [ii].3949–4904 + [ii].4405–5256.323. [50,000.]

cv. 1934. pp.[v].2080+[ii].2081–3956+[ii]. 3957–4440 + [v].2160 + [ii].2161–4060 + [ii].4061–4550 + [ii].4551–5436.350. [50,000.]

— Generalregister VIII über die jahrgänge 1930–1934. 1935–1938. pp.iv.1538+[ii]. 1541–2887.312 + vi.1564 + iv.1536 + [ii]. 1537–3288.

cvi. 1935. pp.[v].2124+[ii].2125–4012+[ii]. 4013–4470 + [v].2168 + [ii].2169–4016 + [ii].4017–4512 + [ii].4513–5420.357. [50,000.]

cvii. 1936. pp.[v].2884+[ii].2885–5024+[ii]. 5025–5602 + [v].2660 + [ii].2661–4272 + [ii].4273–4801 + [ii].4804–5990.397. [60,000.]

cviii. 1937. pp.[v].2080+[ii].2081–3756+

[ii].3757–5164 + [ii].511.43 + [vi].1500 + [ii].1501–3120 + [ii].3121–4482 + [ii].542. 38+[ii].1310.460. [70,000.]

cix. 1938. pp.[vi].1928+[ii].1929–3420+[ii]. 3421–4794 + [ii].492.32 + [vi].1720 + [ii].1721–3048 + [ii].3049–4400 + [ii].468. 32 + [ii].1044.379. [80,000.]

cx. 1939. pp.[vi].1920+[ii].1921–3676+[ii]. 3677–5152 + [ii].502.34 + [ii].2008 + 2009–3008 + 4690 + 468.30 + [ii].1004. 414. [80,000.]

— Generalregister IX über die jahrgänge 1935–1939. 1941– . pp.vi.1912+[ii]. 1913–3803.345 + vi.1894 + [ii].1400 + [ii].1401–2328+ .

in progress.

cxi. 1940. pp.[v].1460+[ii].1461–2756+[ii]. 2757–4050 + [ii].421.26 + [ii].1244 + [ii]. 1245–2572 + [ii].2573–3774 + [ii].398.24. [60,000.]

cxii. 1941. pp.v.1252+[ii].1253–2492+[ii]. 2493–3180 + [ii].412.27 + [ii].1240 + [ii]. 1241–2284 + [ii].2285–3306 + [ii].352.21 +[ii].904.252. [50,000.]

cxiii. 1942.

cxiv. 1943. 1943–1947. pp.[ii].1444+[ii]. 1445–2776 + [vii].268.19 + [ii].1256 +

[ii].1257–2350 + [vii].228.16 + 179. [25,000.]

cxv. 1944. pp.[ii].1468+[ii].1356. [30,000.]

— Generalregister x über die jahrgänge 1940–1944. [By Mary Wittenberg and Günter Schmidt]. 1951– . pp.viii.944+[ii]. 945–1600+[ii].1601–2414+ .
in progress.

cxvi. 1945. 1948. pp.[v].744+[ii].745–1574. 148.10+[ii].932+[ii].933–1876. [30,000.]

cxvii. 1946. 1949–1958. pp.[ii].570+[ii].571–1166 + [ii].1167–1806 + [ii].1807–2476 + [ii].696 + [ii].697–1320 + [ii].1321–1944 + .

cxviii. 1947. pp.[ii].1156+[ii].1157–1916. [20,000.]

— Ergänzungsband. [vol.iii: Chefredakteure: Eugen Klever und Heinz Walter]. 1951–1960. pp.[ii].648 + [ii].649–1388.xii + [ii].1389–2044. [10,000.]

cxix. 1948. pp.[ii].772+[ii].773–1468+[iv]. 147.9 + [ii].680 + [ii].681–1476 + [iv]. 148.9. [20,000.]

— Ergänzungsband. 1951–1957. pp.[ii].692 + [ii].693–1368 + [ii].1369–2232 + [ii]. 2233–2900+[ii].2901–3244. [15,000.]

cxx. 1949. pp.[ii].652+[ii].653–1444+[ii].

136.8 + [ii].720 + [ii].721–1472 + vi. 136.8. [20,000.]

cxxi. 1950. pp.[ii].940+[ii].941–2060+iv. 214.12 + [ii].2061–2436 + [ii].1084 + [ii]. 1085–2028 + [ii].2029–3000 + [ii].290.16 +[iii].847+[ii].371. [30,000.]

cxxii. 1951. pp.[ii].1260+[ii].1261–2388+ [ii].2389–3596 + viii.365.18 + [ii].1384 + [ii].1385–2688 + [ii].2689–3684 + viii.423. 21 + [ii].624 + [ii].625–1250 + [ii].527. [45,000.]

cxxiii. 1952. pp.[ii].1436+[ii].1437–2952+ [ii].2953–4248 + [ii].4249–5680 + [ii]. 5681–7120 + [ii].7121–8080 + iv.971 + [ii].712+[ii].713–1388+[ii].588. [50,000.]

cxxiv. 1953. pp.[ii].1280+[ii].1281–2704+ [ii].2705–3988 + [ii].3989–5444 + [ii]. 5445–6874 + [ii].6875–8270 + [ii].8271–9918 + [ii].9919–11614 + iv.1370 + [ii]. 816+[ii].817–1657+[ii].877. [75,000.]

— [supplement]. 1950–1954. Sonderband . . . 1[–14]. 1957–1959. pp.[iv].656.xiv + [iii]. 657–1404.xvi + [iii].1405–2220.xviii + [iii].2221–2956.xvi + [iii].2957–3676.xvi + [iii].3677–4376.xv + [iii].4377–5104.xvi + [iii].5105–5812 + [iii].5813–6588 + [iii].6589–7224 + [ii].7225–7716 + [iii].

7717–8276 + [iii].8277–8852 + [iii].8853–9396. [60,000.]

cxxvi. 1955. pp.[ii].1428 + [ii].1429–2820 + [ii].2821–4264 + [ii].4265–5704 + [ii].5705–7132 + [ii].7133–8536 + [ii].8537–10180 + [ii].10181–11749 + iv.624 + [ii].625–1348 + [iv].824 + [ii].825–1622 + [ii].999. [75,000.]

cxxvii. 1956. pp.[ii].1204+[ii].1205–2348+[ii].2349–3742 + [ii].3743–4894 + [ii].4895–6002 + [ii].6003–7424 + [ii].7425–8520 + [ii].8521–9612 + [ii].9613–10836 + [ii].10837–11884 + [ii].11995–13016 + [ii].13017–14576 + iv.905 + iv.870 + [ii].1040+[ii].1041–2064+1212. [90,000.]

cxxviii. 1957. pp.[ii].1084+[ii].1085–2100+[ii].2101–3436 + [ii].3437–4580 + [ii].4581–5748 + [ii].5749–7252 + [ii].7253–8404 + [ii].8405–9556 + [ii].9557–10660 + [ii].10661–11764 + [ii].11765–12944 + [ii].12945–14228 + iv.878 + iv.869 + [ii].1032+[ii].1133–2018+1240. [85,000.]

cxxix. 1958. pp.[ii].1192+[ii].1193–2324+[ii].2325–3496 + [ii].3497–4976 + [ii].4977–6140 + [ii].6141–7308 [*sic*, 7356] + [ii].7357–8808 + [ii].8809–9960 + [ii].

9961–11128 + [ii].11129–12580 + [ii].12581–13676 + [ii].13677–14548 + iv.909 + iv.861 + [ii].1040 + [ii].1041–2042 + 1258. [90,000.]

— Das system. 1958. pp.95.

cxxx. 1959. Herausgegeben von Heinrich Bertsch, Wilhelm Klemm, Maximilian Pflücke [July 1959–   : H. Bertsch und W. Klemm], pp.[ii].1332+[ii].1333–2672 + [ii].2673–4024 + [ii].4025–5680 + [ii].5681–7012 + [ii].7013–8440 + [ii].8441–9784 + [ii].9785–11122 + [ii].11123–12746 + [ii].12747–13990 + [ii].13991–15238 + [ii].15239–16886 + iv.989 + iv.1110 + [ii].1290+[ii].1291–2517+639. [105,000.]

cxxxi. 1960. pp.[ii].1384 + [ii].1385–2748 + [ii].2749–4404 + [ii].4405–5704 + [ii].5705–7060 + [ii].7061–8744 + [ii].8745–10116 + [ii].10117–11880 + [ii].11881–13252 + [ii].13253–14580 + [ii].14581–16316 + [ii].16317–17672 + [iv].1056 + [iv].1057–2164 + [iv].1240 + [ii].1241–+836+[ii].837–1555. [110,000.]

cxxxii. 1961. pp.[iv].1440+[iv].1441–2860+[iv].2861–4224 + [iv].4225–5672 + [ii].5673–7100 + [ii].7101–8536 + [ii].8537–10048 + [ii].10049–11592 + [ii].11593–

13064 + [ii].13065–14524 + [ii].14525–15988 + [ii].15989–17456 + [ii].17457–18940 + [iv].1111 + [ii].1112– + . [120,000.]

cxxxiii. 1962. pp.[iv].1484+[iv].1485–2976 + [iv].2977–4408 + [iv].4409–5912 + [iv].5913–7396+[iv].7937–8856+

cxxxiv. 1963. pp.[iv].1528+

in progress.

JAHRESBERICHT über die fortschritte der reinen, pharmaceutischen und technischen chemie, physik, mineralogie und geologie. Giessen [1886 &c.: Braunschweig].

1847–1848. Edited by Justus [von] Liebig and Hermann Kopp. 1849. pp.xxiv.1368. [2000.]

1849. 1850. pp.xvi.868. [1500.]

1850. 1851. pp.xvi.866. [1500.]

1851. 1852. pp.xvi.928. [1500.]

1852. 1853. pp.xx.1036. [1500.]

1853. 1854. pp.xx.980. [1500.]

1854. 1855. pp.xx.960. [1500.]

1855. 1856. pp.xxi.1078. [1500.]

1856. 1857. pp.xx.960. [1500.]

[*continued as:*]

Jahresbericht über die fortschritte der chemie und verwandter theile anderer wissenschaften.

1857. By H. Kopp and Heinrich Will. 1858. pp.xviii.776. [1000.]

1858. 1859. pp.xx.860. [1250.]

1859. 1860. pp.xx.904. [1250.]

1860. 1861. pp.xxii.906. [1250.]

1861. 1863. pp.xxiv.1196. [1500.]

1862. 1863. pp.xxiv.888. [1250.]

1863. Edited by H. Will. 1864. pp.xxvi.970. [1500.]

1864. 1865. pp.xxvii.964. [1750.]

1865. 1866. pp.xxvii.1008. [1500.]

1866. 1868. pp.xxvii.1076. [1500.]

— Register zu den berichten für 1857 bis 1866. 1868. pp.[iv].544.

1867. 1869. pp.xxvii.1118. [1750.]

1868. Edited by Adolph Strecker. 1870. pp. xxxi.1372. [1500.]

1869. 1871. pp.xxxvii.1096. [1500.]

1870. Edited by Alexander Naumann. 1872. pp.xlvii.1482. [1500.]

1871. 1873. pp.xliii.1314. [1500.]

1872. Edited by A. Naumann and August Laubenheimer. 1874. pp.xliv.1276. [1500.]

1873. Edited by A. Naumann. 1875. pp.xl. 1336. [1750.]

1874. 1876. pp.xl.1432. [1750.]

1875. Edited by F. Fittica. 1877. pp.xl.1396. [2250.]
1876. 1878. pp.xliv.1407. [2250.]
— Register zu den berichten für 1867 bis 1876. [1879–]1880. pp.[iii].728.
1877. 1879. pp.lii.1504. [2750.]
1878. 1880. pp.xlviii.1524. [3000.]
1879. 1881. pp.xliv.1420. [2750.]
1880. 1882. pp.xlviii.1716. [3500.]
1881. 1883. pp.lvi.1622. [3750.]
1882. 1884. pp.lxiii.1890. [4500.]
1883. 1885. pp.lxvii.2274. [5000.]
1884. 1885–1887. pp.lvi.136+xliv.1365–2382. [5000.]
1885. 1886–1888. pp.lxxvi.1675+xliv.1677–2708. [5500.]
1886. 1889–1890. pp.li.1682+xxxvi.1683–2654. [4500.]
— General-register für die berichte von 1877 bis 1886. 1898. pp.[iv].616+[iv].768+[iv].769–1620.
1887. 1889–1891. pp.lviii.1860+xliv.1861–3107. [5000.]
1888. 1890–1893. pp.lxvii.1932+xlviii.1933–3371. [5500.]
1889. 1892–1895. pp.lxiii.1858+xlviii.1859–3219. [5000.]

1890. 1894–1896. pp.lxiii.1752+lviii.1753–3357. [5500.]

1891. 1896–1898. pp.lxxiv.1812+lx.1813–3160. [4000.]

1892. 1898–1900. pp.c.1864+lxxviii.1865–3347. [5500.]

1893. Edited by G[uido] Bodländer. 1900–1901. pp.xli.1015 + lxxxviii.1017–2607. [3500.]

1894. Edited by G. Bodländer, W. Kerp and G[aetano] Minunni. 1901–1903. pp.xxxii.723 + [iv].xxxiii–lxvi.725–1716 + [iv].lxvii–cii.1717–3190. [4000.]

1895. 1902–1904. pp.xxxviii.938 + [iv].xxxix–lxxiv.939–2012 + [iv].lxxv–cxiii.2013–3543. [4500.]

1896. Edited by K[arl] von Buchka and G. Bodländer. 1897–1901. pp.xxx.616+[iv].xxxi–lviii.615–1584 + [iv].lix–xcii.1585–2701. [3500.]

— General-register für die berichte von 1887 bis 1896. Edited by G. Bodländer. 1904–1907. pp.[iv].873 + [iv].800 + [iv].801–1580.

1897. Edited by G. Bodländer. 1901–1902. pp.xlviii.1042+[iv].xlix–xcii.1043–2300+[iv].xciii–cxii.2301–3344. [5000.]

1898. Edited by G. Bodländer, W. Kerp and G. Minunni. 1903–1905. pp.xxxvi.816+[iv].xxxvii–lxxvi.817–2044 + [iv].lxxvii–xcvii.2045–3091. [4500.]

1899. Edited by G. Bodländer and W. Kerp. 1904–1905. pp.xxxiii.808+[iv].xxxv–lxx.809–1992 + [iv].lxxi–lxxxix.1993–2888. [4000.]

1900. Edited by G. Bodländer, W. Kerp and J. Troeger. 1907. pp.xxxii.518 + [iv].xxxiii–lxxxi.519–2022. [4500.]

1901. Edited by W. Kerp and J. Troeger. 1907–1908. pp.xxxvii.544 + [iv].xxxix–xcii.545–2118. [5000.]

1902. 1908–1909. pp.xlii.714 + [iv].xliii–xcviii.715–2360. [5500.]

1903. Edited by G. Bodländer and W. Kerp. 1904–1905. pp.xxxii.722 + [iv].xxxiii–lxxv.viii.723–2224. [5000.]

1904. 1906[–1907]. pp.xlvi.800+[iv].xlvii–cii.801–2521. [6000.]

— General-register für die berichte von 1897 bis 1904. Edited by E[mil] Fromm. 1910–1912. pp.[iv].847+[iv].1304.

1905–1908. Edited by J. Tröger and E. Baur. 1909–1912. pp.xiii.1440+[iv].1441–2880

+ [iv].2881–3672 + xvi.1440 + [iv].1441–2880 + [iv].2881–4320 + [iv].4321–5760 + [iv].5761–6094. [25,000.]

1909. 1911–1913. pp.x.949+x.960+[iv].961–2176. [7500.]

1910. 1912–1913. pp.x.996+x.960+[iv].961–2242. [10,000.]

no more published; the issue for 1857 is in two parts, the second of which is entered unter Physics, below.

— Annual report of the progress of chemistry, and the Allied Sciences.

i. 1847–1848. Edited by A[ugust] W[ilhelm] Hofmann and W. De La Rue. 1849. pp. xv.516. [1000.]

ii. 1847–1848. 1850. pp.xviii.586. [1500.]

iii. 1849. Edited by A. W. Hofmann and H[enry] Bence Jones. 1852. pp.xx.628. [1250.]

iv. 1850. 1852. pp.xvi.624. [1250.]

CATALOGUE of books in the Department of chemistry. Queen's college library: Belfast 1858. pp.10. [150.]

RÉPERTOIRE de chimie pure et appliquée. Société chimique de Paris.

[A]. Compte rendu des progrès de la chimie pure en France et à l'étranger. By A[dolphe]

Wurtz [1862: and Félix Le Blanc].

i. 1858–1859. pp.626. [5000.]
ii. 1860. pp.[iii].497. [450.]
iii. 1861. pp.[iii].508. [550.]
iv. 1862. pp.[iii].484. [500.]

[B]. Compte rendu des applications de la chimie en France et à l'étranger. By Ch[arles Louis] Barreswil.

i. 1858–1859. pp.viii.538. [550.]
ii. 1860. pp.[iii].416. [500.]
iii. 1861. pp.[iii].492. [550.]
iv. 1862. pp.[iii].492. [550.]
v. 1863. pp.[iii].500. [600.]

after 1862 and 1863 respectively these two series were incorporated with the Bulletin de la Société chimique de Paris, *of which the abstracts formed part until 1891; after that year each annual issue of the Bulletin forms two volumes, of which the second contains the abstracts of foreign publications, those of french publications continuing to form part of the main work.*

[*continued as:*]

Bulletin de la Société chimique de Paris. [Analyse des travaux étrangers.]

3rd series

viii. 1892. pp.[iv].1480. [1000.]
x. 1893. pp.[iv].1376. [1000.]

xii. 1894. pp.[iv].1668. [1100.]
xiv. 1895. pp.[iv].1689. [1250.]
xvi. 1896. pp.[iv].2289. [1500.]
xviii. 1897. pp.[iv].1524. [1500.]
xx. 1898. pp.[iv].1111. [1500.]
xxii. 1899. pp.[iv].1284. [1500.]
xxiv. 1900. pp.[iv].1315. [1250.]
xxvi. 1901. pp.[iv].1444. [1750.]
xxviii. 1902. pp.[iv].1224. [1750.]
xxx. 1903. pp.[iv].1602. [2000.]
xxxii. 1904. pp.[iv].1596. [2000.]
xxxiv. 1905. pp.[iv].1864. [2000.]
xxxvi. 1906. pp.[iv].1762. [2000.]

4th series

ii. 1907. pp.[iv].1815. [2250.]
iv. 1908. pp.[iv].2007. [2750.]
vi. 1909. pp.[iv].1747. [3060.]
viii. 1910. pp.[iv].1920. [3500.]
x. 1911. pp.[iv].1200+[iv].1201–2164. [4000.]
xii. 1912. pp.[iv].1024 + [iv].1025–2062. [4000.]
xiv. 1913. pp.[iv].1056 + [iv].1057–2022. [4000.]
xvi. 1914. pp.[iv].1228. [2500.]
xviii. 1915. pp.[iv].1071. [2000.]
xx. 1916. pp.[iv].1048. [1750.]

xxii. 1917. pp.[iv].1115. [1500.]
xxiv. 1918. pp.[iv].770. [1250.]
xxvi. 1919. pp.[iv].717. [1250.]
xxviii. 1920. pp.[iv].1268. [2250.]
xxx. 1921. pp.[iv].2082. [3000.]
xxxii. 1922. pp.[iv].2819. [4500.]
xxxiv. 1923. pp.[iv].1488+[iv].1489–2853. [5000.]
xxxvi. 1924. pp.[iii].1416+[iii].1417–2828. [6500.]
xxxviii. 1925. pp.[iii].1008+[iii].1009–2324. [5500.]
xl. 1926. pp.[iii].1132 + [iii].1133–2216. [5500.]
xlii. 1927. pp.[iii].1360 + [iii].1361–2479. [6500.]
xliv. 1928. pp.[iii].1048 + [iii].1049–2209. [5500.]
xlvi. 1929. pp.[iii].1004 + [iii].1005–2147. [6000.]
xlvii[*sic*, xlviii]. 1930. pp.[iii].1200 + [iii]. 1201–5228[*sic*, 2528]. [6500.]
l. 1931. pp.[iii].1380[*sic*, 1280] + [iii].1281–2591. [7000.]
lii. 1932. pp.[iii].1124 + [iii].1125–2323. [7000.]

liv. January–September 1933. pp.[iii].1328+[iii].1329–2712. [10,000.]

[*continued as:*]

Bulletin de la Société chimique de France. Documentation.

October 1933–1934. pp.656. [7000.]

1935. pp.636. [7000.]

1936. pp.720. [7500.]

1937. pp.668. [7000.]

1938. pp.800. [8000.]

1939. pp.592. [6000.]

1940. pp.480. [5000.]

1941. pp.513. [5000.]

1942. pp.503. [5000.]

1943. pp.411. [5000.]

1944. pp.240. [2500.]

1945. pp.305. [3000.]

indexes to the Bulletin *as a whole are now published separately; subsequent issues form part of the* Bulletin.

ERNST AMANDUS ZUCHOLD, Bibliotheca chemica. Verzeichniss der auf dem gebiete der reinen, pharmaceutischen, physiologischen und technischen chemie in den jahren 1840 bis mitte 1858 in Deutschland und im auslande erschienenen schriften. Göttingen 1859. pp.viii.342. [5000.]

— — 1858–1870. By Rud[olf] Ruprecht. 1872. pp.[ii].126. [2250.]

[FRANÇOIS JOSEPH] J[ÉRÔME] NICKLÈS, Revue des travaux de chimie publiés à l'étranger. . . . Année 1868. 1869. pp.111. [125.]

JAHRESBERICHT über die fortschritte auf dem gebiete der reinen chemie. Tübingen.

i. 1873. Herausgegeben von Wilh[elm] Staedel. pp.[ii].xxxviii.558. [1500.]
ii. 1874. pp.xii.631. [1500.]
iii. 1875. pp.xxvi.530. [1500.]
iv. 1876. pp.xx.487. [1750.]
v. 1877. pp.x.642. [1500.]
vi. 1878. Herausgegeben von Ludwig Medicus and W. Staedel. pp.vi.605. [1750.]
vii. 1879. pp.viii.623. [1750.]
viii. 1880. Herausgegeben von W. Staedel. pp.viii.644. [1500.]
ix. 1881. pp.viii.612. [1500.]

JOURNAL of the Chemical society. . . . Abstracts [*afterwards:* of papers].

xxxiv. 1878. pp.xxxvi.1093. [1500.]
xxxvi. 1879. pp.xl.1168. [1750.]
xxxviii. 1880. pp.xli.1021. [1750.]
xl. 1881. pp.xlvii.1283. [2250.]
xlii. 1882. pp.l.1444. [2250.]
xliv. 1883. pp.xliii.1267. [2000.]
xlvi. 1884. pp.liii.1562. [2500.]

xlviii. 1885. pp.lvii.1398. [2750.]
l. 1886. pp.xlix.1190. [2250.]
lii. 1887. pp.xlviii.1256. [2250.]
liv. 1888. pp.liii.1471. [2500.]
lvi. 1889. pp.lv.1374. [2250.]
lviii. 1890. pp.lix.1616. [2500.]
lx. 1891. pp.lxi.1704. [2500.]
lxii. 1892. pp.lix.1674. [2500.]
lxiv. 1893. pp.xxxvii.744 + xxxiv.770. [2750.]
lxvi. 1894. pp.xxxiii.631 + xxviii.642. [2500.]
lxviii. 1895. pp.xxxvii.692 + xxxiii.720. [3000.]
lxx. 1896. pp.xl.716+xlii.1053. [3500.]
lxxii. 1897. pp.xxxv.648 + xxxviii.915. [3000.]
lxxiv. 1898. pp.xxxviii.720+xli.965. [3250.]
lxxvi. 1899. pp.[ii].968+[ii].1228. [3500.]
lxxviii. 1900. pp.[ii].712+vii.1080. [3500.]
lxxx. 1901. pp.x.784+[ii].1039. [3500.]
lxxxii. 1902. pp.[ii].iv.8.852 + [ii].1039. [3500.]
lxxxiv. 1903. pp.[iii].v.8.872 + [ii].1066. [4000.]

lxxxvi. 1904. pp.[ii].v.9.1072 + [ii].1187. [4750.]
lxxxviii. 1905. pp.[ii].vii.8.956 + [ii].1204. [4250.]
xc. 1906. pp.[ii].vi.8.1000+[ii].1266. [5000.]
xcii. 1907. pp.[ii].v.8.1104+[ii].1383. [5250.]
xciv. 1908. pp.[ii].vi.8.1032 + [ii].1479. [5500.]
xcvi. 1909. pp.[ii].8.980+[ii].1470. [5500.]
xcviii. 1910. pp.[ii].8.908+[ii].1492. [5750.]
c. 1911. pp.[ii].9.1056+[ii].1531. [6000.]
cii. 1912. pp.[ii].9.1044+[ii].1614. [6250.]
civ. 1913. pp.[ii].8.1432+1528. [7000.]
cvi. 1914. pp.[ii].1212+[ii].1200.8. [5750.]
cviii. 1915. pp.[ii].1092+[ii].1138.8. [5000.]
cx. 1916. pp.[ii].876+[ii].885.8. [4000.]
cxii. 1917. pp.xv.724+[ii].785.5. [3250.]
cxiv. 1918. pp.xv.564+[ii].648.5. [2750.]
cxvi. 1919. pp.xv.616+[ii].711.5. [3000.]
cxviii. 1920. pp.xiii.920+1061.5. [4500.]
cxx. 1921. pp.[ii].916+[ii].1009.20. [4750.]
cxxii. 1922. pp.[ii].1228 + [ii].1288.20. [6500.]
cxxiv. 1923. pp.[ii].1276 + [ii].1313.20. [6750.]

[*continued as:*]

Abstracts of chemical papers . . . A. Pure chemistry. Bureau of chemical abstracts.

cxxvi. 1924. pp.[ii].1396 + [ii].1304.20. [6750.]

cxxviii. 1925. pp.[ii].1528 + [ii].1733.24. [8750.]

the continuation of this work is entered separately below under Abstracts of chemical papers; *before 1878 the abstracts appeared as part of the text of the* Journal; *from 1878 the abstracts formed the second volume of the* Journal *for each year; during 1893–1913 this volume appeared in two parts, the former devoted to organic chemistry, the latter to physical, inorganic, mineralogical, physiological, agricultural and analytical chemistry; during 1914–1925 physical and agricultural chemistry (biochemistry) were transferred to the first part.*

[OSKAR FRANK], Frank's führer durch die chemische, chemisch-technische und pharmaceutische literatur vom jahre 1850–1882. Wien 1883. pp. xxxi.244. [3750.]

WILLIAM A[UGUSTUS] TILDEN, Books on chemistry. Birmingham Reference library: Lectures: [1886]. pp.[ii].42–55.ii. [75.]

A CATALOGUE of the library of the Chemical society. 1886. pp.viii.327. [4000.]

— [another edition]. 1903. pp.8.324. [6000.]

H[ENRY] CARRINGTON BOLTON, A bibliography of chemistry for the year 1887. Smithsonian institution: Miscellaneous collections (no.665): Washington 1888. pp.[ii].13. [350.]

CARL SCHAEDLER, Biographisch-litterarisches handwörterbuch der wissenschaftlich bedeutenden chemiker. Berlin 1891. pp.vi.162. [750.]

[HENRY CARRINGTON BOLTON], Catalogue of works on alchemy and chemistry exhibited at the Grolier club. New York 1891. pp.32. [110.]
a catalogue of works from the author's collection.

JAHRBUCH der chemie. Bericht über die wichtigsten fortschritte der reinen und angewandten chemie. Frankfurt a.M.

i. 1891. Herausgegeben von Richard Meyer. pp.x.544. [500.]

BIBLIOGRAPHIA hopkinsiensis, 1876–1893. Parts II and III. Chemistry, geology and mineralogy. Johns Hopkins university: Baltimore 1893. pp.vi.50. [1000.]

HENRY CARRINGTON BOLTON, A select bibliography of chemistry, 1492–1892. Smithsonian institution: Smithsonian miscellaneous collections (no.850): Washington 1893. pp.xiii.1212. [12,031.]

— — First supplement . . . (no.1170): 1899. pp.ix.489. [5554.]

— — Section VIII.—Academic dissertations . . . (no.1253): 1901. pp.iv.534. [7500.]

— — Second supplement . . . (no.1440): 1904. pp.[iii].462. [5000.]

— — [supplement]. Русская часть химической библіографіи Ал. Кур. Крупскаго [Aleksandr Kirillovich Krupsky]. Санктпетербургъ 1900. pp.[iii].62. [1000.]

HENRY CARRINGTON BOLTON, A short list of books on chemistry. [New York 1895]. pp.18. [125.]

SEYDEL's führer durch die neuere litteratur der chemie und chemischen technologie. . . . 8. Verbesserte auflage. Berlin 1898. pp.iv.85. [1000.]

limited to works in german; p.85 is the inner page of the back cover.

SAMMEL-KATALOG der in hamburger öffentlichen bibliotheken vorhandenen litteratur aus der chemie und aus verwandten wissenschaften. Chemiker-verein [&c.]: Hamburg 1901. pp.[vi].108. [3000.]

SUBJECT list of works on chemistry and chemical technology in the library of the Patent office.

Patent office library series (no.6 = Bibliographical series, no.3): 1901. pp.106. [885.]

— [second edition]. Subject list of works on chemistry (including alchemy, electrochemistry and radioactivity) . . . (n.s. ZC–ZQ): 1911. pp.iv. 214. [2000.]

CHEMISTRY. International catalogue of scientific literature (section D): Royal society.

i. 1902–1903. pp.xiv.468+xiv.671. [5990.]
ii. 1904. pp.viii.1002. [5669.]
iii. 1905. pp.viii.1071. [6614.]
iv. 1906. pp.xi.431+[iv].433–1283. [7917.]
v. 1908. pp.viii.1202. [9527.]
vi. 1909. pp.viii.1078. [7879.]
vii. 1910. pp.viii.1137. [10,059.]
viii. 1911. pp.viii.1067. [10,067.]
ix. 1912. pp.viii.960. [8909.]
x. 1913. pp.viii.934. [9182.]
xi. 1914. pp.viii.966. [9715.]
xii. 1915. pp.viii.910. [8986.]
xiii. 1916. pp.viii.836. [8480.]
xiv. 1919. pp.viii.761. [7336.]

no more published.

CHEMISCHE novitäten. Bibliographische monatsschrift für die neu erscheinende litteratur auf dem gesamtgebiet der reinen und angewandten chemie und der chemischen technologie. Leipzig.

i. 1904–1905. pp.[iii].284. [2873.]
ii. 1905–1906. pp.[iii].188. [2467.]
iii. 1906–1907. pp.240. [2717.]
iv. 1907–1908. pp.226. [2428.]
v. 1808 [*sic*]–1909. pp.[iii].264. [2954.]
vi. 1909–1910. pp.[iii].208. [2434.]
vii. 1910–1911. pp.[iii].238. [2456.]
viii. 1911–1912. pp.[iii].164. [2144.]
ix. 1912–1913. pp.[iii].132. [1694.]
x. 1913–1914. pp.[iii].152. [2010.]
xi. 1915–1916. pp.[iii].92. [1209.]
xii. 1917–1918. pp.[iii].84. [1364.]
xiii. 1918–1919. pp.[iii].84. [1281.]
xiv. 1919–1921. pp.[iii].136. [1739.]
xv. 1922–1924.
xvi. 1924–1925. pp.116. [1957.]
xvii. 1925–1926.
xviii. 1926. pp.112. [2710.]
xix. 1927–1928. pp.214. [3389.]
xx. 1928–1929. pp.208. [2984.]
xxi. 1929–1931. pp.328. [4441.]
xxii. 1931–1932. pp.200. [3008.]
xxiii. 1932–1933.
xxiv. 1933–1934. pp.192. [3129.]
xxv. 1934–1936. pp.348. [5249.]
xxvi. 1936–1937. pp.208. [3995.]

xxvii. 1937–1938. pp.232. [3854.]
xxviii. 1938–1939. pp.216. [2932.]

no more published?; the subtitle varies.

JOHN FERGUSON, Bibliotheca chemica: a catalogue of the alchemical, chemical and pharmaceutical books in the collection of the late James Young. Glasgow 1906. pp.xxiii.488+[iii].598. [2500.]

privately printed; a facsimile was issued London 1954; the collection is now in the Royal technical college, Glasgow.

CHEMICAL abstracts. American chemical society: Easton, Pa. [vol.xxiii &c.: Ohio state university: Columbus].

i. 1907. Editor, William A[lbert] Noyes. pp.[ii].1086 + [ii].1087–2206 + [ii].2207–3074+[ii].3075–3348. [12,000.]
ii. 1908. pp.[ii].1216+[ii].1217–2318+[ii].2319–3878. [14,000.]
iii. 1909. pp.[ii].1706 + [ii].1707–3562. [14,000.]
iv. 1910. Editor, Austin M. Patterson. pp.[ii].1916+[ii].1917–4028. [15,000.]
v. 1911. pp.[ii].1540+[ii].1541–3182+[ii].3183–4741. [20,000.]

vi. 1912. pp.[ii].1076+[ii].1077–2700+[ii]. 2701–4331. [20,000.]

vii. 1913. pp.[ii].1428+[ii].1429–3252+[ii]. 3253–4926. [25,000.]

viii. 1914. pp.[ii].1888 + [ii].1889–3516 + [ii].3517–4613. [22,500.]

ix. 1915. Editor, E. J. Crane. pp.[ii].xix.1408 +[ii].1409–2824+[ii].2825–3961.[20,000.]

x. 1916. pp.[ii].1446+[ii].1447–2650+[ii]. 2651–3673. [17,500.]

— Decennial index to . . . volumes 1–10. pp. viii.1006 + [ii].1007–1980 + [ii].xvi.1981–3449+[ii].3451–4823.

xi. 1917. pp.[ii].xxv.1056+[ii].1057–2418+ [ii].2149–3995. [20,000.]

xii. 1918. pp.[ii].xxi.1136+[ii].1137–2144+ [ii].2145–3194. [17,500.]

xiii. 1919. pp.[ii].1168+[ii].1169–2616+[ii]. 2617–3928. [20,000.]

xiv. 1920. pp.[ii].xxxii.1620+[ii].1621–3170 +[ii].3171–4642. [22,500.]

xv. 1921. pp.[ii].1640+[ii].1641–3568+[ii]. 3569–4951. [25,000.]

xvi. 1922. pp.[ii].1682 + [ii].1683–3558 + 3559–5463.lx. [27,500.]

xvii. 1923. pp.[ii].1904 + [ii].1905–3430 + [ii].3431–4943. [30,000.]

xviii. 1924. pp.[ii].1590 + [ii].1591–3294 + [ii].3295–4876. [30,000.]

xix. 1925. pp.[ii].1640 + [ii].1641–3174 + [ii].3175–4773. [30,000.]

xx. 1926. pp.[ii].1730+[ii].1731–3362+[ii]. 3363–5160. [30,000.]

— Decennial index to . . . volumes 11–20. pp. vi.1256 + [ii].1257–2452 + [ii].xix.2453–3801+[ii].3803–5196+[ii].5197–6591.

xxi. 1927. pp.[ii].1724+[ii].1725–3490+[ii]. 3491–5512. [32,500.]

xxii. 1928. pp.[ii].1874 + [ii].1875–4008 + [ii].4009–6631. [40,000.]

xxiii. 1929. pp.[ii].2072 + [ii].2073–4104 + [ii].4105–5164+[ii].5165–7349. [45,000.]

xxiv. 1930. pp.[ii].2008 + [ii].2009–4186 + [ii].4187–6066+[ii].6067–8211. [50,000.]

xxv. 1931. pp.[ii].2022 + [ii].2023–4150 + [ii].4151–6160+6161–8304. [50,000.]

xxvi. 1932. pp.[ii].2086 + [ii].2087–4208 + [ii].4209–6184+[ii].6185–8455. [50,000.]

xxvii. 1933. pp.[ii].2064 + [ii].2065–4134 + [ii].4135–6024+[ii].6025–8197. [50,000.]

xxviii. 1934. pp.[ii].coll.3272 + pp.[ii].coll. 3273–6592 + pp.[ii].coll.6593–7596.pp. 7597–8754. [60,000.]

xxix. 1935. pp.[ii].coll.3552 + pp.[ii].coll. 3553–7124 + pp.[ii].coll.7125–8408.pp. 8409–9739. [65,000.]

xxx. 1936. pp.[ii].coll.3680 + pp.[ii].coll. 3681–7388 + pp.[ii].coll.7389–8692.pp. 8693–10,097.cxxvi. [67,500.]

— Decennial index to . . . volumes 21–30. pp.[iv].1572 + [ii].1573–3095 + [ii].3097–4680+[ii].4681–6357+[ii].6359–7980.

— Patents index . . . 1907–36. 1944. pp.v.479.

xxxi. 1937. pp.[ii].coll.4160 + pp.[ii].coll. 4161–8256 + pp.[ii].coll.8257–8996.pp. 8997–10,484. [72,500.]

xxxii. 1938. pp.[ii].coll.4828 + pp.[ii].coll. 4829–9564 + pp.[ii].coll.9565–11,111. [75,000.]

xxxiii. 1939. pp.[ii].coll.4464 + pp.[ii].coll. 4465–9048 + pp.[ii].coll.9049–9720.pp. 9721–11,330. [80,000.]

xxxiv. 1940. pp.[ii].coll.3948 + pp.[ii].coll. 3949–7760 + pp.[ii].coll.7661–8340.pp. 8341–9726. [65,000.]

xxxv. 1941. pp.[ii].coll.3848 + pp.[ii].coll. 3849–7760 + pp.[ii].coll.7761–8368.pp. 8369–9735. [65,000.]

xxxvi. 1942. pp.[ii].coll.3396 + pp.[ii].coll. 3397–6840 + pp.[ii].coll.6841–7368.pp.

7369–8569. [60,000.]

xxxvii. 1943. pp.[ii].coll.3300+pp.[ii].coll. 3301–6496 + pp.[ii].6497–6940.pp.6941–8144. [55,000.]

xxxviii. 1944. coll.7790. [52,500.]

xxxix. 1945. coll.6705. [45,000.]

xl. 1946. coll.8890. [60,000.]

xli. 1947. coll.9318. [62,500.]

xlii. 1948. coll.10,990. [75,000.]

xliii. 1949. pp.[ii].coll.3244 + pp.[ii].coll. 3245–6468 + pp.[ii].coll.6469–9540.pp. 9541–11,534. [90,000.]

xliv. 1950. pp.[ii].coll.3744 + pp.[ii].coll. 3745–7588 + pp.[ii].coll.7589–11,184.pp. 11,185–13,534. [100,000.]

xlv. 1951. pp.[ii].coll.3656 + pp.[ii].coll. 3657–7380 + pp.[ii].coll.7381–10,680 + pp.10,681–12,959.cclv. [100,000.]

xlvi. 1952. coll.12,406. [100,000.]

xlvii. 1953. coll.12,888+pp.685+1733+387. 15. [78,000.]

xlviii. 1954. coll.14,302+pp.757.14+2023+ 444. [90,000.]

xlix. 1955. coll.16,528+pp.826.13+2267+ 613. [100,000.]

l. 1956. coll.17,536+pp.898.16+314+2604 +690. [110,000.]

li. 1957. coll.18,706+pp.996.21.27+2593+734. [120,000.]

lii. 1958. coll.21,256+pp.1176.27.40+2895+718. [130,000.]

liii. 1959. Editor: Charles L. Bernier. coll. 23,114 + pp.1280.32.32 + 3042 + 1066. [140,000.]

liiii. 1960. coll.26,028+pp.1382.34.48+1927+679. [160,000.]

in progress.

CHEMISTRY. Selected list of books. Public library: Detroit 1915. pp.24. [350.]

CATALOGUS van de boeken en tijdschriften geplaatst in de laboratoria voor scheikunde en toegepaste scheikunde, en voor mineralogie en geologie. Universiteit: Bibliotheek: Amsterdam 1919. pp.[iii].95. [chemistry: 1250.]

BIBLIOGRAPHIA chimica. Internationaler literatur-anzeiger für chemie, chemische technologie und alle grenzgebiete. Leipzig &c.

i. 1922. Herausgegeben . . . von Hermann A[ugust] L[udwig] Degener. pp.xxiv.244. [5000.]

ii. 1923. pp.xxvi [*sic*, xxvii].180. [3500.]

iii. 1924. pp.xxiv.116. [2000.]

iv. 1925. pp.xxxii.152. [3000.]

v. 1926. pp.xxxii.140. [2500.]
no more published.

JOAQUÍN DIAZ DE LA JARA, Bibliografía química de autores españoles. . . . Tomo I. Madrid 1924. pp.48. [500.]
no more published.

ABSTRACTS of chemical papers. Bureau of chemical abstracts.
1924. pp.1030. [7500.]
1925. pp.1016. [7500.]
[*continued as:*]
British chemical [and physiological] abstracts [1946 &c.: British abstracts].
1926 pp.[iv].1284+[iv].1030+430. [15,000.]
1927. pp.[iv].1228+[iv].974+514. [17,500.]
1928. pp.[iv].1408+[iv].944+576. [20,000.]
1929. pp.[iv].1500+[iv].1050+642. [22,500.]
1930. pp.[iv].1628+[iv].1170+705. [25,000.]
1931. pp.[iv].1466 + [iv].1156 + [ii].592. [22,500.]
1932. pp.[iv].1296 + [iv].1138 + [ii].578. [25,000.]
— Collective index, 1923–1932. Part I. Index of authors. [1934]. pp.[ii].1092+[ii].1093–2101.
1933. pp.[iv].1344+[iv].1088+640. [27,500.]

1934. pp.[iv].1422+[iv].1118+692. [27,000.]
1935. pp.[iv].1552+[iv].1168+754. [30,000.]
1935. pp.[iv].1572+[iv].1238+794. [30,000.]
1937. pp.[iv].636+[iv].530+[iv].504+[iv].1414+823. [32,500.]
— Collective index, 1933–1937. pp.[ii].1964+[ii].1584.
1938. pp.[iv].644+[iv].518+[iv].1066+[iv].1512+973. [35,000.]
1939. pp.[iv].628+[iv].566+[iv].1114+[iv].1304+646+302. [35,000.]
1940. pp.[iv].448+[iv].386+[iv].946+512+228. [27,500.]
1941. pp.[iv].coll.492+pp.[iv].388+pp.[iv].coll.1088 + pp.392 + pp.[iv].coll.564 + pp.[iv].coll.452 + pp.[iv].coll.352 + pp.245. [17,500.]
1942. pp.[iv].coll.412+pp.[iv].coll.432+pp.[iv].coll.952 + pp.[iv].coll.540 + pp.[iv].coll.452 + pp.[iv].coll.280 + pp.336 + 222. [15,000.]
1943. pp.[iv].coll.324+pp.[iv].coll.404+pp.[iv].coll.928 + pp.[iv].coll.508 + pp.[iv].400 + pp.[iv].coll.308 + pp.333 + 233. [15,000.]
1944. pp.[iv].coll.296+pp.[iv].coll.384+pp.[iv].coll.856 + pp.[iv].coll.420 + pp.[iv].

coll.384 + pp.[iv].coll.268 + pp.[iv].coll. 268+pp.[iv].coll.208+pp.351. [12,500.]

1945. pp.[iv].coll.344+pp.[iv].coll.420+pp. [iv].coll.904 + pp.[iv].coll.404 + pp.[iv]. coll.404 + pp.[iv].coll.264 + pp.[iv].coll. 300+pp.402+264+61. [15,000.]

1946. pp.[iv].coll.400+pp.[iv].coll.708+pp. [iv].coll.1100 + pp.[iv].coll.432 + pp.[iv]. coll.480 + pp.[iv].coll.268 + pp.[iv].coll. 316+pp.682+302. [17,500.]

1947. pp.[iv].coll.352+pp.[iv].coll.528+pp. [iv].coll.912 + pp.[iv].coll.340 + pp.[iv]. coll.348 + pp.[iv].coll.244 + pp.[iv].coll. 244+pp.588+246. [15,000.]

1948. pp.[iv].coll.424+pp.[iv].coll.948+pp. [iv].coll.1316 + pp.[iv].coll.732 + pp.[iv]. coll.636 + pp.[iv].coll.460 + pp.[iv].coll. 312 + pp.412 + 413–904 + 905–1197. [25,000.]

1949. pp.[iv].coll.536+pp.[iv].coll.916+pp. [iv].coll.1304 + pp.[iv].coll.956 + pp.[iv]. coll.1216 + pp.[iv].coll.532 + pp.[iv].coll. 490+pp.612+613–1227. [35,000.]

1950. pp.[iv].coll.804 + pp.[iv].coll.1324 + pp.[iv].coll.2036 + pp.[iv].coll.1124 + pp. [iv].1246 + pp.[iv].coll.532 + pp.[iv].coll. 520+pp.836. [40,000.]

1951. pp.[iv].coll.1068+pp.[iv].coll.1360+pp.[iv].coll.1640 + pp.[iv].coll.1312 + pp.[iv].coll.1176 + pp.[iv].coll.660 + pp.[iv].488+pp.826+93. [40,000.]

1952. pp.[iv].coll.1008+pp.[iv].coll.1752+pp.[iv].coll.2100 + pp.[iv].coll.1724 + pp.[iv].coll.1292 + pp.[iv].coll.712 + pp.[iv].coll.584. [45,000.]

1953. pp.[iv].coll.968 + pp.[iv].coll.1552 + pp.[iv].coll.1480 + pp.[iv].coll.1260 + pp.[iv].coll.996 + pp.[iv].coll.504 + pp.[iv].528 + pp.[ii].804 + [ii].567. [50,000.]

for 1924–1936 each issue consists of two sections, A. Pure chemistry and B. Applied chemistry, and of a collective index; this index first appeared for 1926; for 1924–1925 only B is shown above; this section also forms a supplement to the Journal of the Society of chemical industry, *in which abstracts appeared before 1924 as part of the text; section A for 1924–1925 is entered above under* Journal of the Chemical society. . . . Abstracts; *for 1937–1939 there are five sections: A 1. General, physical and inorganic chemistry, A 2. Organic chemistry, A 3. Biochemistry [Physiology and biochemistry], B. Applied chemistry, and index; from 1939 A and B have separate indexes; B for 1940 appeared with the* Transactions *of the Society of chemical industry; from 1941 B is in three parts: i. General and inorganic, ii. Industrial organic, iii. Agriculture, foods, sanitation &c.; from 1944 there is a section C. Analysis and apparatus.*

F. A. MASON, An introduction to the literature of chemistry for senior students and research chemists. Oxford 1925. pp.41. [300.]

PHYSIK, bearbeitet von Ludwig Schiller. Chemie, bearbeitet von Walter Becker. Jahresberichte des Literarischen zentralblattes über die wichtigsten wissenschaftlichen neuerscheinungen des gesamten deutschen sprachgebietes (1924, vol.xx): Leipzig 1925. pp.104. [Chemistry: 150.]

A MAGYAR kémiai irodalom bibliográfiája.
1901–1925. Ősszeállította: Gáspár Margít. Országos könyvforgalmi es bibliográfiai központ (ser. B, vol.x, no.1): Budapest 1931. pp.xii.334. [6000.]
1926–1945. 1957. pp.319. [8118.]
1946–1955. 1958. pp.518. [12,343.]

MELVIN GUY MELLON, Chemical publications, their nature and use. International chemical series: New York 1926. pp.viii.253. [1000.]
— — Third edition. 1958. pp.x.327. [1500.]

E[VAN] J[AY] CRANE and AUSTIN M[CDOWELL]

PATTERSON, A guide to the literature of chemistry. New York 1927. pp.[ix].438. [3750.]

— — Second edition. [By] E. J. Crane . . . A. M. Patterson . . . Eleanor B. Marr. [1957]. pp.xv.397. [3000.]

SELECT list of books on chemistry and chemical technology in the . . . libraries. Public libraries committee: Newcastle-upon-Tyne 1927. pp.32. [425.]

INDEX of chemical literature in Indianapolis. American chemical society: Indiana section: [Indianapolis] 1931. pp.35. [600.]

A MAGYAR kémiai irodalom bibliográfiája. Budapest.

1901–1925. [By] Margit Gáspár. Orszagos könyvforgalmi: A magyar tudományos irodalom bibliográfiája (vol.x): 1931.

1926–1945. Akadémiai kiadó: 1957. pp.xii. 319. [8118.]

1946–1955. 1958. pp.518. [12,343.]

FRITZ FERCHL-MITTENWALD, Chemisch-pharmazeutisches bio- und bibliographikon. Gesellschaft für geschichte der pharmazie [Berlin]: Mittenwald 1937–1938. pp.[iv].603. [25,000.]

limited to the period 1500–1850.

READERS' guide to books on mathematics, physics and chemistry. Library association: County libraries section [no.11]: 1938. pp.[iv].36. [900.]

BYRON A[VERY] SOULE, Library guide for the chemist. International chemical series: New York &c. 1938. pp.xiii.302. [500.]

ביבליוגרפיה של הספרות הכימית .... Bibliography of chemical literature available in Palestine. Compiled by the Central committee for the development of chemical industry in Palestine. Edited by the Jewish agency for Palestine, Department for trade and industry. Jerusalem 1942. pp. [209]. [4000.]*

[KATHERINE R. THOMSON and MARY MARGARET SERVICE], Catalogue of the [John] Ferguson collection of books, mainly relating to alchemy, chemistry, witchcraft and gipsies, in the library of the university of Glasgow. [Edited by William Ross Cunningham]. Glasgow university publications: [Glasgow] 1943. pp.xv.384+[ii].385–820. [6000.]

CHEMIE a chemicka technologie. Přehled technické a hospodářské literatury: Praha.

i.
ii.
iii.

iv.
v.
vi.
vii.
viii.
ix. 1952. pp.864. [10,000.]
x. 1953. pp.519. [7500.]
xi. 1954.
xii. 1955.
xiii. 1956. pp.528. [7500.]
xiv. 1957. pp.642. [9000.]
xv. 1958. pp.576. [9000.]
xvi. 1959. pp.644. [9000.]

V. P. ALEKSEEV [*and others*], Химия в изданиях Академии наук СССР. Академия наук СССР: Библиотека: Химическая литература СССР (vol.i): Москва.

i. 1728–1930. 1947. pp.332. [2400.]
ii. 1931–1935. 1951. pp.246. [1276.]

in progress.

CHEMISTRY in wartime in the Netherlands. A review of scientific work done by dutch chemists in the years 1940–1945. Amsterdam 1947.

СИСТЕМАТИЧЕСКИЙ указатель статей в иностранных журналах.... Химические науки. Всесоюзная государственная библиотека иностранной литературы. Москва.

i–ii. 1948. 1950. pp.332+332. [5732.]

iii–iv. 1949. 1951–1952. pp.408 + 468. [8730.]

BILTEN sadržaja stranih časopisa. Dodatak "Zelenom biltenu". D: Hemija, hemiska industrija. Savezna uprava za unapredenje proizvodnje: [Belgrade].

i. 1950. pp.340. [1700.]

ii. 1951. pp.342. [1700.]

iii. 1952. pp.490. [2500.]

iv. 1953. pp.800. [4000.]

v. 1954. pp.660. [3300.]

no more published.

G[EORGE] MALCOLM DYSON, A short guide to chemical literature. 1951. pp.[iii].144. [200.]

— — Second edition. 1958 [1959]. pp.[vii].157. [750.]

NERIO GAUDENZI, Guida bibliografica internazionale per il chimico. Libri e reviste. Firenze [1952]. pp.[v].514. [4393.]

ХИМИЯ. Систематический указатель статей в иностранных журналах. Всесоюзная государственная библиотека иностранных журналов. [&c.]: Москва.

1952. pp.96+88+80+59+80+72+96+88 +132. [14,589.]
1953. pp.117+100+92+96+68+76+76+ 60+57+56+49 [15,000.]
1954. pp.104+92+96+88+81+76+61+ 64. [12,500.]
1955. pp.77+84+81+89+92+96+72+84 76+104. [20,000.]
1956. pp.112+90+81+88+72+72+77+ 68+88+84. [17,500.]
1957.
1958. pp.128+100+116+112+116+128+ 130+120+140+104. [25,000.]
1959. pp.101+100+92+84+88+108+ 105+124+112+84. [25,000.]
1960. pp.132+108+112+128+104+92 +112+100+104+108. [25,000.]
1961. pp.132+140+127+108+120+108 +124+128+86+92. [30,000.]
1962. pp.83+92+112+108+104+108+ 92+90+96+140. [25,000.]
1963. pp.124+100+124+92+114+100 +120+103+116+160. [30,000.]

in progress.

1. A. BAITIN, Химическая литература. Библиографический справочник (1920–1951 гг.). Москва 1953. pp.564. [6000.]

Chemistry

TITLES of theses submitted for doctoral degrees in chemistry and chemical engineering at american educational institutions October 1, 1951 to October 30, 1952. American chemical society: Committee on professional training: Washington [1953]. pp.[iv].55. [600.]

FACULTIES, publications and doctoral theses in chemistry and chemical engineering at United States universities. American chemical society: Committee on professional training: [Washington] 1953 &c.*

details of this work are entered under Academies: Chemistry, above.

РЕФЕРАТИВНЫЙ журнал. Химия. Академия наук СССР: Институт научной информации: Москва.

1953. coll.3732. pp.xii.xii.xiv.xvi.xx.xxii. [10,042.]

1954. pp.392.xx + 352.xviii + 320.xvi + 338.xiv + 352.xvi + 356.xvii + 326.xvi + 304.xvi + 322.xvi + 308.xvi + 324.xvi + 346.xvii + 333.xvi + 327.xvii + 338.xvii + 336.xvii + 320.xvi + 318.xviii + 31c. xvi + 301.xviii + 281.xvi + 295.xvi + 308.xvi. [51,047.]

— Авторский и патентный указатели за 1953–1954 гг. pp.1406.

— Предметный указатель за 1953–1954 гг. 1958–1959. pp.1656+384.

1955. pp.312.xiv + 298.xv + 334.xvi + 355. xvi + 347.xviii + 392.xix + 434.xxii + 455.xxi + 478.xxiii + 456.xxiv + 468. xxiv + 480.xxiv + 468.xxiv + 430.xxiii + 449.xxiv + 452.xxv + 450.xxvi + 440. xxv + 443.xxv + 473.xxvi + 467.xxviii. 509.xxviii. [57,470.]

— Авторский и патентный указатели. pp.1584.

1956. pp.471.xii + 479.xvi + 454.xviii + 477.xvi + 502.xviii + 505 + 507 + 495 + 529 + 519 + 477 + 523 + 520 + 516 + 549 + 556 + 517 + 508 + 504 + 491 + 473+473+465+504. [80,462.]

— Авторский указатель. 1959. pp.1844.*

— Патентный указатель. 1959. pp.1845–2032.*

— Предметный указатель за 1955–1956 гг. 1960–1961. pp.716+608+663+737+534 + 364 + 618 +   +   + 604 + 367+440+478.

— Формульный указатель за 1955–1956 гг. 1960. pp.1428.

1957. pp.496 + 576 + 568 + 536 + 584 + 596 + 616 + 612 + 629 + 509 + 524 + 525 + 528 + 505 + 504 + 497 + 485 + 453 + 421 + 413 + 412 + 432 + 420 + 424. [79,039.]

1958. pp.543 + 503 + 512 + 484 + 520 + 555 + 548 + 576 + 565 + 557 + 541 + 569 + 528 + 556 + 529 + 607 + 560 + 552 + 480 + 496 + 502 + 496 + 624 + 608. [84,041.]

1959. pp.524 + 557 + 563 + 549 + 621 + 632 + 636 + 672 + 633 + 609 + 581 + 567 + 601 + 628 + 576 + 581 + 616 + 571 + 541 + 576 + 593 + 592 + 608 + 680. [89,013.]

1960. pp.540 + 644 + 661 + 649 + 713 + 701 + 657 + 621 + 625 + 631 + 640 + 832 + 815 + 712 + 668 + 609 + 644 + 656 + 608 + 619 + 585 + 604 + 629 + 640. [99,365.]

— Авторский указатель. pp.768+641+561+626.

1961. pp.612 + 640 + 656 + 696 + 736 + 744 + 692 + 732 + 672 + 663 + 715 + 744 + 752 + 764 + 684 + 683 + 680 + 624 + 592 + 532 + 532 + 548 + 632 + 668. [120,000.]

— Авторский указатель. pp.776+

1962. pp.596 + 688 + 676 + 700 + 728 + 772 + 736 + 676 + 736 + 740 + 712 + 696 + 728 + 744 + 664 + 636 + 640 + 640 + 672 + 616 + 556 + 640 + 848 + 1108+504+408+664+680. [130,000.]

1963. pp.616 + 636 + 696 + 708 + 684 + 736 + 688 + 800 + 776 + 768 + 784 + 788 + 800 + 772 + 728 + 724 + 676 + 700 + 664 + 616 + 576 + 716 + 716 + 768+ + + . [140,000.]

in progress.

CURRENT chemical papers. A classified world list of new papers in pure chemistry. Chemical society.

1954. pp.634. [20,000.]
1955. pp.718. [20,000.]
1956. pp.731. [22,500.]
1957. pp.820. [25,000.]
1958. pp.825. [25,000.]
1959. pp.908. [27,500.]
1960. pp.914. [27,500.]
1961. pp.1091. [30,000.]
1962. pp.1090. [30,000.]
1963. pp.1115. [32,500.]

in progress.

S. R. SERGIENKO and S[EMEN] I[SAKOVICH] VOLFKOVICH, Русские химики. Аннотированный указатель литературы. Библиотека СССР имени В. И. Ленина: Центральная политехническая библиотека: Что читать о выдающихся деятелях отечественной науки и техники (no.5): Москва 1954. pp.146. [500.]

OLDŘICH HANČ, VRATISLAV HUMMEL and BŘETISLAV HLAVICA, Chemická literatura její dokumentace a použití. Československá akademie věd: Sekce chemická: Praha 1954. pp.424. [2500.]

— — 2 . . . vydání. 1961. pp.473. [4000.]

HALINA ANGERSTEIN [*and others*], Piśmiennictwo chemiczne w Polsce ludowej. Warszawa 1955. pp.208. [3539.]

BIBLIOGRAFIA brasileira de química. Instituto brasileiro de bibliografia e documentação: Rio de Janeiro.

i. 1950–1955. pp.185. [957.]
ii. 1956–1958. pp.159. [879.]
iii. 1959–1960. pp.84. [359.]

in progress.

ANDRÁS BELICZAY, TAMÁS WINDHOLZ and JENŐ GALÁNTAY, A vegyi irodalom útmutatója. Budapest 1956. pp.172. [1000.]

INHALTSVERZEICHNISSE sowjetischer fachzeitschriften. Reihe III B. Chemie, biologie. Zentralstelle für wissenschaftliche literatur: Berlin.

v. 1956. pp.1578. [4500.]
vi. 1957. pp.2144.181. [6000.]
vii. 1958. pp.2281. [6000.]
viii. 1959. pp.2476. [6500.]

in progress; the earlier and parallel volumes are entered under Science, above.

CHEMISTRY (excluding chemical technology). Linda hall library: Bibliographic bulletin (no.14): Kansas City 1957. pp.[iv].100. [1406.]*

ABSTRACTS of bulgarian scientific literature. Chemistry and chemical technology. Bulgarian academy of sciences: Department for scientific and technical information and documentation: Sofia.*

i. 1957. pp.[iv].67. [103.]
ii. 1958–1959. pp.ii.84+[ii].89. [236.]
iii. 1959–1960. pp.108+67. [260.]
iv. 1960–1961. pp.46+22. [180.]

HANS BERNDT and STEGFRIED GÜNTHER, Bibliographie der deutschen hochschulschriften zur chemie. Ein systematisches verzeichnis der in den jahren 1957–1958 an deutschen universitäten und hochschulen eingereichten dissertationen und habilitationsschriften. Deutsche bücherei: Sonderbibliographien (no.23): Leipzig 1960. pp.viii.264. [2631.]

ST. JANISZEWSKA-DRABAREK, J. SUROWIŃSKI and A. SZUCHNIK, Literatura chemiczna, technika posługivania się. Warszawa 1960. pp.163. [200.]

RUSSIAN chemical reviews. Uspekhi khimii. Chemical society.*

1960. pp.[v].689. [10,000.]
1961. pp.[vii].671. [10,000.]
1962. pp.[vii].736. [10,000.]

in progress?

V[LADIMIR] V[ASILEVICH] FELDT and A[NASTASIA] T[ERENTEVNA] FELDT, Аннотированный указатель литературы по химии. Москва 1961. pp.160. [300.]

CHEMICAL titles. American chemical society: [Easton, Pa.] 1961 &c.*

in progress; six preliminary numbers were issued in 1960.

ХИМИЯ. Систематический указатель статей в иностранных журналах. Всесоюзная государственная библиотека иностранной литературы: Москва 1961 &c.*

in progress.

R. T. BOTTLE, Use of the chemical literature. 1962. pp.x.231.

ÍNDICE bibliográfico del Centro de investigación y de estudios avanzados. Sección 3ª. Química. Instituto politécnico nacional: México 1962 &c.
in progress.

Agricultural chemistry. see **Agriculture: Chemistry.**

6. *Analytic chemistry*

BRIEF list of books on chemical analysis. Library of Congress: Washington 1918. ff.2. [19.]*

TH[EODOR] DÖRING, Analytische chemie. Wissenschaftliche forschungsberichte: Naturwissenschaftliche reihe (vol.i): Dresden &c. 1921. pp. [vii].98. [600.]

HERMINE and SÉBASTIEN SABETAY, Les travaux récents d'analyse et de synthèse organiques et la chimie des parfums, de 1935 à 1938. Monographies de chimie pure et appliquée (new ser.): 1941. pp.823. [5000.]

WILHELM KLEMM [*and others*], Analytical chemistry of inorganic substances. Office of military government for Germany: Field information

agencies technical: Fiat review of german science, 1939–1946: Wiesbaden 1948. pp.[viii].326. [2000.]
the text is in german.

BIBLIOGRAPHY of standard, tentative and recommended or recognised methods of analysis. Society of public analysis and other analytical chemists: Analytical methods committee: Cambridge [1951]. pp.vii.225. [500.]

BIBLIOGRAFICKÝ přehled analytické chemie anorganické. Výběrovy seznam čs. a cizojazyčne časopisecké literatury ze zvláštním zaměřením k rozborům metalurgickým. Státní vědecka knihovna v Ostravě: Příručka (no.41 &c.) [Ústredná technická knižnica]: Ostrava [Bratislava].*

1945–1952. . . . (no.41). [By] Vilem Mayer. pp.xi.ff.347+257. [5210.]
1953. . . . (no.51). ff.157. [1131.]
1954. . . . (no.59). pp.219. [1114.]
1955.
1956. pp.299. [2098.]
1957. pp.339. [2460.]
1958. pp.321. [2008.]
1959. pp.335. [2263.]
1960. pp.[ii].290. [2055.]

in progress.

ANALYTICAL abstracts. Society for analytical chemistry: Cambridge.

i. 1954. [Edited by Norman Evers]. pp.[370]. cxxxv. [3190.]
ii. 1955. pp.[484].clix. [3556.]
iii. 1956. pp.[560].clxiii. [3820.]
iv. 1957. pp.[576].clxvi. [4223.]
v. 1958. pp.[568].clxxxviii. [4412.]
vi. 1959. pp.[680].ccxviii. [5073.]
vii. 1960. pp.[720].ccxxxii. [5522.]
viii. 1961. pp.[704].ccxxiv. [5334.]
ix. 1962. pp.[672]. [5564.]
x. 1963. pp.[680]. [5503.]

in progress.

[WIKTOR KEMULA *and others*], Polska bibliografia analityczna. Chemia. Polska akademia nauk: Ośredek bibliografii i dokumentacji naukowej: Warszawa.

i. 1955. pp. . [500.]
ii. 1956. pp.98.106.174. [1100.]
iii. 1957. pp.213. [500.]

no more published.

A[LEKSEI] I[VANOVICH] BUSEV, Аналитическая химия. Литература на русском языке (1941–1952 гг.). Академия наук СССР: Институт научной информации: Москва 1956. pp.404. [8495.]

7. *Chemical warfare*

LIST of recent references on chemical warfare. Library of Congress: [Washington] 1923. ff.6. [58.]*

BOOKS and pamphlets, published since 1919, on chemical warfare and defence. Science library: Bibliographical series (no.292): 1937. ff.5. [80.]*

— Supplement . . . (no.342): 1937. ff.2. [30.]*

— Supplement . . . (no.500): 1939. ff.2. [30.]*

CHARLES H[ANCOCK] WENTZ, A check list, civilian defense collection, section D, subsection 5. Chemical warfare, decontamination. Library of Congress: Reference department: Washington 1943. ff.[i].17. [201.]*

— — [another edition]. Chemical warfare. A check list. 1944. pp.[ii].19.38. [544.]*

[CLARENCE JAY WEST], Chemical warfare. Bibliographic series (no.1): Cambridge, Mass. 1949. pp.12. [450.]

CHEMICAL and biological warfare (URSS). Bibliography. Library of Congress: Aerospace information division: Report (B-63-52): [Washington] 1963. ff.[ii].34. [278.]*

8. *Inorganic chemistry*

GUSTAV THEODOR FECHNER, Repertorium der neuen entdeckungen in der unorganischen chemie. Lehrbuch der theoretischen und praktischen chemie (vol.vii): Leipzig 1830–1833. pp.viii.768+iv.435+[ii].581. [4000.]

ABRIDGMENTS of specifications. Class 1, acids, alkalies, oxides, and salts, inorganic. Patent office.

1855–1866. pp.xx.166. [1000.]
1867–1876. pp.xxiii.224. [1000.]
1877–1883. pp.xxii.220. [1000.]
1884–1888. pp.xviii.192. [1000.]
1889–1892. pp.xviii.148. [750.]
1893–1896. pp.xx.164. [750.]
1897–1900. pp.xxii.186. [1000.]
1901–1904. pp.xxiv.208. [1000.]
1905–1908. pp.xxvi.268. [1250.]
[*continued as:*]

1. i. Chemical processes and apparatus.
1909–1915. pp.xi.186. [750.]
1916–1920. pp.x.115. [400.]
1921–1925. pp.xiii.187. [750.]
1926–1930. pp.xvii.353. [1500.]

1. ii. Inorganic compounds other than metallic oxides, hydrates, oxyacids, and salts (in-

cluding alkali manufacture and cyanogen compounds).

1909–1915. pp.xiv.144. [600.]
1916–1920. pp.xiv.104. [500.]
1921–1925. pp.xv.124. [500.]
1926–1930. pp.xvii.167. [750.]

I. iii. Oxides, hydrates, oxyacids, and salts (other than alkali manufacture and cyanogen compounds).

1909–1915. pp.xx.157. [750.]
1916–1920. pp.xvi.112. [500.]
1921–1925. pp.xix.146. [750.]
1926–1930. pp.xxiv.216. [860.]

no more published.

PATENTS for inventions. Fifty years subject index, 1861–1910. Class I. Patent office.

i. Chemical processes and apparatus. 1915. pp.7. [750.]

ii. Inorganic compounds other than metallic oxides, hydrates, oxyacids, and salts. 1914. pp.21. [3000.]

iii. Oxides, hydrates, oxyacids, and salts, metallic. 1913. pp.37. [5000.]

ROBERT SCHWARZ, Anorganische chemie. Wissenschaftliche forschungsberichte: Naturwissenschaftliche reihe (vol.xvi): Dresden &c. 1927. pp. xi.140. [600.]

LIST of books on inorganic chemistry published in America. Science library: Bibliographical series (no.30): [1932]. ff.[i].3. [35.]*

BIBLIOGRAFICKÝ přehled analytické chemie anorganické. Ostrava 1945 &c.

details of this work are entered under Chemistry: Analytical, above.

WILHELM KLEMM [*and others*], Inorganic chemistry. Office of military government for Germany: Field information agencies technical: Fiat review of german science, 1939–1946: Wiesbaden.

- i. [Elements, nuclear chemistry, compounds]. 1948. pp.[xii].278. [600.]
- ii. [Metal halides and chalcogenides, additional compounds]. 1948. pp.[iv].179. [400.]
- iii. [Compounds of higher order, intermetallic systems]. 1948. pp.[iv].275. [600.]
- iv. [Structure, physical measurement, thermochemistry, electrochemistry]. 1948. pp. [iv].260. [600.]
- v. [Equilibria, scientific bases of metallurgical processes]. 1948. pp.[iv].315. [600.]
- vi. [Active states and reactions in the solid state, constitution and catalysis, solutions, new methods, bibliography, history, in-

dexes]. 1949. pp.[iv].434. [750.]
the text is in german.

S. V. ROTKOVA and I. H. METSATUNYAN, Химия редких элементов. Библиографический указатель отечественной и зарубежной литературы. Академия наук СССР [&c.]. Москва.

i. 1951–1954. 1960. pp.420. [7000.]
ii. 1955–1956. 1964. pp.264. [4500.]

in progress.

9. *Medical and physiological chemistry*

ANNUAIRE de chimie, comprenant les applications de cette science à la médecine et à la pharmacie, ou répertoire des découvertes et des nouveaux travaux en chimie faits dans les diverses parties de l'Europe.

1845. Par E[ugène Auguste Nicolas] Millon et J[ules] Reiset. pp.
1846. pp.xxii.874. [433.]
1847. pp.xiv[*sic*, xxiv].818. [526.]
1848. pp.xxiii.531. [518.]
1849. pp.xix.630. [375.]
1850. pp.xviii.765. [417.]
1851. Par E. Millon et [François Joseph] J[érôme] Nicklès. pp.xvi.576. [329.]

no more published.

Chemistry

JAHRESBERICHT über die fortschritte der thierchemie [*afterwards:* oder der physiologischen und pathologischen chemie; *afterwards:* oder der physiologischen, pathologischen und immunochemie und der pharmakologie]. Wien [vols.iii–xlvii. Wiesbaden; xlviii–xlix: München].

i. 1871. Herausgegeben von Richard Maly. pp.vii.345. [300.]

ii. 1872. pp.[v].377. [300.]

iii. 1873. pp.[vii].336. [250.]

iv. 1874. pp.viii.483. [300.]

v. 1875. pp.viii.344. [300.]

vi. 1876. Herausgegeben von Felix Hoppe-Seyler. pp.vii.295. [300.]

vii. 1877. Herausgegeben von R. Maly. pp. [vii].397. [400.]

viii. 1878. pp.[vii].400. [400.]

ix. 1879. Herausgegeben von Richard Přibram. pp.[vii].432. [500.]

x. 1880. Herausgegeben von R. Maly, pp. [vii].506. [500.]

— Sach- und autoren-register über die ersten zehn jahrgänge. . . . Von Rudolf Andreasch. 1881. pp.v.154.

xi. 1881. pp.[v].494. [500.]

xii. 1882. Herausgegeben von R. Přibram. pp.[vii].544. [600.]

xiii. 1883. Herausgegeben von R. Maly. pp. [vii].436. [500.]
xiv. 1884. pp.[iii].558. [750.]
xv. 1885. pp.[ix].542. [600.]
xvi. 1886. Redigirt von R. Andreasch. pp. [vii].552. [750.]
xvii. 1887. pp.[vii].512. [750.]
xviii. 1888. pp.[v].378. [750.]
xix. 1889. pp.[v].542. [1000.]
xx. 1890. pp.[vii].xiv.494. [1000.]
— Autoren- und sach-register zu den bänden XI–XX. . . . Von R. Andreasch. 1892. pp. [iv].198.
xxi. 1891. Herausgegeben . . . von M. von Nencki und R. Andreasch. pp.[vii].518. [1000.]
xxii. 1892. pp.[vii].694. [1250.]
xxiii. 1893. pp.[v].751. [1500.]
xxiv. 1894. pp.[v].911. [1500.]
xxv. 1895. pp.[v].717. [1500.]
xxvi. 1896. pp.[v].1038. [2000.]
xxvii. 1897. pp.[vii].979. [2000.]
xxviii. 1898. pp.[vii].888. [2000.]
xxix. 1899. pp.[vii].1058. [2250.]
xxx. 1900. pp.[ix].1140. [2500.]
— Autoren- und sachregister zu den bänden

xxi–xxx. . . . Von R. Andreasch. 1903. pp.[vi].460.

xxxi. 1901. Herausgegeben . . . von R. Andreasch und Karl Spiro. pp.xxxix.1054. [2250.]

xxxii. 1902, pp.[vii].1142. [2500.]

xxxiii. 1903. pp.[v].1298. [3000.]

xxxiv. 1904. pp.[vii].1240. [3000.]

xxxv. 1905. pp.viii.1120. [3000.]

xxxvi. 1906. pp.viii.1062. [2750.]

xxxvii. 1907. pp.xi.1276. [3500.]

xxxviii. 1908. pp.[viii].1412. [4000.]

xxxix. 1909. pp.[vi].1476. [4500.]

xl. 1910. pp.[vi].1528. [5000.]

— Namen- und sachverzeichnis zu den bänden xxxi–xl. . . . Von R. Andreasch. 1924. pp.iv.836.

xli. 1911. pp.[viii].1424. [5000.]

xlii. 1912. pp.[viii].1474. [5000.]

xliii. 1913. pp.[viii].1714. [6000.]

xliv. 1914. pp.[viii].1296. [5000.]

xlv. 1915. pp.[x].917. [4000.]

xlvi. 1916. pp.[ix].922. [3500.]

xlvii. 1917. pp.[viii].884. [3000.]

xlviii. 1918. pp.[viii].998. [3000.]

xlix. 1919. pp.[viii].1079. [3500.]

amalgamated with the Jahresbericht über die fortschritte der anatomie und physiologie, *to form the* Jahresbericht über die gesamte physiologie und experimentelle pharmakologie, *which is entered under Physiology, below.*

LIST of references on chemistry in relation to health and disease. Library of Congress: [Washington] 1925. ff.6. [48.]*

FORTSCHRITTE der physiologischen chemie, 1929–1934. Berlin [1934]. pp.[iv].311.

FRITZ SCHÖNHÖFER [*and others*], Chemotherapy. Office of military government for Germany: Field information agencies technical: Fiat review of german science, 1939–1946: Wiesbaden 1948. pp.[viii].304. [1250.]
the text is in german.

EMIL LEHNARTZ [*and others*], Physiological and pathological chemistry, including nutrition. Office of military government for Germany: Field information agencies technical: Fiat review of german science, 1939–1946: Wiesbaden 1948. pp. [viii].348. [2000.]
the text is in german.

10. *Organic chemistry*

EMIL TH. WOLFF, Quellen-literatur der theoretisch-organischen chemie, oder verzeichniss der vom anfang des letzten viertheils des vorigen jahrhunderts bis zum schluss des jahres 1844 ausgeführten chemischen untersuchungen über die eigenschaften und die constitution der organischen substanzen, ihrer verbindungen und zersetzungsproducte. Halle 1845. pp.xii.coll.808. [5000.]

ABRIDGMENTS of specifications. Class 2. Acids and salts, organic, and other carbon compounds (including dyes). Patent office.

1855–1866. pp.xiii.70. [300.]
1867–1876. pp.xii.42. [200.]
1877–1883. pp.xiv.72. [400.]
1884–1888. pp.xii.66. [300.]
1889–1892. pp.xv.100. [500.]
1893–1896. pp.xxiv.202. [1000.]
1897–1900. pp.xxxvi.402. [1500.]
1901–1904. pp.xxx.274. [1000.]
1905–1908. pp.xxvi.262. [1000.]

M[AX] M[ORITZ] RICHTER, Lexikon der kohlenstoff-verbindungen. Hamburg &c. 1900. pp.xvi. 1264+[vii].1265–2482. [125,000.]

first published in 1883 as Tabellen der kohlenstoffverbindungen.

ROBERT STELZNER, Literatur-register der organischen chemie, geordnet nach M. M. Richters formelsystem. Deutsche chemische gesellschaft [Berlin]: Braunschweig [Berlin].

i. 1910–1911. 1913. pp.xli.1286. [60,000.]
ii. 1912–1913. 1919. pp.xl.1672. [75,000.]
iii. 1914–1915. 1921. pp.xii.100.1462. [65,000.]
iv. 1916–1918. 1923. pp.viii.1381. [60,000.]
v. 1919–1921. 1926. pp.xv.1773. [80,000.]

BEILSTEINS handbuch der organischen chemie. Vierte auflage, die literatur bis 1. Januar 1910 [&c.] umfassend. Herausgegeben von der Deutschen chemischen gesellschaft. Bearbeitet von Bernhard Prager [*and others*]. Berlin 1918– . pp.xxxv.984+viii.920+x.938+ .
in progress.

RUDOLF PUMMERER, Organische chemie. . . . Zweite . . . auflage. Wissenschaftliche forschungsberichte: Naturwissenschaftliche reihe (vol.iii): Dresden &c. 1923. pp.xii.210. [500.]

CATALOGUS van de laboratorium-bibliotheken. I. . . . Van't Hoff laboratorium, Organisch-chemisch lab., Pharmaceutisch laborat., Physisch laboratorium. Rijksuniversiteit: Bibliotheek: Utrecht 1933. pp.[iv].100. [1100.]

ABSTRACTS of chemical papers. . . . A 2. Organic

chemistry. Bureau of chemical abstracts: 1937 &c.
details of this work are entered under Chemistry: General, above.

WALTER HÜCKEL [*and others*], Theoretical organic chemistry. Office of military government for Germany: Field information agencies technical: Fiat review of german science, 1939–1946: Wiesbaden 1948. pp.[viii].145+[ii].160. [1250.]
the text is in german.

SHERLOCK SWANN, Bibliography of electroorganic chemistry. Part I. University of Illinois: Engineering experiment station: Circular series (no.50): [Urbana 1948]. pp.114. [1569.]
no more published.

KARL ZIEGLER [*and others*], Preparative organic chemistry. Office of military government for Germany: Field information agencies technical: Fiat review of german science, 1939–1946: Wiesbaden 1948. pp.[viii].432 + [iv].300 + [iv].352. [3000.]
the text is in german.

SELECT bibliography on quantitative organic analysis by micro- and semi-micro-methods. Central library: Research bibliographies (new ser., no.1): Sheffield [1948]. ff.6. [50.]*

MÁRIA JUST, A szegedi tudományegyetem Szerves kémiai intézete dolgozóinak szakirodalmi munkássága az 1945–1955 években. A Szegedi egyetemi könyvtár kiadványai (no.35): Szeged 1955. pp.15. [102.]

T[ATYANA] I[VANOVNA] TEMNIKOVA and V. A. CHERKASOVA, Справочная литература по органической химии. Ленинградский . . . государственный университет имени А. А. Жданова: [Leningrad] 1961. pp.92. [100.]

ORGANOMETALIC compounds. Translation and technical information services.*

i. 1961. pp.356.53. [2631.]
ii. 1962. pp.362. [2779.]
iii. 1963. pp.398. [3788.]

NORMAN KHARASCH, WALTER WOLF and ELAINE C. P. HARRISON, Index to reviews, symposia volumes and monographs in organic chemistry for the period 1940–1960. New York &c. 1962. pp. [ii].ix.345. [5000.]*

11. *Physical chemistry*

PHYSIKALISCH-chemisches centralblatt. Vollständiges internationales referatenorgan für die physikalische chemie und die angrenzenden gebiete der chemie und physik. Leipzig.

i. 1904. pp.[ii].807. [1631.]
ii. 1905. pp.[ii].823. [1644.]
iii. 1906. pp.[ii].807. [1574.]
iv. 1907–1908. pp.li.720. [1519.]
v.
vi. 1908–1909. pp.xxxviii.592. [1222.]

ALFRED BENRATH, Physikalische chemie. Wissenschaftliche forschungsberichte: Naturwissenschaftliche reihe (vols.viii, xiv): Dresden &c. 1923–1925. pp.viii.108+x.192. [2000.]

READING list of text-books on physical chemistry published 1940–1945. Science library: Bibliographical series (no.628): 1946. ff.3. [45.]*

KLAUS CLUSIUS [*and others*], Physical chemistry. Office of military government for Germany: Field information agencies technical: Fiat review of german science, 1939–1946: Wiesbaden 1948. pp.[viii].270. [1500.]

the text is in german.

12. *Technical and industrial chemistry*

JAHRES-BERICHT über die fortschritte der chemischen technologie für fabrikanten, hütten- und forstleute, cameralisten, chemiker und pharmaceuten. Leipzig.

i. 1855. Edited by Johannes Rudolf Wagner. pp.viii.508. [1000.]
ii. 1856. pp.viii.472. [1000.]
iii. 1857. pp.viii.520. [1000.]
iv. 1858. pp.xvi.648. [1250.]

[*continued as:*]

Jahres-bericht über die fortschritte und leistungen der chemischen technologie und technischen chemie.

v. 1859. pp.xii.708. [1500.]
vi. 1860. pp.xi.668. [1500.]
vii. 1861. pp.xvi.738. [1500.]
viii. 1862. pp.xvi.736. [1500.]
ix. 1863. pp.xii.776. [1500.]
x. 1864. pp.xii.736. [1500.]
— Generalregister über band 1 bis x, bearbeitet von Friedrich Gottschalk. 1866. pp.[iii]. 308.

[*continued as:*]

Jahres-bericht über die leistungen der chemischen technologie.

xi. 1865. pp.xvi.784. [1500.]
xii. 1866. pp.xvi.756. [1500.]
xiii. 1867. pp.xvi.800. [1500.]
xiv. 1868. pp.xvi.836. [1750.]
xv. 1869. pp.xv.832. [1750.]
xvi. 1870. pp.xv.824. [1750.]

xvii. 1871. pp.xv.952. [2000.]
xviii. 1872. pp.xiv.936. [2000.]
xix. 1873. pp.xvi.996. [2000.]
xx. 1874. pp.xviii.1962. [2500.]
— Generalregister über band XI bis XX. bearbeitet von F. Gottschalk. 1876. pp.[ii].374.
xxi. 1875. pp.xviii.1138. [2500.]
xxii. 1876. pp.xix.1200. [2500.]
xxiii. 1877. pp.xx.1144. [2500.]
xxiv. 1878. pp.xxiv.1260. [2500.]
xxv. 1879. pp.xxvi.1286. [3000.]
xxvi. 1880. Edited by Ferdinand Fischer. pp. xxxii.980. [2000.]
xxvii. 1881. pp.xxvi.1082. [2000.]
xxviii. 1882. pp.xxiv.1200. [2500.]
xxix. 1883. pp.xx.1332. [2500.]
xxx. 1884. pp.xxiv.1360. [3000.]
— Generalregister über band XXI bis XXX, herausgegeben von F. Fischer. 1889. pp.[ii]. 134.
xxxi. 1885. pp.xxiv.1376. [3000.]
xxxii. 1886. pp.xxiv.1211. [3000.]
xxxiii. 1887. pp.xxviii.1262. [3000.]
xxxiv. 1888. pp.xxvi.1214. [3000.]
xxxv. 1889. pp.xxxi.1256. [3000.]
xxxvi. 1890. pp.xxxvi.1248. [3000.]
xxxvii. 1891. pp.xxxvi.1240. [3000.]

xxxviii. 1892. pp.xxxv.1180. [3000.]

xxxix. 1893. pp.xl.1272. [3000.]

xl. 1894. pp.xxxiv.1208. [3000.]

— Generalregister über band XXXI bis XL bearbeitet von [Richard] Guericke, Unger und Wildt. 1898. pp.[ii].646.

xli. 1895. pp.xxxix.1208. [3000.]

xlii. 1896. pp.xxxvi.1280. [3000.]

xliii. 1897. pp.xxxiv.1222. [3000.]

xliv. 1898. pp.xxxvi.1260. [3000.]

xlv. 1899. pp.xxxiv.1256. [3000.]

xlvi. 1900. pp.xix.604+xxvi.629. [3000.]

xlvii. 1901. pp.xix.604+xxiv.663. [3000.]

xlviii. 1902. pp.xxiii.564+xxiv.656. [3000.]

xlix. 1903. pp.xxiii.612+xxiii.636. [3000.]

l. 1904. pp.xxvii.628+xxvii.580. [3000.]

— Generalregister über band XLI bis L. 1908. pp.[iii].260.

li. 1905. pp.xxviii.640+xxiv.636. [3000.]

lii. 1906. pp.xxxi.646+xxxii.599. [3000.]

liii. 1907. pp.xxxv.658+xxvii.612. [3000.]

liv. 1908. pp.xxxiv.665+xxviii.619. [3000.]

lv. 1909. pp.xxviii.669+xxvi.620. [3000.]

lvi. 1910. pp.xxxv.700+xxxi.648. [4000.]

lvii. 1911. Edited by Paul F. Schmidt and B[erthold Leopold Peter] Rassow. 1912. pp.xxiv.736+xxv.615. [4000.]

lviii. 1912. pp.xxviii.734+xxiv.691. [4000.]

lix. 1913. pp.xxviii.759+xxv.668. [5000.]

lx. 1914. Edited by B. Rassow, P. F. Schmidt and W. Everding. 1915. pp.xxiv.678+xviii.530. [4000.]

lxi. 1915. pp.xix.520+xiv.445. [3000.]

lxii. 1916. pp.xvi.451+xiv.344. [2500.]

lxiii. 1917. pp.xv.454+xii.316. [2500.]

lxiv. 1918. pp.xvi.390+xiii.259. [2500.]

lxv. 1919. Edited by B. Rassow and P. F. Schmidt. pp.xxvi.584+xix.457. [4000.]

lxvi. 1920. pp.xxxii.672+xxiii.549. [5000.]

lxvii. 1921. pp.xxxiv.710+xxiii.520. [5000.]

lxviii. 1922. pp.xxxvii.740+xxvi.658. [6000.]

lxix. 1923. Edited by B. Rassow and A. Loesche. pp.vi.684+iv.611. [6000.]

lxx. 1924. pp.vi.735+iv.706. [6000.]

— Generalregister über die jahrgänge 51–70. Bearbeitet von Adolf Loesche. 1928. pp. [iii].491.

— Patentregister zum jahresbericht . . . 1878–1924. Bearbeitet von A. Loesche. 1925. pp. [iii].398. [65,000.]

lxxi. 1925. pp.vi.781+iv.595. [6000.]

lxxii. 1926. pp.vi.592+iv.649. [5000.]

lxxiii. 1927. pp.viii.565+iv.640. [5000.]

lxxiv. 1928. pp.viii.514+iv.612. [4000.]
lxxv. 1929. pp.vi.731+iv.796. [6000.]
lxxvi. 1930. pp.vi.787+iv.1027. [8000.]
lxxvii. 1931. pp.vi.696+iv.874. [7000.]
lxxviii. 1932. pp.viii.806+viii.1027. [8000.]
lxxix. 1933. pp.vi.536+iv.488. [6000.]
lxxx. 1934. pp.vi.499+iv.431. [5000.]
lxxxi. 1935. pp.vi.567 + iv[*sic,* vi].468. [6000.]
— Generalregister der jahrgänge 71–81. . . . Bearbeitet von A. Loesche. 1938. pp.439.
— Patentregister . . . 1925–1935. Bearbeitet von A. Loesche. 1936. pp.[iii].190. [30,000.]
lxxxii. 1936. pp.vi.614+iv.443. [6000.]
lxxxiii. 1937. pp.vi.346+iv.341. [4000.].

no more published; there have been minor variations in the title; since 1900 the first volume in each issue has been devoted to inorganic chemistry and the second to organic.

CHEMISCH-TECHNISCHES repertorium. Uebersichtlich geordnete mittheilungen der neuesten erfindungen, fortschritte und verbesserungen auf dem gebiete der technischen und industriellen chemie. Herausgegeben von Emil Jacobsen. Berlin.
[i]. 1862. pp.xii.72+xii.112. [1000.]
ii. 1863. pp.[ii].viii.112+[ii].viii.122. [1000.]

iii. 1864. pp.[ii].viii.120+x.122. [1000.]
[iv]. 1865. pp.viii.124+viii.126. [1000.]
[v]. 1866. pp.viii.104+viii.112. [1000.]
— Generalregister zu jahrgang I–V. 1867. pp.48.
[vi]. 1867. pp.viii.128+viii.137. [1250.]
[vii]. 1868. pp.viii.133+viii.169. [1500.]
[viii]. 1869. pp.viii.149+x.172. [1500.]
[ix]. 1870. pp.x.161+ix.183. [1500.]
[x]. 1871. pp.viii.184+viii.204. [2000.]
— General-register zu jahrgang VI–X. 1873. pp.[iv].114.
[xi]. 1872. pp.x.216+viii.256. [2500.]
[xii]. 1873. pp.x.248+viii.276. [2500.]
[xiii]. 1874. pp.viii.270+x.318. [3000.]
[xiv]. 1875. pp.ix.332+ix.352. [3000.]
[xv]. 1876. pp.xvi.603. [3000.]
— General-register zu jahrgang XI–XV. 1897. pp.[ii].168.
[xvi]. 1877. pp.xx.768. [3000.]
[xvii]. 1878. pp.xvi.496+ix.319. [3000.]
[xviii]. 1879. pp.xv.424+xv.376. [3000.]
[xix]. 1880. pp.xx.503. [2500.]
[xx]. 1881. pp.[ii].216+[ii].224. [2500.]
— Generalregister zu jahrgang XVI–XX. 1883. pp.[ii].236.
[xxi]. 1882. pp.200+268. [2500.]

[xxii]. 1883. pp.292+298. [3000.]
[xxiii]. 1884. pp.288+264. [3000.]
[xxiv]. 1885. pp.308+318. [3000.]
[xxv]. 1886. pp.260+444. [3000.]
— General-register zu jahrgang XXI–XXV. 1889. pp.313.
[xxvi]. 1887. pp.356+416. [3500.]
[xxvii]. 1888. pp.300+370. [3000.]
[xxviii]. 1889. pp.304+295. [2500.]
[xxix]. 1890. pp.348+252. [2500.]
[xxx]. 1891. pp.308+384. [3000.]
— General-register zu jahrgang XXVI–XXX. 1893. pp.339.
[xxxi]. 1892. pp.314–306. [2500.]
[xxxii]. 1893. pp.336+438. [3000.]
[xxxiii]. 1894. pp.368+350. [2500.]
xxxiv. 1895. pp.344+456. [3000.]
xxxv. 1896. pp.692. [2500.]
— General-register zu jahrgang XXXI–XXXV. 1898. pp.292.
xxxvi. 1897. pp.678. [2500.]
xxxvii. 1898. pp.708. [2500.]
xxxviii. 1899. pp.692. [2500.]
xxxix. 1900. pp.910. [3500.]
xl. 1901. pp.724. [3000.]

no more published.

ABRIDGMENTS of specifications relating to acids,

alkalies, oxides, and salts, A.D. 1622–1866. Commissioners of patents for inventions: 1869. pp. xix.1096. [1500.]

— — Part II. A.D. 1867–1876.

i. Acids, chlorine, sulphur, &c. 1887. pp.[ii]. x.234. [500.]

ii. Alkalies, oxides, and salts. 1888. pp.xviii. 588. [1000.]

iii. Benzene derivatives and other carbon compounds. 1888. pp.viii.120. [300.]

— — Part III. A.D. 1877–1883.

i. 1893. pp.xviii.296. [600.]

ii. 1894. pp.xxxii.634. [1250.]

iii. 1894. pp.xviii.264. [750.]

no more published.

TECHNISCH-CHEMISCHES jahrbuch. Herausgegeben von Rudolf Biedermann. Berlin.

i. 1880. pp.419–465. [500.]

xxxiii. 1905. pp.570–594. [300.]

SUBJECT list of works on certain chemical industries, including destructive distillation, mineral oils and waxes, gaslighting, acetylene; oils, fats, soaps, candles and perfumery; paints, varnishes, gums, resins; paper and leather industries, in the library of the Patent office. Patent office: Library series (no.7 = Bibliographical series, no.4): 1901. pp.100. [1003.]

— [second edition]. 1911.

[i]. Subject list of works on chemical technology (including oils, fats, soaps, candles, and perfumery; paints, varnishes, gums, resins, india-rubber; paper and leather industries) . . . (n.s. YN-ZB). pp.iv.172. [1500.]

[ii]. Subject list of works on peat, destructive distillation, artificial lighting, mineral oil and waxes, gaslighting and acetylene. Patent office library: Subject lists (n.s. YK-YM). pp.104. [1000.]

FORTSCHRITTE der technik. . . . Jahrbuch des Internationalen institutes für techno-bibliographie. . . . Band v: chemische technik. Berlin.

ii. 1910. pp.[vi].12.xvi.779. [18,500.]
iii. 1911. pp.548.28.27. [12,500.]
iv. 1912. pp.344. [7500.]

no more published; for earlier issues of this work see Technology, below.

LIST of reference on engineering chemistry. Library of Congress: Washington 1917. ff.2. [20.]*

SELECTED list of books on technological chemistry and chemical industries. Public library of New South Wales: Sydney 1919. pp.18. [450.]

Chemistry

УКАЗАТЕЛЬ изданий Научного химико-технического издательства Н.-Т.О.В.С.Н.Х. [Научно-технический отдел, Высший совет народного хозяйства]. Петроград 1920. pp.21. [100.]

КРАТКИЙ отчет о деятельности. Научное химико-техническое издательство: Ленинград 1920.

— 5 . . . издание. 1925. pp.112. [400.]

EDWARD CHAUNCEY WORDEN, Chemical patents index. A comprehensive and detailed index of . . . United States patents and patent reissues granted . . . 1915–1924 inclusive, covering the entire field of chemical technology. New York 1927–1934. pp.904+1190+1004+1132+1091. [12,500.]

[ALLAN J. HOLDEN and J. DAVIDSON PRATT], An index to acts of parliament and statutory rules and others affecting the chemical industry. Association of british chemical manufacturers [London]: Cambridge 1931. pp.4.ff.5–25. [300.]

[—] — Second edition. 1937. pp.5.ff.6–38. [600.]

[—] — — Supplement.

i. 1938. ff.8. [50.]

ii. 1939. ff.8. [60.]

iii. 1940. pp.18. [200.]

DECHEMA werkstoff blätter [Dechema-werkstoff-bericht]. Kurzreferate über werkstoff-fragen in der chemischen technik aus dem schrifttum des in- und auslandes. Beilage zu den zeitschriften des Vereins deutscher chemiker. Berlin.

[i]. Bearbeitet von E. Rabald. 1935. pp.viii. 105. [4004.]
ii. 1936. pp.iii.104. [3565.]
iii. 1937. pp.x.104. [3492.]
iv. 1938. pp.x.104. [3453.]
v. 1939. pp.x.94. [3152.]
vi. 1940.
vii. 1941. pp.[ii].106. [3074.]
viii. 1942. pp.[iv].140. [2972.]

ABSTRACTS of chemical papers. . . . B. Applied chemistry. Bureau of chemical abstracts: 1937 &c.

details of this work are entered unter Chemistry: General, above.

READERS' guide to books on chemical technology. Library association: County libraries section [no.22]: 1939. pp.16. [300.]

— — [another edition]. Readers' guide to chemical technology. . . . (new series, no.9): 1950. pp.40. [600.]

C[LIFFORD] C[OOK] FURNAS, A bibliography of unit operations of chemical engineering. [University of Louisville: Department of chemical engineering: Louisville 1939]. ff.[iii].64. [2000.]*

MINIMUM list of recommended books for a chemical engineering library. American institute of chemical engineers: Book list (no.1): New York 1939. pp.8. [170.]

— Supplement . . . (no.2): 1940. pp.24. [600.]

JOAQUÍN NAVARRO SAGRISTA, Manuel bibliográfico sobre industrias químicas, 1920–1940. Madrid 1940. pp.195. [2500.]

INDUSTRIAL chemistry. Compiled by . . . the Society of chemical industry. National book council: Book list (no.177): 1941. pp.8. [280.]

AN ABSTRACT of available literature on the occupation of the industrial chemist. New York 1941. pp.6 (folder).

ABSTRACTS of chemical patents vested in the Alien property custodian. Section 1[–33]. Chicago 1944.

CHEMICAL engineering. Library of Congress: Washington 1944. ff.5. [44.]*

MARIA SERRALLACHI, Bibliografía química. Documentación científico-industrial. Barcelona 1946. pp.xvi.360. [1000.]

AUSZÜGE deutscher patent-anmeldungen. Chemie. pp. +888.7+638.2. [15,000.]
covers the period 1939–1945; issued by the allied occupation forces.

CATALOGUE des documents sur [Documents sur] l'industrie chimique allemande. No.[1]–18. Direction des industries chimiques: 1951–1952.*

CURRENT literature abstracts bulletin. Olim Mathiesen chemical corporation: Niagara Falls, N.Y.*

i. 1952–1953. ff.15.28. [100.]
ii. 1953–1954. ff.200. [750.]
iii. 1954. ff.226. [897.]
iv. 1955. ff.290. [1126.]
v. 1956. ff.339. [1235.]
vi. 1957. ff.336. [1048.]
vii. 1958. ff.347. [1158.]

ÍNDICE de revistas científicas y técnicas. Serie A. Química [industrial]. Patronato "Juan de la Cierva" de investigación técnica: [Madrid].

i. 1953.
ii. 1954.
iii. 1955.
iv. 1956. pp.1382. [30,000.]
v. 1957.
vi. 1958.

vii. 1959. pp.820. [13,994.]
viii. 1960. pp.818. [13,992.]

CHEMICAL products and processes. Government-owned inventions. Department of commerce: Small business administration: Patents abstract series (no.2): Washington [1954]. pp.iii. 101. [1350.]*

A[LEKSEI] N[IKOLAEVICH] BUCHENKOV, Химизация народного хозяйства СССР. Рекомендательный указатель литературы. Государственная . . . библиотека СССР имени В. И. Ленина: Центральная политехническая библиотека: Москва 1956. pp.59. [125.]

CHEMICAL technology. Linda hall library: Bibliographic bulletin (no.15): Kansas City 1958. pp. [v].94. [1233.]*

AZ INTÉZET tudományos munkatársai műveinek bibliográfiája 1949–1958. Nehézvegyipari kutató intézet: Dokumentációs közleményei (no.12): Veszprém 1959. pp.36. [300.]*

SOVIET research and development in mass transfer chemical engineering unit operations. Department of commerce: Office of technical services: Washington [1960]. pp.iii.109. [550.]*

KATALOG książek chemicznych (1950–1960). Państwowe wydawnictwa techniczne: Warszawa [1960]. pp.168. [500.]

L. R. DAY, Books on the chemical and allied industries. A subject catalogue of books in the Science library . . . published 1930 and onwards. 1961. pp.ix.118. [3000.]*

CHEMICAL engineering library book list. American institute of chemical engineers: New York [1961]. pp.29. [1000.]*

BOOKS on the chemical and allied industries . . . received . . . in 1960. Science library: Bibliographical series (781): [1961]. pp.10. [250.]*

THE NATIONAL catalog of patents. A guide to the . . . patents . . . as described in the Official gazette, grouped in the classes and subclasses of the Manual of classification of the United States patent office. Chemical. New York.

1961. pp.1568. [11,000.]
1962. pp.1777. [12,500.]

in progress.

M[ARK] S[OLOMONOVICH] SHULMAN, Авторефераты выполненных исследований. Центральный научно-исследовательский институт спиртовой и ликерно-водочной промышленности при ВСНХ: Москва 1962. pp.36. [150.]

13. *Miscellaneous*

[PIERRE JOSEPH PELLETIER], Notice sur les recherches chimiques publiées par J. Pelletier. 1829. pp.8. [20.]

— — [another edition]. Analyse succincte des mémoires et observations chimiques [&c.]. [1837]. pp.32. [35.]

[EDMOND FREMY], Analyse succincte des mémoires de chimie publiés per m. Fremy. 1844. pp.[ii].25. [22.]

— — [another edition]. Publications de chimie générale. 1890. pp.56.

[ALEXANDRE ANTOINE BRUTUS BUSSY], Notice sur les principaux ouvrages chimiques de m. Bussy. [1850]. pp.11. [20.]

[AMÉDÉE GÉLIS], Notice sur les travaux de chimie de m. A. Gélis. [1862]. pp.12. [35.]

VERZEICHNISS der während eines zeitraums von 25 Jahren aus dem Chemischen universitätslaboratorium zu Jena unter der direction des professor dr. A. Geuther hervorgegangenen arbeiten. Jena 1888. pp.14. [151.]

[GEORGES LEMOINE], Notice sur les travaux de

chimie de m. Georges Lemoine. 1891. pp.30. [100.]

— — [another edition]. 1899. pp.[ii].42. [150.]

ENRIQUE HERRERO DUCLOUX, Los estudios químicos en la republica Argentina (1810–1910). Buenos Aires 1912. pp.431. [3000.]

TITLES of published papers from the Wellcome chemical research laboratories. [1921]. pp.[6]. [191.]

— [another edition]. [1933]. pp.12. [300.]

REFERENCES, 1926–1931, on the indexing, &c. of chemical literature. Science library: Bibliographical series (no.22): [1931]. ff.[i].3. [54.]*

F. TWYMAN, Spectrochemical abstracts, 1933–1937. 1938. pp.52. [239.]

F. TWYMAN, Analyse spectrochimique en 1938. 1939. pp.[iii].74. [100.]

LLOYD MCKINLEY, Bibliography of some achievements in chemistry, 1930–1941. [Municipal] university: [Bulletin (vol.xvii, no.6):] Wichita, Kan. 1942. pp.[ii].60.

N. F. KARETNIKOVA, Сергей Семенович Наметкин. Академия наук Союза ССР: Материалы к биобиблиографии ученых ССР: Серия химических наук (vol.2): Москва &c. 1946. pp.51. [250.]

HAMILTON MCCOMBIE and A. E. CUMMINS, List of german books on chemical & cognate subjects published 1939–1946. Chemical council: [1947]. pp.19. [500.]

HERBERT COBLANS, A bibliography of chemical research in South Africa, 1910–1939. Natal university college: Pietermaritzburg 1947. pp.ix.25. [368.]

G. C. AKERLOF and E. WILLS, Bibliography of chemical reactions in electric discharges. Department of commerce: Office of technical services: Washington 1951. pp.[v].249. [750.]*

BIBLIOGRAFIA chemii, ze szczególnym uwzględnieniem potrzeb nauczyciela i szkoły. Pedagogiczna biblioteka wojewódzka: Poznán.

1952.

1953. Opracowali: Ignacy Eichstaedt, Zofia Grotowa. 1956. pp.116. [239.]

1954. 1957. pp.80. [179.]

TITLES of papers delivered before the Division of chemical literature of the American chemical society, 1943–1953. [New York 1953]. pp.[vi].46. [470.]*

ERIK LARSSON, Kemiska institutionen i Lund. Lärare och bibliografi 1897–1956. Lund 1958. pp. 41. [1000.]

Chemistry

BIBLIOGRAPHY of papers published by László Zechmeister and co-authors in the fields of chemistry and biochemistry, 1913–1958. Wien 1958. pp.[vii].22. [257.]

DAGMAR KLESKEŇOVÁ, Metodika chemie. Bibliografie knižních publikací a časopiseckých článků. Státní pedagogická knihovna: Publikace (no.49): Brně 1959. pp.[iv].82. [900.]*

RALPH L. WENTWORTH and MARIE P. CANFIELD, Patent literature in radiation chemistry. Cambridge, Mass. [1964]. pp.[iii].81. [500.]*

Electricity and magnetism.

1. *Bibliographies*

KATHARINE MAYNARD, A bibliography of bibliographies in electrical engineering, 1918–1929. Special libraries association: Providence, R.I. 1931. pp.[iii].156. [2250.]

LIST of bibliographies on automatisation in electrical engineering. Science library: Bibliographical series (no.46): [1932]. single sheet. [11.]*

Electricity

2. *Periodicals*

A BIBLIOGRAPHY of electrical literature. Current sources and reference books. Special libraries association: Information bulletin (no.6): Providence, R.I. 1928. pp.62. [600.]

JAPANESE periodicals containing electrical papers. Science library: Bibliographical series (no.241): 1936. single sheet. [17.]*

CATALOGUE of text books, reference books and periodicals. British electrical and allied industries research association: Library: 1938. pp.61. [500.]

LIST of current periodicals in the Science museum library, dealing with electrical engineering, published in the United States. Science library: Bibliographical series (no.642): 1947. ff.2. [46.]*

GEORGE M. BERNARD, Les sources de documentation technique aux États-Unis dans les domaines de l'électricité et de la mécanique. [*c*.1950]. pp.11. [80.]

JULES CLÉMENT, Catalogue collectif des périodiques relatifs à l'électrotechnique existant dans les principales bibliothèques spécialisées de Belgique. Commission belge de bibliographie: Bibliographia belgica (no.33): Bruxelles 1958. pp.43. [879.]

RIVISTE di elettronica: televisione, radiotecnica, telecomunicazioni, ecc. i cui articoli possono essere forniti dal servizio fotoriproduzione documentaria. Consiglio nazionale delle ricerche: Roma 1954. pp.15.

ZEITSCHRIFTEN-BESTANDSVERZEICHNIS der hochschulbibliothek. Hochschule für elektrotechnik: [Ilmenau 1959]. pp.[v].119. [1108.]*

3. *General*

FR[IEDRICH] WILH[ELM] AUG[UST] MURHARD, Versuch einer historisch-chronologischen bibliographie des magnetismus. Kassel 1797. pp.166. [750.]

ABRIDGMENTS of specifications relating to electricity and magnetism, their generation and applications. Patent office: 1859. pp.xcv.772. [1000.]

— 1858–1866. 1870. pp.xxiv.864. [1000.]

— — Second edition. 1874. pp.xxiv.900. [1000.]

— [supplements].

i. Generation of electricity and magnetism ... 1867–1876. 1882. pp.7.152. [200.]

— 1877–1883. 1894. pp.xxviii.568. [1000.]

i. Conducting and insulating ... 1867–1876. 1882. pp.vii.162. [250.]

— 1877–1883. 1894. pp.xvi.224. [500.]

iii. Transmitting and receiving signals, controlling mechanical action, and exhibiting electrical effects . . . 1867–1876. 1883. pp. xii.400. [600.]
— 1877–1883. 1894. pp.xxxi.954. [1250.]
iv. Electric lighting, igniting, and heating . . . 1839–1876. 1880. pp.vii.125. [200.]
— 1877–1883. 1893. pp.xxvi.664. [1000.]
v. Electro-deposition and electrolysis . . . 1805–1876. 1882. pp.xi.276. [400.]
— 1877–1883. 1894. pp.xiii.96. [200.]
vi. Electric motive power engines and similar apparatus . . . 1837–1876. 1883. pp.vii.100. [200.]
— 1877–1883. 1894. pp.x.114. [200.]
no more published.

SIR FRANCIS RONALDS, Catalogue of books and papers relating to electricity, magnetism, the electric telegraph, &c. including the Ronalds library. . . . . Edited by Alfred J[ames] Frost. Society of telegraph engineers: 1880. pp.xxvii.564. [11,500.]

F[RANCESCO] ROSSETTI and G[IOVANNI] CANTONI, Bibliografia italiana di elettricità e magnetismo. Padova 1881. pp.117. [2750.]

LIST of a selection of works relating to electricity

& magnetism exhibited by Latimer Clark . . . at the Exposition internationale d'électricité. Paris 1881. pp.10. [100.]

GUSTAV MAY, Die weltliteratur der elektricität und des magnetismus von 1860–1883, mit besonderer berücksichtigung der elektro-technik. . . . Mit sachregister von . . . O. Salle. Elektrotechnische bibliothek (vol.xx): Wien &c. 1884. pp.xv. 203. [2000.]

— — A bibliography of electricity and magnetism, 1860 to 1883, with special reference to electrotechnics. 1884. pp.viii.203. [2000.]

SUBJECT catalogue of the memorial library of the International electrical exhibition. Franklin institute: Philadelphia [1884]. pp.xcvii. [1349.]

— [another edition]. 1885 [on cover: 1884]. pp.clxv. [2500.]

FORTSCHRITTE der elektrotechnik. Vierteljährliche berichte über die neueren erscheinungen auf dem gesammtgebiete der angewandten elektricitätslehre. Berlin.

i. 1887. Herausgegeben von Karl Strecker. pp.[xi].840.xl. [4493.]
ii. 1888. pp.[viii].764.lxviii. [4672.]
iii. 1889. pp.[viii].798.liv. [5213.]

iv. 1890. pp.viii.894.lviii. [6000.]
v. 1891. pp.viii.858.xxxviii. [5987.]
vi. 1892. pp.viii.878.xxx. [7213.]
vii. 1893. Herausgegeben von K. Strecker und Karl Kahle. pp.viii.941.xxx. [7593.]
viii. 1894. Herausgegeben von K. Kahle. pp. viii.647.xix.177. [8106.]
ix. 1895. Herausgegeben von K. Strecker. pp. viii.764.xii.160. [7201.]
x. 1896. pp.viii.827.xxiv.xvi.174. [7910.]
xi. 1897. Herausgegeben von K. Kahle. pp. viii.1024. [8178.]
xii. 1898. pp.viii.1076.xviii. [8012.]
xiii. 1899. pp.vii.1044.xiv. [8294.]
xiv. 1900. Herausgegeben von K. Strecker. pp.viii.1259.xii. [10,659.]
xv. 1901. pp.viii.1031.xii. [8191.]
xvi. 1902. pp.viii.1206.xii. [9683.]
xvii. 1903. pp.viii.1264.viii. [10,494.]
xviii. 1904. pp.viii.1185.vi. [10,470.]
xix. 1905. pp.viii.1164.vi. [10,766.]
xx. 1906. pp.viii.1266.v. [11,157.]
xxi. 1907. pp.viii.1450.v. [12,696.]
xxii. 1908. pp.viii.1504.iv. [13,323.]
xxiii. 1909. pp.viii.1478.iv. [13,770.]
xxiv. 1910. pp.viii.1474.iv. [13,327.]
xxv. 1911. pp.x.1410.iv. [13,063.]

no more published; vols.viii–xi include a separate 'patent-heft' by — Born.

FRITZ VON SZCZEPANSKI, Bibliotheca electrotechnica. Wissenschaftlich mit autorenregister versehenes repertorium der neueren deutschen, französischen und englischen elektrotechnischen litteratur. St. Petersburg &c. 1892. pp.75. [1250.]

ELEKTROTECHNISCHE bibliographie. Monatliche rundschau über die literarischen erscheinungen des in- und auslandes, einschliesslich der zeitschriftenliteratur.... Unter ständiger mitwirkung der Elektrotechnischen gesellschaft zu Leipzig zusammengestellt von Georg Maas. I (einziger) band. Leipzig 1893. pp.vi.154. [2250.]

MAX OSTERBERG, Synopsis of current electrical literature compiled from the ... technical journals and magazines during 1895. New York 1896. pp.xv.143. [1250.]

SEYDEL's führer durch die litteratur der theoretischen physik und elektrotechnik.... 6.... auflage. Berlin 1898. pp.[iv].91. [1400.]
limited to works in german.

SCIENCE ABSTRACTS. Physics and electrical engineering. Issued under the direction of the Institution of electrical engineers [and] the Physical society of London.

i. 1898. pp.viii.834. [1423.]
ii. 1899. pp.xi.961. [2000.]
iii. 1900. pp.xii.1100. [2525.]
iv. 1901. pp.xii.1192. [2506.]
v. 1902. pp.xii.1056. [2362.]

[*continued as:*]

Science abstracts. Section B. Electrical engineering. Institution of electrical engineers.

vi. 1903. pp.xv.503. [1069.]
vii. 1904. pp.xvi.1288. [2965.]
viii. 1905. pp.xvi.719. [1710.]
ix. 1906. pp.xvi.616. [1481.]
x. 1907. pp.xvi.664. [1474.]
xi. 1908. pp.xvi.655. [1367.]
xii. 1909. pp.viii.584. [1191.]
xiii. 1910. pp.iv.520. [1104.]
xiv. 1911. pp.iv.608. [1281.]
xv. 1912. pp.xvi.672. [1356.]
xvi. 1913. pp.iv.675. [1294.]
xvii. 1914. pp.iv.628. [1208.]
xviii. 1915. pp.iv.624. [1152.]
xix. 1916. pp.iv.496. [949.]
xx. 1917. pp.xi.492. [864.]
xxi. 1918. pp.iv.493. [886.]
xxii. 1919. pp.iv.482. [942.]

xxiii. 1920. pp.vii.636. [1117.]
xxiv. 1921. pp.viii.656. [1197.]
xxv. 1922. pp.viii.652. [1248.]
xxvi. 1923. pp.viii.651. [1574.]
xxvii. 1924. pp.viii.654. [1650.]
xxviii. 1925. pp.viii.692. [1811.]
xxix. 1926. pp.viii.668. [1831.]
xxx. 1927. pp.xii.724. [1985.]
xxxi. 1928. pp.xii.776. [2210.]
xxxii. 1929. pp.xii.787. [2250.]
xxxiii. 1930. pp.xii.851. [2357.]
xxxiv. 1931. pp.xii.820. [2607.]
xxxv. 1932. pp.xvi.911. [2926.]
xxxvi. 1933. pp.xvi.948. [3078.]
xxxvii. 1934. pp.xvi.906. [2992.]
xxxviii. 1935. pp.xvi.899. [3080.]
xxxix. 1936. pp.xviii.1007. [3525.]
xl. 1937. pp.xviii.934. [3252.]
xli. 1938. pp.xx.875. [3622.]
xlii. 1939. pp.xx.728. [2811.]
xliii. 1940. pp.xv.584. [2155.]
xliv. 1941. pp.ix.290. [1818.]
xlv. 1942. pp.vi.219. [1672.]
xlvi. 1943. pp.xii.272. [2600.]
xlvii. 1944. pp.xiv.243. [2197.]
xlviii. 1945. pp.xiv.292. [2744.]
xlix. 1946. pp.xv.416. [2977.]

ment of electrical engineering, no.65:] Cambridg[e]
1929. pp.108. [1280.]

ENGINEERING index . . . of library extracts Electricity supply commission: Johannesburg.*

i. 1929. ff.[i].8.pp.86. [1000.]
ii. 1930. ff.[i].8.pp.87–174. [1000.]
iii. 1932. pp.[iii].175–319. [1250.]
iv. 1932. pp.[iii].320–458. [750.]
v. 1933. pp.[ii].320–464. [750.]
vi. 1934. pp.[ii].155. [750.]
vii. 1935. pp.[ii].108. [1000.]
viii. 1936. pp.[ii].96. [1000.]
ix. 1937. pp.[ii].94. [1000.]
x. 1938. ff.[i].113. [1000.]
xi. 1939. pp.[ii].134. [1000.]
xii. 1940. pp.61. [750.]
xiii. 1941. pp.89. [750.]
xiv. 1942. pp.83. [750.]
xv. 1943. pp.78. [750.]
xvi. 1944. pp.85. [750.]
xvii. 1945. pp.113. [1000.]
xviii. 1946. pp.142. [1250.]
xix. 1947.
xx. 1948. pp.[ii].102. [1000.]
xxi. 1949. pp.[ii].96. [800.]
xxii. 1950. pp.[ii].107. [900.]

l. 1947. pp.x.406. [3233.]
li. 1948. pp.xv.392. [3119.]
lii. 1949. pp.xx.616. [4151.]
liii. 1950. pp.xii.717. [4366.]
liv. 1951. pp.xii.737. [4650.]
lv. 1952. pp.xii.797. [5132.]
lvi. 1953. pp.xvi.824. [5144.]
lvii. 1954.
lviii. 1955.
lix. 1956. pp.viii.678.
lx. 1957. pp.viii.764. [6451.]*
lxi. 1958. pp.xxiv.787. [6434.]*
lxii. 1959. pp.xxiv.944. [7718.]*
lxiii. 1960. pp.xxvi.1078. [8537.]*
lxiv. 1961. pp. .95 [7110.]*
lxv. 1962. pp.xxx.1909. [15,038].*
lxvi. 1963. pp. . 256 .]16,300.]*

in progress; section A is entered under Physics, below.

LIST of works on electricity in the New York public library. [New York 1902]. pp.100. [3500.]

BOOKS and pamphlets in [Schuyler Skaats] Wheeler gift published prior to 1826. American institute of electrical engineers: New York 1903. ff.[i].43. [900.]

Electricity

SUBJECT list of works on electricity, magnetism, and electro-technics, in the library of the Patent office. Patent office library series (no.14 = Bibliographical series, no.11): 1904. pp.288. [2374.]

[HERBERT L. COWING], Books on electricity. An annotated list. Pratt institute free library: Brooklyn 1906. pp.33. [300.]

BOOKS on electricity and allied subjects in the Carnegie free library. Braddock, Pa. 1907. pp. [33]. [300.]

WILLIAM D[IXON] WEAVER, Catalogue of the [Schuyler Skaats] Wheeler gift of books, pamphlets and periodicals in the library of the American institute of electrical engineers. . . . With . . . descriptive and critical notes by [Michael Francis O'Reilly, *in religion*] brother Potamian. New York 1909. pp.vii.504+475. [6000.]

FORTSCHRITTE der technik. . . . Jahrbuch des Internationalen institutes für techno-bibliographie. . . . Band II: Elektrotechnik. Berlin.

ii. 1910. pp.[vi].30.xviii.636. [11,500.]
iii. 1911. pp.556.17.46. [10,000.]
iv. 1912. pp.480. [10,000.]

no more published; for the earlier issues of this work see Technology, below.

JAHRBUCH der elektrotechnik. Übersicht über die wichtigeren erscheinungen auf dem gesamtgebiet der elektrotechnik. Erster [–Sechzehnter] jahrgang, das jahr 1912[–1927]. Berlin.

[S. P. THOMPSON], Hand list of the magnetic and electrical books in the library of Silvanus Phillips Thompson. 1914. pp.[v].120. [1000.]
privately printed.

PAUL FLEURY MOTTELAY, Bibliographical history of electricity & magnetism. 1922. pp.xx.673. [5000.]

A BIBLIOGRAPHY of electrical literature. Current sources and reference books. Special libraries association: Information bulletin (no.6): Providence, R.I. 1928. pp.62. [600.]

ELECTRICITY. A selected list of books . . . compiled . . . by the British electrical development association. Third edition. [National book council:] Bibliography (no.26): 1928. pp.[3]. [60.]

KATHARINE MAYNARD and MURRAY F[RANK] GARDNER, A classified list of theses in electrical engineering presented at Massachusetts institute of technology, 1902–1929, including also recent research reports. Massachusetts institute of technology: Publication [vol.lxv, no.46 = Depart-

xxiii. 1951. pp.100. [900.]
xxiv. 1952.
xxv. 1953. ff.[i].77. [750.]
xxvi. 1954. ff.[i].87. [750.]
xxvii. 1955. ff.[ii].85. [750.]
xxviii. 1956. ff.[ii].84. [750.]

READERS' guide to books on electrical engineering. Library association: County libraries section [no.4]: 1937. pp.[iii].16. [500.]

ELEKTROTECHNISCHE berichte. Unter mitwirkung des Verbandes deutscher elektrotechniker e.v. herausgegeben. Berlin.
i. Herausgegeben von Franz Moeller. 1937. pp.v.463. [1250.]
ii. 1937. pp.v.465. [1250.]
iii. 1937. pp.v.466. [1500.]
iv. 1937. pp.v.467. [1500.]
v. 1938. pp.v.468. [1500.]
vi. 1938. pp.v.466. [1500.]
vii. 1938. pp.v.467. [1750.]
viii. 1938. pp.v.470. [1750.]
ix. 1939. pp.v.469. [1750.]
x. 1939. pp.v.559. [1750.]
xi. 1939. pp.v.471. [1750.]
xii. 1939. pp.v.474. [1750.]
xiii. 1940. pp.v.472. [1750.]

xiv. 1940. pp.v.459. [1750.]
xv. 1940. pp.v.475. [1750.]
xvi. 1940. pp.v.546. [1750.]
xvii. 1941. pp.v.443. [1500.]
xviii. 1941. pp.v.472. [1750.]
xix. 1941. pp.v.463. [1750.]
xx. 1942. pp.v.581. [1750.]
xxi. 1942. pp.v.453. [1500.]
xxii. 1943. pp.v.450. [1500.]

vol.x includes a subject index to vols.i–x, vol.xx one to vols.xi–xx.

CATALOGUE of text books, reference books and periodicals. British electrical and allied industries research association: Library: 1938. pp.61. [500.]

— Supplement.
i. 1939. pp.20. [100.]
ii. 1940. pp.16. [80.]

A SELECTED list of recent references on electrical engineering. Library of Congress: Washington 1939. ff.8. [89.]*

ELECTRICAL engineering. Library of Congress: Washington 1944. ff.5. [36.]*

BULLETIN of the 'Centro Volpi di elettrologia'. English edition. [Venice].
i. 1938. pp.27.98.35.121.23.148.31.124. [1000.]

ii. 1939. pp.13.6.37.289.37. [764.]
iii. 1940. pp.15.7.27.39.343.40. [675.]
iv. 1941. pp.37.15.15.25.356.45. [710.]
v. 1942. pp.320. [640.]
v. January–June 1943. pp.102. [200.]

[*continued as:*]

Bollettino di documentazione elettrotecnica.
i. 1943–1944. pp.104. [200.]
ii. 1944. pp.[88]. [83.]
iii. 1945. pp.58. [55.]
iv. 1946. pp.60. [113.]
v. 1947. pp.[74]. [280.]
vi. 1948. pp.[76]. [287.]
vii. 1949. pp.[73]. [276.]

[*continued as:*]

Elettrotecnica. Bibliografia italiana.
viii. 1950. pp.65. [658.]
ix. 1951. pp.75. [898.]
x. 1952. pp.87. [916.]
xi. 1953. pp.81. [918.]
xii. 1954. pp.56. [623.]
xiii. 1955. pp.52. [542.]
xiv. 1956. pp.53. [566.]
xv.
xvi.
xvii. 1959. pp.65. [589.]
xviii. 1960. pp.72. [704.]

xix. 1961.

xx. 1962. pp.85. [675.]

in progress.

CATALOGUE of books and periodicals in E .R. A. library. British electrical and allied industries research association: 1945. pp.97. [1250.]

p.97 is p.[iii] of the cover.

LIST of publications by members of the E R A staff. British electrical and allied industries research association: 1945. pp.24. [230.]

ANNOTATED list of published reports. British electrical and allied industries research association: [1946]. pp.112. [600.]

BIBLIOGRAFIA elettrotecnica straniera. Anno I [&c.] Gruppo Edison: Giunta tecnica: [Milano] 1947 &c.

AUSZÜGE deutscher patent-anmeldungen. Elektrotechnik. [Berlin 1947–1948]. pp.964+953.2+816.2+992.2. [26,117.]

covers the period 1939–1945; issued by the allied occupation forces.

УКАЗАТЕЛЬ технической литературы. Центральное бюро технической информации Министерства электропромышленности СССР: Москва 1948 &c.

in progress.

CATALOGUE of the lending library. Institution of electrical engineers: 1950. pp.159. [2600.]

L. D. BELKIND, Русские электротехники. Рекомендательный указатель литературы. Гос. библ. СССР имени В. И. Ленина and Центральная политехническая библиотека: Москва 1951. pp.132.

ENERGETIKA a elektrotechnika. Přehled technické literatury a hospodářské literatury. Československý ústav pro technickou a hospodářskou dokumentaci: Praha.

viii. 1951.
ix. 1952. pp.424. [6077.]
x. 1953. pp.492. [6667.]
xi. 1954. pp.476. [7115.]
xii. 1955. pp.537. [6293.]
xiii. 1956. pp.547. [7153.]
xiv. 1957. pp.528. [7505.]
xv. 1958. pp.528. [7505.]
xvi. 1959. pp.570. [7003.]
xvii. 1960. pp.628. [8068.]
xviii. 1961.
xix. 1962. pp.624. [7665.]
xx. 1963. pp.520. [7196.]

in progress; the numbering continues that of the

Prehled techniské literatury, *which is entered under Technology, below.*

ELECTRICAL and electronic apparatus. Government owned-inventions. Department of commerce: Small business administration: Patent abstract series (no.5): Washington [1954]. ff.i.160. [1915.]*

SUMMARIES of reports of the Electrotechnical laboratory. No.1, 1950 [&c.]. Ministry of international trade and industry: Agency of industrial science and technology: Electrotechnical laboratory: Tokyo 1954 &c.

in progress.

CATALOGUE of the lending library. Institution of electrical engineers: 1956. pp.203. [3200.]

РЕФЕРАТИВНЫЙ журнал. Электроника. Академия наук СССР: Институт научной информации: Москва.

1956. pp.294.vi+312+350+323+384+367 + 398 + 368 + 278 + 256 + 256 + 284 + 282. [23,099.]

1957. pp.376 + 378 + 355 + 320 + 327 + 327 + 320 + 344 + 343 + 336 + 344 + 292 + 270 + 286 + 278 + 265 + 276 + 256+214+200+216. [40,089.]

1958. pp.264 + 267 + 279 + 283 + 289 + 284 + 281 + 304 + 316 + 256 + 264 + 275 + 292 + 296 + 280 + 572 + 302 + 311 + 288 + 323 + 325 + 314 + 291 + 290+289. [47,607.]

1959. pp.332 + 299 + 310 + 292 + 288 + 315 + 302 + 316 + 320 + 318 + 304 + 316 + 328 + 308 + 314 + 295 + 281 + 244 + 230 + 228 + 231 + 259 + 251 + 276. [51,289.]

1960. pp.368 + 392 + 508 + 512 + 536 + 536 + 516 + 502 + 460 + 360 + [340] + [366] + [266] + [420] + [404] + [416] + [324] + [316] + [370] + [320] + [316] + [240] + [268] + [267] + 1232 + 618. [83,000.]

continued in two parts, one of which is entered under Automation, above, the other being:

Электроника и энергетика.

1961. [12 volumes, each containing eight separately paginated parts].

in progress.

V[IKTOR] I[SIDOROVICH] GRANSKY, Что читать электромонтеру. Рекомендательный указатель литературы. Государственная... библиотека имени Салтыкова-Щедрина: Ленинград 1957. pp.88. [357.]

LIBRARY services and technical information for the radio and electronic engineer. British institution of radio engineers: [1958]. pp.72. [1000.]

CATALOGUE of the lending library. Institution of electrical engineers: 1959. pp.224. [3250.]

ELLEN M[ABEL] CODLIN, Handlist of basic reference material for librarians and information officers in electrical and electronic engineering. Aslib electronic group: 1959. pp.[ii].37. [250.]*

ELECTRICAL engineering. Public libraries: [Nottingham] 1959. pp.12. [250.]

ELECTRICS current. Survey of current electrical & electronics literature in journals at the Central library. [Acton] 1961 &c.*
in progress.

INFORMACJA bibliograficzna wydziału elektrycznego za okres 1950–1960. Politechnika śląska: Wydział elektryczny: Informacija dla przemysłu: Gliwice 1962. pp.70. [300.]*

4. *Conducting and insulating electricity*

ABRIDGMENTS of specifications. Class 36. Electricity, conducting and insulating. Patent office. 1855–1866. pp.xii.132. [650.]

1867–1876. pp.ix.70. [350.]
1877–1883. pp.xi.94. [450.]
1884–1888. pp.ix.72. [350.]
1889–1892. pp.x.78. [400.]
1893–1896. pp.x.76. [350.]
1897–1900. pp.xii.92. [400.]
1901–1904. pp.xiv.146. [700.]
1905–1908. pp.xiv.176. [850.]
1909–1915. pp.xiii.158. [850.]
1916–1920. pp.x.64. [300.]
1921–1925. pp.xiii.141. [700.]
1926–1930. pp.xviii.256. [1250.]

no more published.

PATENTS for inventions. Fifty years subject index, 1861–1910. Class 35. Electricity, conducting and insulating. Patent office: 1914. pp.18. [2500.]

CERAMIC materials as dielectrics. Science library: Bibliographical series (no.140): 1934. ff.[i].5. [86.]*

BIBLIOGRAPHY, vibration in electrical conductors including allied researches. [Hydro-electric power commission: Toronto 1936]. ff.[371]. [2700.]*

DIGEST of literature on dielectrics. National research council: Division of engineering and industrial research: Conference on electrical insulation: Washington.

i–x. 1936–1946. pp.704. [4000.]
xi. 1947. pp.94. [370.]
xii. 1948. pp.150. [750.]
xiii. 1949. pp.162. [960.]
xiv. 1950. pp.230. [1400.]
xv. 1951. pp.203. [1200.]
xvi. 1952. pp.179. [850.]
xvii. 1953. Edited by A. E. Middleton and H. M. Philofsky. pp.[ii].x.177. [1000.]
xviii. 1954. Edited by H. M. Philofsky and R. W. Crowe. pp.[ii].ix.176. [1000.]
xix. 1955. Edited by R. W. Crowe and J[ohn] D. Hoffman. pp.[ii].x.237. [15,000.]
xx. 1956. Edited by J. D. Hoffman and John Hart. pp.[ii].xi.239. [1500.]
xxi. 1957. Edited by J. Hart and Robert A. Sodermann. pp.xii.283. [2000.]
xxii. 1958. Edited by R. A. Sodermann and Louis J. Frisco. pp.xiii.293. [2000.]
xxiii. 1959. Edited by L. J. Frisco and Thomas D. Callinan. pp.xiv.423. [3000.]
xxiv. 1960. Edited by T. D. Callinan and Ann M. Parks. pp.vii.303. [2000.]

xxv. 1961. Edited by Ann M. Parks. pp.vii. 419. [3000.]

in progress.

BIBLIOGRAPHY and abstracts on electrical contacts. Prepared by committee B-4 . . . of the American society for testing materials. Philadelphia [1944]. pp.v.137. [837.]

REGINALD B[RODRICK] GORDON, Electrical insulation, 1941–1943. A bibliography. John Crerar library: Reference list (no.53): Chicago 1944. ff. [ii].32. [450.]*

BIBLIOGRAPHY. Vibration in electric conductors including allied researches. Hydro-electric power commission: [Toronto] 1946. ff.[217]. [1750.]*

— Addenda. 1947. ff.[58]. [300.]*

— Final addendum. 1951. ff.[54]. [300.]*

[B. M. TAREEV *and others*], Научная литература по диэлектрикам. Академия наук СССР: Физический институт им. П. Н. Лебедева: Москва 1952. pp.671. [7000.]

L. O. ANDERSON, Ferroelectrics of barium titanate type, a bibliography. Articles appearing in open literature only, with abstracts. U.S. naval ordnance test station [Inyokern]: China Lake, Cal. 1952. pp.iv.79. [250.]*

BIBLIOGRAPHY of semiconductors, 1946–Sept. 1952. Science library: Bibliographical series (no. 711): 1952. ff.21. [400.]*

A BIBLIOGRAPHY on testing of insulating materials and systems for thermal degradation. American institute of electrical engineers: New York 1956. pp.28. [500.]*

A[LEKSANDRA] N[IKODIMOVNA] GRUZDEVA, Научная литература по диэлектронам и полупроводникам. Библиография 1953–1955. Электротехнический институт: Ленинград 1957. pp.148. [1750.]

[V. P. ZHUZE *and others*], Научная литература по полупроводниковым электронным приборам (детекторы и транзисторы). Библиография 1945–1955. Академия наук СССР: Институт полупроводников: Москва &c. 1959. pp.328. [3500.]

JUDY HAYES and ESTHER PEREIRA, Meissner effect, penetration depth, and residual magnetism in superconductors. California institute of technology: Jet propulsion laboratory: Astronautics information literature search (no.149): Pasadena 1959. pp.v.85.

SUPERCONDUCTIVITY 1959–1961. An annotated

bibliography. Department of commerce: Office of technical services: [Washington 1961]. pp.iii.187. [355.]*

5. Electric generators and motors

ABRIDGMENTS of specifications. Class 35. Dynamo-electric generators and motors (including frictional and influence machines, magnets, and the like). Patent office.

1855–1866. pp.viii.50. [150.]
1867–1876. pp.vii.64. [200.]
1877–1883. pp.xv.294. [1000.]
1884–1888. pp.xi.176. [500.]
1889–1892. pp.xii.158. [500.]
1893–1896. pp.xii[*sic*, xiii].156. [500.]
1897–1900. pp.xvii.288. [750.]
1901–1904. pp.xxvi.556. [1500.]
1905–1908. pp.xx.646. [2000.]
1909–1915. pp.xx.562. [1500.]
1916–1920. pp.xvii.387. [1000.]
1921–1925. pp.xxi.494. [1500.]
1926–1930. pp.xxi.499. [1500.]

no more published.

PATENTS for inventions. Fifty years subject index, 1861–1910. Class 35. Dynamo-electric generators and motors (including frictional and influence machines, magnets, and the like). Patent office: 1916. pp.23. [3000.]

READING list on fractional horse-power electric motors, 1935–1939 inclusive. Science library: Bibliographical series (no.546): 1941. ff.2. [30.]*

READING list on electric motors, especially capacitor, polyphase and repulsion motors. 1935 to 1939 inclusive. Science library: Bibliographical series (no.549): 1941. ff.2. [30.]*

BATTERY electric road vehicles, 1946–1953. Science library: Bibliographical series (no.726): 1954. ff.2. [34.]*

SOME references to I.C. engine-electric road vehicles. Science library: Bibliographical series (no.729): 1954. ff.2. [41.]*

ELECTRIC motors — fractional horse power. Public library of South Australia: Research service: [Adelaide] 1958. ff.7. [111.]*

6. *Electric lighting and heating*

ABRIDGMENTS of specifications. Class 39. Electric lamps and furnaces. Patent office.

1855–1866. pp.v.10. [20.]
1867–1876. pp.v.12. [35.]
1877–1883. pp.xv.326. [1000.]
1884–1888. pp.ix.116. [350.]

1889–1892. pp.xii.162. [500.]
1893–1896. pp.xiv.186. [550.]
1897–1900. pp.xx.250. [750.]
1901–1904. pp.xx.334. [1000.]
1905–1908. pp.xxii.424. [1250.]
[*continued as:*]

i. Electric lamps, arc and incandescent-arc, and vacuum or low-pressure apparatus for electric discharges through gases or vapours.

1909–1915. pp.x.159. [500.]
1916–1920. pp.viii.97. [300.]
1921–1925. pp.xii.270. [750.]
1926–1930. pp.xix.387. [1250.]

ii. Electric lamps, incandescent.

1909–1915. pp.xii.214. [800.]
1916–1920. pp.x.106. [400.]
1921–1925. pp.xi.156. [600.]
1926–1930. pp.xi.155. [600.]

iii. Heating by electricity (including electric furnaces and ovens).

1909–1915. pp.xii.290. [800.]
1916–1920. pp.ix.200. [600.]
1921–1925. pp.xiii.269. [800.]
1926–1930. pp.xv.286. [800.]

no more published.

PATENTS for inventions. Fifty years subject

index, 1861–1910. Class 39. Patent office.

i. Electric lamps, arc and incandescent-arc, and vacuum or low-pressure apparatus for electric discharges through gases or vapours. 1913. pp.14. [1750.]

ii. Electric lamps, incandescent. 1915. pp.10. [1250.]

iii. Heating by electricity (including electric furnaces and ovens). 1914. pp.10. [1250.]

LIST of references on thermo-electricity. Library of Congress: Washington 1920. ff.4. [40.]*

BELLE BOONE BEARD [*and others*], Electricity in the home. Being a list of books and articles with brief abstracts. New York 1927. pp.[ii].173. [750.]

A BRIEF list of references on government ownership of electric light and power utilities. Library of Congress: Washington 1935. ff.6. [62.]*

— Additional references. 1936. ff.7. [71.]*

INCANDESCENT electric lamps. 1933–1937. Science library: Bibliographical series (no.408): 1938. ff.5. [77.]*

T. H. MESSENGER and D. V. ONSLOW, Capacity current heating. Résumé of published information. British electrical and allied industries research association: Technical report (Y/T6): 1945. pp.19. [241.]

BIBLIOGRAPHY on high frequency and dielectric induction heating. Northwestern university: Library: Evanston, Ill. 1946. ff.[ii].97. [900.]*

ELIZABETH G[OULD] DAVIS, Rural electrification in the United States. A bibliography of selected references. Department of agriculture: Bibliographical bulletin (no.24): Washington 1954. pp.iv.148. [1780.]*

V[LADIMIR] P[ANTELEIMONOVICH] ZHULE and E[LENA] I[VANOVNA] GUSENKOVA, Библиография по термоэлектричеству. Академия наук СССР: Институт полупроводников: Ленинград 1963. pp.252. [3500.]

7. *Hydro-electricity*

SELECT list of references on water power: conservation, electric transmission, etc. Library of Congress: Washington 1909. ff.9. [35.]*

HYDRO-ELECTRIC bibliography and manufacturers' statements. National electric light association: New York 1929. pp.[iii].20. [300.]

— Bibliography of hydro-electric subjects [1930]. pp.[ii].22. [550.]

LA ENERGÍA hidroeléctrica en la república Argentina. Universidad nacional: Instituto de la producción: Series bibliográficas (vol.9): [Buenos Aires] 1947. pp.22. [250.]

BIBLIOGRAPHICAL index of works published on hydro-electric plant construction. United nations: Geneva 1957. pp.[x].90. [300.]*

PUMPED storage for hydroelectric power. A selected list of references in english. Tennessee valley authority: Knoxville 1957. pp.8. [100.]*

— [another edition]. pp.9. [125.]*

8. *Measuring and testing electricity*

ABRIDGMENTS of specifications. Class 37. Electricity, measuring and testing. Patent office.

1855–1866. pp.v.12. [35.]
1867–1876. pp.v.12. [35.]
1877–1883. pp.vii.78. [250.]
1884–1888. pp.viii.64. [200.]
1889–1892. pp.vii.70. [200.]
1893–1896. pp.viii.64. [200.]
1897–1900. pp.xii.134. [400.]
1901–1904. pp.xvi.184. [500.]
1905–1908. pp.xii.194. [600.]
1909–1915. pp.xiv.274. [800.]
1916–1920. pp.xi.152. [500.]

1921–1925. pp.xx.440. [1250.]

1926–1930. pp.xxii.539. [1500.]

no more published.

PATENTS for inventions. Fifty years subject index, 1861–1910. Class 37. Electricity, measuring and testing (including electric resistances and inductances). Patent office: 1913. pp.12. [1500.]

STROBOSCOPIC testing of electricity meters. Science library: Bibliographical series (no.673): 1949. ff.2. [29.]*

9. *Regulating and distributing electricity*

ABRIDGMENTS of specifications. Class 38. Electricity, regulating and distributing. Patent office.

1855–1866. pp.viii.32. [100.]

1867–1876. pp.viii.50. [150.]

1877–1883. pp.xvi.276. [800.]

1884–1888. pp.xv.206. [600.]

1889–1892. pp.xx.236. [750.]

1893–1896. pp.xviii.234. [750.]

1897–1900. pp.xxx.502. [1500.]

1901–1904. pp.xliv.1010. [3000.]

1905–1908. pp.lii.1338. [4000.]

[*continued as:*]

i. Electric couplings, and cut-outs other than electro-magnetic and thermal.

1909–1915. pp.xv.197. [750.]
1916–1920. pp.xii.113. [450.]
1921–1925. pp.xviii.232. [1000.]
1926–1930. pp.xxi.268. [1000.]

ii. Electric currents, converting and transforming other than by rotary converters and rotary transformers, and condensers.

1909–1915. pp.ix.139. [550.]
1916–1920. pp.viii.119. [500.]
1921–1925. pp.xiv.315. [1250.]
1926–1930. pp.xvii.400. [1500.]

iii. Electric motor control systems and motor and like controllers.

1909–1915. pp.xii.415. [1000.]
1916–1920. pp.ix.178. [400.]
1921–1925. pp.xi.297. [600.]
1926–1930. pp.xiii.332. [750.]

iv. Electric supply and transmission systems and apparatus not otherwise provided for.

1909–1915. pp.xiii.374. [1000.]
1916–1920. pp.x.178. [500.]
1921–1925. pp.xiv.306. [1000.]
1926–1930. pp.xvi.343. [1000.]

v. Electric switches and electro-magnetic and thermal cut-outs (other than motor and like controllers).

1909–1915. pp.xliv.914. [3000.]

1916–1920. pp.xxxi.482. [1500.]
1921–1925. pp.xlvi.560+[ii].561–928. [3000.]
1926–1930. pp.xlviii.480 + [ii].481–1011. [3000.]

no more published.

PATENTS for inventions. Fifty years subject index, 1861–1910. Class 38. Patent office.

i. Electric couplings, and cut-outs other than electro-magnetic and thermal. 1913. pp.16. [2000.]
ii. Electric currents, converting and transforming other than by rotary converters and rotary transformers, and condensers. 1913. pp.8. [800.]
iii. Electric motor control systems and motor and like controllers. 1916. pp.10. [1250.]
iv. Electric supply and transmission systems and apparatus not otherwise provided for. 1915. pp.11. [1250.]
v. Electric switches and electro-magnetic and thermal cut-outs (other than motor and like controllers). 1917. pp.64. [10,000.]

LIST of references on central-station electric service for rural communities. Library of Congress: Washington 1920. ff.5. [52.]*

AMPLIFICATION of small currents. Science library:

Bibliographical series (no.343): 1937. ff.5. [94.]*

BIBLIOGRAPHY of relay literature, 1927–1939. American institute of electrical engineers: New York 1941. pp.15. [200.]

BIBLIOGRAPHY on circuit-interrupting devices, 1928–1940. American institute of electrical engineers: New York 1942. pp.26. [256.]

THE ELECTRICITY supply industry. [House of commons: Library:] Bibliography (no.14): [1946]. ff.14. [82.]*

10. *Miscellaneous*

JOHANN GEORG KRÜNITZ, Verzeichniss der vornehmsten schriften von der elektricität und den elektrischen curen. Leipzig 1769. pp.200. [473.]

WILLIAM B[URT] GAMBLE, List of works relating to electric welding. Public library: New York 1913. pp.13. [600.]

HEYWARD SCUDDER, The electrical conductivity and ionization constants of organic compounds. A bibliography of the periodical literature from 1889 to 1910 inclusive, including all important work before 1889 and corrected to the beginning of 1913. 1914. pp.568. [2100.]

BRIEF list of books on electromagnetic theory. Library of Congress: Washington 1917. ff.3. [27.]*

Electricity

BRIEF list of references on the electrification of railroads. Library of Congress: [Washington] 1920. ff.7. [88].*

BRIEF list of United States documents in the field of electro-chemistry. Library of Congress: Washington 1923. ff.2. [14.]*

BIBLIOGRAPHY on core losses in electrical machinery and related subjects, 1885–1924. National research council: Washington 1925. pp.[iii]. vi.138.18.5. [1266.]*

MAX KNOBEL and C[HARLES] J[OSEPH] BROCKMAN, Bibliography of electro-organic chemistry. National research council: Washington 1926. ff.[iii].70.30. [809.]*

HANS PFEIFFER, Elektrizität und eiweisse, insbesondere des zellplasmas. Wissenschaftliche forschungsberichte: Naturwissenschaftliche reihe (vol.xxi): Dresden &c. 1929. pp.xii.150. [1500.]

P[ERCY] D[AVIES] MORGAN, Memorandum and bibliography on corona on overhead lines. British electrical and allied industries research association: Technical report (F/T30): 1930. ff.[i].8. [71.]*

MOTO-ELECTRIC effect. [Science library: Bibliographical series (no.3): 1931]. single sheet. [19.]*

BIBLIOGRAPHY on the effect of electricity on plants. Science library: Bibliographical series (no. 24): [1931]. ff.[i].3.2. [79.]*

P. H. PRAUSNITZ and J. REITSTÖTTER, Elektrophorese, elektroosmose, elektrodialyse in flüssigkeiten. Wissenschaftliche forschungsberichte: Naturwissenschaftliche reihe (vol.xxiv): Dresden &c. 1931. pp.xii.308. [2500.]

NON-METALLIC resistances. Science library: Bibliographical series (no.93): [1933]. single sheet. [10.]*

BIBLIOGRAPHY of the economics of the grid system in Great Britain. Science library: Bibliographical series (no.141): 1934. ff.[i].3. [46.]*

CARBON electrodes, 1930–35. Science library: Bibliographical series (no.215): 1936. ff.3. [54.]*

RELATION between melting and current strength in fuses, with special reference to silver wires. Science library: Bibliographical series (no.281): 1936. ff.4. [65.]*

APPLICATIONS of electricity to railways. Bibliography of periodical articles appearing in a select list of periodicals covering the calendar year 1936

[&c.]. Association of american railroads: Washington [1937 &c.].
in progress.

THE ELECTRICAL resistance of the human body. Science library: Bibliographical series (no.310): 1937. ff.4. [44.]*

REFERENCES to electrodialysis and electroendosmosis (exclusive of patents) covering the period 1931–March 31st 1937. Science library: Bibliographical series (no.319): 1937. ff.12. [225.]*

CONTACT resistance. (1929–1936). Science library: Bibliographical series (no.348): 1937. ff.3. [46.]*

FRITZ VILBIG, Schrifttumsverzeichnis zum lehrbuch der hochfrequenztechnik. Zweite . . . auflage. Leipzig 1939. pp.[iii].172. [3500.]

PIEZOELECTRIC crystals. Science library: Bibliographical series (no.448): 1939. ff.11. [220.]*

DISRUPTIVE atmospheric electrical discharges (refs. from 1900 to date). Science library: Bibliographical series (no.455): 1939. ff.18. [360.]*

BIBLIOGRAPHY of electrical safety, 1930–1941. American institute of electrical engineers: New York 1942. pp.14. [550.]

NEUTRAL earthing and Petersen coils. 1937 to date. Science library: Bibliographical series (no. 597): 1943. ff.2. [40.]*

АННОТИРОВАННЫЙ указатель литературы по радиоэлектронике. Москва 1945 &c.
in progress.

SELECT bibliography on the installation and utilisation of electricity in iron and steel works. Central library: Research bibliographies (new ser., no.57): Sheffield [1948]. ff.3. [40.]*

G. C. AKERLOF and E. WILLS, Bibliography of chemical reactions in electric discharges. Department of commerce: Office of technical services: Washington 1951. pp.[v].249. [750.]*

J. D. PHILIPPOFF, Electric power industry of the U.S.S.R. An annotated bibliography. Library of Congress: Reference department: Washington 1952. pp.xi.154. [1765.]*

FRANCIS D. HORIGAN, Electrical characteristics of snow and ice. A literature search. QM research & development laboratories: Technical library: Bibliographic series (no.28): Philadelphia [1953]. pp. [ii].15. [39.]*

[GERHARD KURT MARTIN PFESTORF], Survey of literature on high voltage engineering and allied subjects (1935–53). Indian institute of science: Memoirs (n.s., no.1): Bangalore [1954]. pp.xi.206. [6000.]

ELIZABETH GOULD DAVIS, Rural electrification in the United States. A bibliography of selected references. Department of agriculture: Bibliographical bulletin (no.24): Washington 1954. pp.iv. 148. [1780.]*

A[LEKSEI] N[IKOLAEVICH] BUCHENKOV, Радиоэлектроника, ее достижения и перспективы развития. Рекомендательный обзор литературы. Государственная . . . библиотека СССР имени В. И. Ленина: Новости техники (no.1): Москва 1956. pp.15. [30.]

A[LEKSEI] N[IKOLAEVICH] BUCHENKOV, Электрификация СССР к 40-летию Великой Октябрьской социалистической революции. Рекомендательный указатель литературы. Государственная . . . библиотека СССР имени В. И. Ленина: Центральная политехническая библиотека: Москва 1957. pp.63. [250.]

L[YUDMILA] M[ARIANOVNA] IVANOVA, В помощь электромонтеру. Государственная... библиотека СССР имени В. И. Ленина: Москва 1957. pp.92. [150.]

THERMOELECTRICITY abstracts. U. S. naval research laboratory: Library branch: Washington 1959. pp.vii.74. [1021.]*

V. N. GLINKINA, Электроискровая обработка токопроводящих материалов. Библиографический указатель (1955–1959). Академия наук СССР: Центральная научно-исследовательская лаборатория электрической обработки материалов: Москва 1960. pp.71. [930.]

N[ATALYA] I[VANOVNA] KRASTING, Единая энергетическая система СССР. Рекомендательный обзор литературы. Государственная . . . библиотека имени В. И. Ленина: Новости науки и техники (no.24): Москва 1960. pp.19. [13.]

L[IDYA] S[ERGEEVNA] SAMOKHVALOVA, Достижения советской энергетики. Рекомендательный ... указатель литературы. Государственная библиотека СССР имени В. И. Ленина [&c.]: Москва 1961. pp.43. [80.]

NUCLEAR effects on the propagation of electromagnetic waves. Annotated bibliography. [Library of Congress:] Aerospace information division: AID report (no.62–155): [Washington] 1962. ff.[ii].30. [114.]*